Hausdorff on Ordered Sets

 Earlier

Felix Hausdorff in Greifswald, 1914–1921
Universitäts-und Landesbibliothek Bonn, Hss.-Abt.
NL Hausdorff: Kapsel 65: Nr. 06

History of Mathematics • *Volume 25*
SOURCES

Hausdorff on Ordered Sets

J. M. Plotkin
Editor

Translated by J. M. Plotkin

American Mathematical Society
London Mathematical Society

Editorial Board

American Mathematical Society
Joseph W. Dauben
Peter Duren
Karen Parshall, Chair
Michael I. Rosen

London Mathematical Society
Alex D. D. Craik
Jeremy J. Gray
Peter Neumann
Robin Wilson, Chair

Original German text translated by J. M. Plotkin

The article *Sum of $\aleph_1$ sets* by Felix Hausdorff, originally published as "Summen von $\aleph_1$-Mengen", Fundamenta Mathematicae XXVI (1936) is reprinted by permission of Instytut Matematyczny Polskiej Akademii Nauk.

2000 *Mathematics Subject Classification.* Primary 01A75, 01A60, 03-03, 06-03, 26-03.

For additional information and updates on this book, visit
www.ams.org/bookpages/hmath-25

Library of Congress Cataloging-in-Publication Data
Hausdorff, Felix, 1868–1942.
 [Selections. English. 2005]
 Hausdorff on ordered sets / J. M. Plotkin, editor ; translated by J. M. Plotkin.
  p. cm. — (History of mathematics, ISSN 0899-2428 ; v. 25)
 Includes bibliographical references.
 ISBN 0-8218-3788-5 (alk. paper)
 1. Ordered sets–History. 2. Hausdorff, Felix, 1868–1942. 3. Logic, Symbolic and mathematical–History. 4. Lattice theory–History. 5. Functions–History. I. Plotkin, J. M. (Jacob M.), 1941– II. Title. III. Series

QA171.48.H38 2005
511.3′2–dc21
                                                                2005045328

Copying and reprinting. Individual readers of this publication, and nonprofit libraries acting for them, are permitted to make fair use of the material, such as to copy a chapter for use in teaching or research. Permission is granted to quote brief passages from this publication in reviews, provided the customary acknowledgment of the source is given.

Republication, systematic copying, or multiple reproduction of any material in this publication is permitted only under license from the American Mathematical Society. Requests for such permission should be addressed to the Acquisitions Department, American Mathematical Society, 201 Charles Street, Providence, Rhode Island 02904-2294, USA. Requests can also be made by e-mail to reprint-permission@ams.org.

© 2005 by the American Mathematical Society. All rights reserved.
Printed in the United States of America.

The American Mathematical Society retains all rights
except those granted to the United States Government.
∞ The paper used in this book is acid-free and falls within the guidelines
established to ensure permanence and durability.
The London Mathematical Society is incorporated under Royal Charter
and is registered with the Charity Commissioners.
Visit the AMS home page at http://www.ams.org/

10 9 8 7 6 5 4 3 2 1   10 09 08 07 06 05

To the memory of Felix Hausdorff

Felix Hausdorff in his study at home, Bonn, June 8–14, 1924
Universitäts-und Landesbibliothek Bonn, Hss.-Abt.
NL Hausdorff: Kapsel 65: Nr. 29

Contents

Preface	ix
Selected Hausdorff Bibliography	1
Introduction to "About a Certain Kind of Ordered Sets"	3
About a Certain Kind of Ordered Sets [H 1901b]	11
Introduction to "The Concept of Power in Set Theory"	23
The Concept of Power in Set Theory [H 1904a]	31
Introduction to "Investigations into Order Types I, II, III"	35
Investigations into Order Types [H 1906b]	45
I. The Powers of Order Types	46
II. The Higher Continua	58
III. Homogeneous Types of the Second Infinite Cardinality	75
Introduction to "Investigations into Order Types IV, V"	97
Investigations into Order Types [H 1907a]	113
IV. Homogeneous Types of Cardinality of the Continuum	113
V. On Pantachie Types	129
Introduction to "About Dense Order Types"	173
About Dense Order Types [H 1907b]	175
Introduction to "The Fundamentals of a Theory of Ordered Sets"	181
The Fundamentals of a Theory of Ordered Sets [H 1908]	197
Introduction to "Graduation by Final Behavior"	259
Graduation by Final Behavior [H 1909a]	271
Appendix. Sums of $\aleph_1$ Sets [H 1936b]	305
Bibliography	317

Preface

Introduction

As the nineteenth century was ending, Georg Cantor (1845–1918) published his own *Summa Theologica* for set theory. This two-part work, entitled *Beiträge zur Begründung der transfiniten Mengenlehre*, appeared in 1895 and 1897 in volumes 46 and 49, respectively, of Mathematische Annalen. The second installment of *Beiträge* turned out to be Cantor's last published work on set theory. Though *Beiträge* could well serve as a primer on basic set theory, it also seemed unfinished. Fundamental questions remained. Are all cardinals comparable? Can all sets be well-ordered? Are there just two possibilities for the cardinalities of infinite sets of reals? Yet in some cosmically-just way this work marked a fitting ending for a journey that began with the bread and butter question of uniqueness of representations of functions by trigonometric series and ended in the realm of pure set theory.

The year 1897 also saw Cantor's contributions gain official and public recognition from his peers. The First International Congress of Mathematicians took place in Zurich in August of 1897, and Cantor was in attendance. In a major address, the Zurich mathematician Adolf Hurwitz (1859–1919), who was one of the congress's organizers, devoted considerable space to an exegesis of Cantor's ordinal numbers and their application to the analysis of closed point sets. Hurwitz propounded Cantor's work as the basis for an investigation of analytic functions in terms of their sets of singularities.[1] At this same congress, Jacques Hadamard (1865-1963) gave a short talk entitled *Sur certaines applications possibles de la théorie des ensembles.*

The next international congress was held in Paris in August 1900. There, David Hilbert (1862–1943) gave the famous address in which, after a provocative meditation on the esthetics of mathematical problems and on the criteria for admissible solutions, he listed twenty-three problems whose solutions would advance the entire enterprise of mathematics.[2] The first three

---

[1] *Über die Entwickelung der allgemeinen Theorie der analytischen Funktionen in neurer Zeit*, Verhandlungen des ersten internationalen Mathematiker-Kongresses in Zürich vom 9. bis 11. August 1897 (Leipzig: Teubner), 1898, 91–112. Hurwitz's discussion of Cantor's work appears on pp. 94–97.

[2] Because of time constraints, Hilbert mentioned only ten problems in his actual address: three from foundations, four from arithmetic and algebra, and three from function theory. All twenty-three problems are in the printed version: *Mathematische Probleme*,

ix

problems dealt with foundations, and problem No. 1 was labelled *Cantors Problem von der Mächtigkeit des Continuums*. As its title indicates, the first problem calls for a proof of what is now termed Cantor's Weak Continuum Hypothesis: there are only two possible cardinalities for the infinite subsets of the continuum, that of the set of natural numbers and that of the entire set of reals. In addition, the first problem also calls for a proof of the well-orderability of the continuum, again as conjectured by Cantor.[3]

Thus as the twentieth century began, Cantor, whose work in set theory was finally receiving its due, had begun a de facto retirement, and David Hilbert, one of the two mathematical greats of the day—the other being Henri Poincaré (1854–1912)—had made the positive resolution of two of Cantor's most important conjectures his own number one problem.[4] It was time for the appearance of the second generation of Cantorians.

The first decade of the new century witnessed a burst of activity in set theory with many of the participants belonging to a group twenty to thirty years younger than Cantor; among them were Felix Bernstein (1878–1956), Gerhard Hessenberg (1874–1925), Phillip Jourdain (1879–1919), Ernst Zermelo (1871–1953), and Felix Hausdorff (1868–1942). Although he was not of this younger generation, the veteran mathematician Arthur Schoenflies (1853–1928) also worked in set theory during this period. He was an ardent disseminator of Cantorian ideas, writing an encyclopedia article and book-length reports on set theory and its application to the study of point sets, i.e., subsets of $\mathbb{R}$ and $\mathbb{R}^n$ ([Sch 1898; 1900; 1908]).

However, in the years 1900–1910, it was Zermelo and Hausdorff who came to dominate the discourse on set theory and who represented the two trends that were to characterize set theory's future: its abstraction and its pursuit as a mathematical theory. Zermelo, in seeking to clarify the existence assumptions in his proof of well-ordering, axiomatized the concept of set based solely on the membership relation and a few simple operations; he became the father of abstract set theory.[5] Hausdorff carried on in Cantor's footsteps, developing set theory as a branch of mathematics worthy of study in its own right and capable of undergirding both general topology and

---

Göttinger Nachrichten (1900), 253–297. His entire text, translated into English by Dr. Mary Winston Newton, appears in the *Bulletin of the American Mathematical Society*, **8** (1901–1902), 437–479.

[3]Hilbert felt that the well-orderability of the continuum had a close connection with the continuum problem and that perhaps it was the key to its solution. He asked that the solver provide a specific (i.e., definable) well-ordering: "Es erscheint mir höchst wünschenswert, *einen direkten Beweis dieser merkwürdigen Behauptung von Cantor zu gewinnen* [his italics], etwa durch wirkliche Angabe einer solchen Ordnung der Zahlen ..." ["It appears highly desirable to me *to produce a direct proof of Cantor's remarkable assertion*, perhaps by actually indicating one such ordering of the [real] numbers ..."]

[4]Actually, their position on Hilbert's list is an artifact of his organizational scheme, but their placement did have a psychological effect.

[5]See [Ka 1996, 9–12] and [Ka 2004]. The latter article is a penetrating examination of Zermelo's place in the history of set theory.

measure theory; he was a founder of the first and a significant contributor to the second.

At least superficially the education and early professional careers of both men seem quite similar: dissertations and habilitations in applied mathematics, an interest in Cantor's set theory as manifested in 1901 by the offering of the first courses devoted entirely to set theory and by the appearance of their own first publications in set theory. We pause to present brief biographical sketches for both (up through the mid-teens of the early nineteen hundreds), asking the reader's indulgence as we maintain the conceit of parallel development.

Ernst Zermelo was born in Berlin in 1871. After studying at its university and then at Halle and Freiburg, he returned to Berlin where he received his doctorate in 1894; his dissertation on the calculus of variations was directed by H. A. Schwarz (1843–1921), Weierstrass's successor. He then became an assistant to Max Planck at Berlin's Institute for Theoretical Physics, staying there from 1894 to 1897. He moved to Göttingen in 1897, where he completed his *Habilitation* in 1899 with a work on hydrodynamics. As a *Privatdozent* at Göttingen, Zermelo was influenced by Hilbert to turn his attention to questions in set theory and foundations.[6]

At Göttingen in the winter semester of 1900–1901, Zermelo offered a course devoted entirely to set theory (in the spirit of Cantor's *Beiträge*). This was the first such course offered anywhere. In 1901, he published his first set theory paper, *Über die Addition transfiniter Cardinalzahlen*, in the Göttinger Nachrichten. It was a modest contribution to the arithmetic of cardinal numbers. However, his second publication in set theory ([Ze 1904]) ensured his immortality: he proved that every set could be well-ordered on the basis of a newly articulated principle, the Axiom of Choice.[7] Initial reactions were less than laudatory; criticism came swiftly and was not *sotto voce*. Zermelo enjoyed the private support of Hilbert, but in public he was largely left to defend himself. And defend himself he did. In [Ze 1908a], he offered a new proof of the Well-Ordering Theorem together with a brilliantly written reply to his critics. In a second, related, paper [Ze 1908b], Zermelo presented his axiomatization for the concept of set, which, with important later contributions by Fraenkel and Skolem, is the widely recognized ZF (ZFC) set theory of today. This first phase of Zermelo's foundational

---

[6] In [Pe 1990], Volker Peckhaus gives a detailed picture of Zermelo's years at Göttingen. Though plagued by financial and health problems, Zermelo enjoyed increasing status with Hilbert and the members of his school. From 1905, he was *Titularprofessor*, a position financed by "soft money." Hilbert's manuevers to regularize his position failed.

[7] At the Third International Congress of Mathematicians, held in Heidelberg in August 1904, Julius König presented a proof—later withdrawn—that the continuum could not be well-ordered. (See the *Introduction to "The Concept of Power in Set Theory,"* p. 24.) A little over a month after this unsettling (for set theorists) meeting, Zermelo discovered his proof that the continuum and any other set could be well-ordered. Zermelo's proof is contained in a brief letter to Hilbert dated September 18, 1904.

research ended in 1910 when he left Göttingen for a Professorship at Zurich. In 1916, Zermelo retired from this position for health reasons.[8]

Felix Hausdorff was born in Breslau (now Wrocław) in 1868 and moved with his parents to Leipzig at an early age. After studying at Leipzig, Freiburg, and Berlin, he returned to Leipzig where he received his doctorate in 1891; his dissertation on the refraction of light in the atmosphere was supervised by Heinrich Bruns (1848–1919), a mathematical astronomer and, from 1882, director of Leipzig's observatory. In 1893–1895, Hausdorff worked at the observatory doing scientific calculations. As a *Privatdozent* for astronomy and mathematics, he completed his *Habilitation* at Leipzig in 1895 with a work on the absorption of light in the atmosphere. By the late 1890s, Hausdorff began to move away from applied mathematics and towards pure mathematics with publications in probability theory and non-Euclidean geometry.

At Leipzig, in the summer semester of 1901, he taught a *Beiträge*-influenced course on set theory, historically the second such course—after Zermelo. In 1901, he also published his first paper on set theory, *Über eine gewisse Art geordneter Mengen*, in the Reports of Leipzig's Königlich Sächsischen Gesellschaft der Wissenschaften. In 1902, he became *aussordentlicher Professor* at Leipzig; shortly thereafter, in what may have been an unwise career move, he turned down the offer of a similar position at Göttingen.[9] Hausdorff was appointed *Extraordinarius* at Bonn in 1910. He became *ordentlicher Professor* at Greifswald in 1913.[10] During the period 1900–1910, he published thirteen articles on mathematics. Seven of them—some 281 journal pages—were in set theory, mostly about ordered sets. In 1914, he published the book *Grundzüge der Mengenlehre*.[11] The first six chapters, approximately half of this classic and highly praised textbook, are devoted to Hausdorff's version of Cantorian set theory as developed through

---

[8] Zermelo's move to Zurich terminated what Peckhaus calls his first *pro*-Hilbert phase [Pe 2002]. After his retirement, he moved back to Germany in 1921 and took up residence in the Black Forest near Freiburg. In 1926, he was appointed to an honorary Professorship at Freiburg, which was an unsalaried position. He returned to foundational research in the period 1927–1935; Peckhaus labels this Zermelo's *contra*-Hilbert phase, because he totally opposed Hilbert's finitistic proof theory. Zermelo lost his position at Freiburg in 1935 for failing to give the Nazi salute properly; he was reinstated in 1946. He died in Freiburg on May 21, 1953.

[9] See [Di 1967, 52] on this episode.

[10] Hausdorff returned to Bonn as *Ordinarius* in 1921. He was forcibly retired by the Nazis in 1935 because he was a Jew. He died in Bonn on January 26, 1942. Hausdorff, his wife, and his wife's sister committed suicide just before they were to be interned at the former cloister *Zur Ewigen Anbetung* in Endenich. Endenich was the first stop before deportation to the East and almost certain death. A photocopy and text transcription of Hausdorff's stoic and poignant farewell letter, dated January 25, 1942, appears in [Br 1996, 263–267]; an English translation of this letter appears in [Ei 1992, 101–102].

[11] Hausdorff's dedication in *Grundzüge* reads: "Dem Schöpfer der Mengenlehre Herrn Georg Cantor in dankbarer verehrung gewidmet" [["Dedicated to the creator of set theory Herr Georg Cantor in grateful admiration"]].

a decade of his own research; the last four chapters form an important founding text for general topology.

This ends our brief attempt at comparative biography. An intriguing question remains unanswered: for Zermelo, it was Hilbert's influence that brought him to set theory and foundations, but what led Hausdorff to set theory? Both Cantor and Hausdorff attended the 1897 Zurich congress, and the mathematicians at Halle and Leipzig met frequently. These gatherings, known as *Kränzchen*, alternated between the two locations, and some at Halle were even held at Cantor's home ([Kow 1950, 106]). But social interaction alone does not account for intellectual affinity. For some important clues, we need to consider the works of the writer Paul Mongré.

Mongré's first book, which showed the influence of both Nietzsche and Schopenhauer, was the aphoristic *Sant'Ilario–Gedanken aus der Landschaft Zarathustras*; it appeared in 1897. A second book followed in 1898. Entitled *Das Chaos in kosmischer Auslese–Ein erkenntniskritischer Versuch*, it was critical of metaphysics and disparaged the ability of the empirical world as recognized by our consciousness to reveal anything about the transcendental world.[12] In 1900, Mongré published *Ekstasen*, a book of poetry, and in 1904 his one act play *Der Arzt seiner Ehre*, a comedy, was performed in Berlin and Hamburg. In the period 1897–1910, Paul Mongré, the friend of avant garde artists and writers and the quasi-disciple of Nietzsche, also published essays and reviews in various cultural periodicals. The variety and quantity of Mongré's output alone might pique our interest, but for us his most remarkable attribute is his true identity—he was Felix Hausdorff.[13]

It seems that it was Hausdorff's philosophical-literary alter ego, Paul Mongré, who inspired his interest in set theory. Walter Purkert, in the first section of his very informative introduction to the second volume of Hausdorff's Collected Works [Pu 2002, 2–7], presents strong internal evidence that Hausdorff became acquainted with Cantorian set theory somewhere in the period 1897–1898. In particular, Purkert establishes that between the writing of *Sant'Ilario* and *Das Chaos* Hausdorff-Mongré learned of Cantor's

---

[12]We refer to these books as *Sant'Ilario* and *Das Chaos* for short. The English translations of their titles are: *Saint Ilario–Thoughts from Zarathustra's Countryside* and *Chaos in Cosmic Selection–A Critical Epistemological Essay*, respectively.

[13]A complete Hausdorff-Mongré bibliography is available from
   http://www.aic.uni-wuppertal.de/fb7/hausdorff/schriver.htm.
All our references to Hausdorff's publications follow the designations assigned by this bibliography.

A reprint of *Das Chaos* appeared in 1976 with an altered title: *Zwischen Chaos und Kosmos oder Vom Ende der Metaphysik*, with an introduction by Max Bense (Baden-Baden: Agis Verlag). Both *Sant'Ilario* and *Das Chaos* are reprinted in *Felix Hausdorff, Gesammelte Werke, Band VII, Philosophisches Werk*, W. Stegmaier (ed.) (Berlin-Heidelberg-New York: Springer-Verlag), 2004. This volume also includes line-by-line commentaries for *Sant'Ilario* and *Das Chaos*. Werner Stegmaier's introduction describes Hausdorff's philosophical antecedents and the main ideas in his philosophical works. In addition, three short articles by Hausdorff on Nietzsche's theory of *return* [[*Wiederkunft des Gleichen*]] and his *Will to Power* [[*Der Wille zur Macht*]] are reprinted.

1878 result that all the $\mathbb{R}^n$ have the same cardinality. He also cites several places in *Das Chaos* where connections between philosophical argumentation and set theoretic concepts and theorems are made.[14]

Philosophical interests, as expressed in his writings as Paul Mongré, may have brought Hausdorff to Cantor, but several commentators have seen the solution of Cantor's continuum problem as the driving force behind Hausdorff's work on ordered sets in the years 1901–1909.[15] It is not hard to imagine that any mathematician who was worth his or her salt and who had more than a passing acquaintance with Cantor's work would be tantalized by the challenge set by Hilbert and would take up the continuum problem. With respect to this challenge, starting in 1904, Hausdorff, the skilled mathematician, extended Cantor's concept of cardinal exponentiation to create the tool of *ordered powers* of ordered sets; he discovered the fundamental relation of *cofinality*, which led him to isolate an important class of invariants, the regular cardinals and their inverses. He then used ordered products and his invariants to develop a representation theory, *ab initio*, for ordered sets in general and the important subfamily of densely ordered sets in particular. His representation theory often bumped into the continuum problem, but he seemed to use such occurrences to direct his attention to more realistic goals or to measure the difficulty of a specific problem.

For the interested scholar, there is no better place to begin to seek answers to meta-questions, such as the role played by the continuum problem in Hausdorff's research—motivational goal or cautionary symbol of intractability—than in the articles themselves. These same articles are also a treasure trove for those wanting to find the sources of concepts and methods that have hitherto been seen as originating in *Grundzüge der Mengenlehre*. These range from the perhaps mundane identification of sets with functions via characteristic functions to the famous back-and-forth construction of isomorphisms, from the extremely fruitful concept of η-set to what we now call Hausdorff's maximal principle. Of course, we also find things that did not make it into *Grundzüge*: worthwhile ideas, such as the concept of *type ring* and important results, such as the calculation of a sharp lower bound for the cardinality of irreducible, continuous $c_{\sigma\sigma}$-sets and the proof that Hausdorff gaps exist in certain partially ordered sets. Then there are the *firsts*: for example, the first elimination of CH from a proof ([H 1907a]), the first statement of GCH ([H 1908]), and the first use of a maximal principle in algebra ([H 1909a]). Finally, through these articles we are afforded glimpses of Hausdorff's opinions on foundations as he tries to do battle against the ineffability of the uncountable.

---

[14] *Das Chaos* is discussed in connection with Hausdorff's mathematical work in [Ei 1992, 85–91; 1994, 3–10]; Erhard Scholz echoes this title in his article *Logischen Ordnungen im Chaos* ([Scho 1996]) and uses it thematically in commenting on Hausdorff's work up to 1919.

[15] In particular, see [Mo 1989, 109; 1996, 127, 130], [Br 1996, 5], [Koe 1996, 86], and [Scho 1996, 109, 114].

Our main aim in translating Hausdorff's papers on ordered sets is to make them available to a larger audience for whatever purpose. At the same time, we hope to expose more people to the early work of a creative and productive mathematician of the first rank. The period covered is truly an amazing one in the history of set theory: "a hundred flowers bloomed," and many wilted in the intense light. However, under the aegis of Hilbert's school, Zermelo created the new branch of abstract set theory, while separately the inner-directed Hausdorff exploded upon the scene and usurped the position of the era's number one Cantorian.[16]

The Translations and Introductory Essays

The articles translated in this volume are: *Über eine gewisse Art geordneter Mengen* [[*About a Certain Kind of Ordered Sets*]] [H 1901b]; *Der Potenzbegriff in der Mengenlehre* [[*The Concept of Power in Set Theory*]] [H 1904a]; *Untersuchungen über Ordnungstypen I, II, III* [[*Investigations into Order Types I, II, III*]] [H 1906b]; *Untersuchungen über Ordnungstypen IV, V* [[*Investigations into Order Types IV, V*]] [H 1907a]; *Über dichte Ordnungstypen* [[*On Dense Order Types*]] [H 1907b]; *Grundzüge einer Theorie der geordneten Mengen* [[*The Fundamentals of a Theory of Ordered Sets*]] [H 1908]; *Die Graduierung nach dem Endverlauf* [[*Graduation by Final Behavior*]] [H 1909a]; and in the Appendix, *Summen von $\aleph_1$ Mengen* [[*Sums of $\aleph_1$ Sets*]] [H 1936b]. Further bibliographical data for these articles appears in our *Selected Hausdorff Bibliography*.

The provenance of these translations requires some explanation. I was led to Hausdorff's *Grundzüge der Mengenlehre* and then to his earlier papers on ordered sets in the early 1990s as I tried to discover the origins of the *back-and-forth* construction. This resulted in [Pl 1993] and in my realization of the wealth of material in [H 1906b; 1907a]. I was determined to produce translations of both. In the summer of 1995, I finished a first version of [H 1906b]. The project lay dormant until 1997, when at the AMS winter meetings in San Diego, I spoke about Hausdorff's work on order types. Afterwards, Marion Scheepers introduced himself, and I learned of his interest in these same papers. Marion had done some partial translations of Hausdorff while pursuing his study of gap theorems (see [Sc 1993]).

Shortly after our meeting in San Diego, we began an extensive e-mail correspondence in which we decided to work together on translations of the remaining papers on ordered sets. As amateur translators we were both feeling our way. We agreed that the translations were to be fairly literal and that anachronistic terminology was to be avoided. Each of us had primary responsibility for certain papers: I took [H 1901b; 1907b; 1908; 1909a §1] and Marion took [H 1904a; 1907a; 1909a §§2-5]. Translations were passed

---

[16] The creative outpouring from Hausdorff-Mongré in the span 1897–1910 is truly remarkable, as is its intellectual scope. The Hausdorff-Mongré bibliography lists twenty-two items for Hausdorff and twenty-one items for Mongré during this era; for Mongré, this includes three books and for Hausdorff, three memoir-length articles.

back and forth electronically, and the task of smoothing rough spots and maintaining uniformity across various iterations was my responsibility. By September 1, 1998, the first versions of all these papers had been completed. In the spring of 1999, Marion decided to add a translation of [H 1936b] because of its connection with [H 1909a] and its relevance to the gap literature; a first version was finished by June. The joint phase of our translating work ended in the fall of 1999. Since then, I have gone back to the originals and reworked the translations. Of course, having the versions from our joint efforts made the resulting iterations much less burdensome, and it allowed me to concentrate on passages that had earlier caused us difficulty. Multiple passes of proofreading helped to increase the accuracy of the translations. This reworking took place from the winter of 2000 through the fall of 2002.

In January of 2003, I began writing introductory essays for the papers. Each translation, except that of [H 1936b] in the Appendix, is preceded by an essay, prosaically entitled *Introduction to —*. These essays, each equipped with its own end notes, are "programs" that take the reader through the highlights of each article, make connections among articles, and offer commentary on relationships to the work of others. (All translations from German sources appearing in the introductions were done by me.) The introductions can be read before, during, or after the relevant article—or they can be ignored. If *before*, I hope that they inspire a possibly wavering reader to take up the actual work.

I wish to offer further comments on the particulars of the translations and introductions. The numbers appearing in the margins of the pages of the translations refer to the corresponding pages in the originals. All page references to Hausdorff's articles that appear in the introductions are to the relevant page or pages in the original article, which can be easily found in the translation by using the numbers in the margins. Usually the first occurrence of a translated technical term in a given article is accompanied by the original German term in double brackets, for example: initial segment [[Abschnitt]]. This is not done for certain ubiquitous terms. The "set" words always receive the same translation: *Menge = set*, *Inbegriff = aggregate*, and *Gesamtheit = totality*; the term *Mächtigkeit*, which Cantor appropriated from the geometer Jacob Steiner, is always translated as *cardinality*. Double brackets are also used to enclose clarifying information. In the introductions, I have freely used standard set theory abbreviations: CH = continuum hypothesis; GCH = generalized continuum hypothesis; AC = axiom of choice; ZF (ZFC) = Zermelo-Fraenkel set theory (Zermelo-Fraenkel set theory plus the Axiom of Choice).

Through the miracle of LaTeX 2_ε, I have attempted to maintain the look of the originals, even making the font sizes in proper names a little larger or adding space between their letters as was the custom of the times. I have duplicated some of Hausdorff's one of a kind notations and have continued his practice of eliding the letter "i" or the letter "j" (but not both) when itemizing a list by letters. The organization into paragraphs is exactly as in the originals; however, sentences have been punctuated to reflect English usage. Minor and obvious misprints, which were amazingly very few in number, are corrected without comment. I have kept the original indexing style for footnotes (numerals or asterisks) and, where possible, the same values for indices; any deviations from this are noted by placing the original designation in double brackets within the current footnote. Notes from the translator are rare, but when they occur, they are indexed by small roman letters.

Acknowledgements

In preparing my essays, I have benefited from scholarly works emanating from Germany, in particular the books edited by Eugen Eichorn and Ernst-Jochen Thiele ([EiT 1994]) and by Egbert Brieskorn ([Br 1996]).[17] These volumes are still both important and useful. But currently the greatest boon to Hausdorff scholarship is (or will be) the Hausdorff Edition, a project undertaken by a consortium of scholars to publish the collected works of Hausdorff-Mongré. This nine-volume edition is also to include excerpts from his *Nachlaß*, along with editorial essays and commentary. Three volumes have already appeared. Of particular importance to me during the preparation of this book was the publication of the Hausdorff Edition's volume containing *Grundzüge der Mengenlehre*. It was the first volume published (designated officially as volume II), and I profited greatly from reading Walter Purkert's introductory essay; his essay is available on-line ([Pu 2002]). It was also very helpful to have the catalog (Findbuch) of Hausdorff's *Nachlaß* prepared by Walter Purkert; it too is available on-line ([Pu1995]).[18]

Sources in English that mention Hausdorff, other than in passing, are rather sparse. An exception is [Mo 1982], in which the works in the period 1904–1909 that we have translated are covered. I found Gregory Moore's book very helpful, not only for his commentary on specific articles, but also for the rich portrait that he paints of an important era. Moore has also written a short article on Hausdorff highlighting the years 1901–1908 ([Mo 1996]).

In a more personal vein, I want to thank Marion Scheepers whose participation came at a crucial time. His enthusiasm for this project provided the will to pursue what seemed like a daunting task. I also want to thank

---

[17] The first appraisals of Hausdorff's life and work appeared in Germany twenty-five years after his death with the traditional, but belated, essays of remembrance in the *Jahresbericht* of the **DMV** [Di 1967], [Lo 1967]. No doubt the necessity to acknowledge the circumstances that forced Hausdorff from his professorship and eventually led him to choose suicide had been a difficult psychological barrier to societal recognition of this great German mathematician.

[18] The miraculous survival of the Second World War by Hausdorff's scientific *Nachlaß* is recounted in [Bo 1967] and [Ber 1967]; also see Bergmann's article in [Br 1996, 271–281]. The forward to [Pu 1995] provides a complete history of the *Nachlaß*.

my colleagues at Michigan State University: Susan Schuur, Christel Rotthaus and Martin Fuchs in the Department of Mathematics; and Elizabeth Mittman in the German Department. Professor Schuur read my essays and made helpful comments and suggestions. Professors Rotthaus and Fuchs provided emergency help when I was overcome by a German language conundrum, and Professor Mittman allowed me to attend her classes and be a student of German once more.

As with any undertaking of long duration, there were peaks and valleys, and I gratefully acknowledge the debt I owe to friends who offered their encouragement and provided a sympathetic ear when I needed it. Foremost among these is my wife and colleague Susan Schuur.

It goes without saying that any of this volume's shortcomings are my responsibility and no one else's.

J.M. Plotkin
East Lansing, 2004

Selected Hausdorff Bibliography

[H 1901b] *Über eine gewisse Art geordneter Mengen*, Berichte über die Verhandlungen der Königlich Sächsischen Gesellschaft der Wissenschaften zu Leipzig, Mathematisch-Physische Classe **53** (1901), 460–475.

[H 1904a] *Der Potenzbegriff in der Mengenlehre*, Jahresbericht der Deutschen Mathematiker-Vereinigung **13** (1904), 569–571.

[H 1906b] *Untersuchungen über Ornungstypen I, II, III*, Berichte über die Verhandlungen der Königlich Sächsischen Gesellschaft der Wissenschaften zu Leipzig, Mathematisch-Physische Klasse **58** (1906), 106–169.

[H 1907a] *Untersuchungen über Ornungstypen IV, V*, Berichte über die Verhandlungen der Königlich Sächsischen Gesellschaft der Wissenschaften zu Leipzig, Mathematisch-Physische Klasse **59** (1907), 84–159.

[H 1907b] *Über dichte Ordnungstypen*, Jahresbericht der Deutschen Mathematiker-Vereinigung **16** (1907), 541–546.

[H 1908] *Grundzüge einer Theorie der geordneten Mengen*, Mathematische Annalen **65** (1908), 435–505.

[H 1909a] *Die Graduierung nach dem Endverlauf*, Abhandlungen der Königlich Sächsischen Gesellschaft der Wissenschaften zu Leipzig, Mathematisch-Physische Klasse **31** (1909), 295–334.

[H 1914a] *Grundzüge der Mengenlehre* (Leipzig: Veit); reprinted (New York: Chelsea) 1949, 1965, 1978; reprinted in *Felix Hausdorff, Gesammelte Werke, Band II, Grundzüge der Mengenlehre*, E. Brieskorn et al. (eds.) (Berlin-Heidelberg-New York: Springer-Verlag), 2002.

[H 1927a] *Mengenlehre* (Leipzig: de Gruyter); reprinted (New York: Dover), 1944.

[H 1936b] *Summen von $\aleph_1$ Mengen*, Fundamenta Mathematicae **26** (1936), 241–255.

Introduction to "About a Certain Kind of Ordered Sets"

In the summer semester of 1901 at the University of Leipzig, Hausdorff gave one of the first courses devoted entirely to set theory. Officially, he had three students.[1] The handwritten notes for his lectures on *Mengenlehre* are found in his *Nachlaß* [Pu 1995, 18]. In these notes, an annotation written in red pencil appears in the top margin of page 37:

> (a) by oral communication from Cantor, (b) given by me 27.6.1901. Dissertation of F. Bernstein received 29.6.1901.
> [Pu 2002, 8]

The designations (a) and (b) refer to the two directions that result in the calculation of the cardinality of the set of all countable order types. In his oral communication, Cantor had (most likely) told Hausdorff of his construction of a family of countable order types of cardinality $2^{\aleph_0}$, showing that the set of countable order types has at least the cardinality of the continuum (part (a)). On June 27, 1901, Hausdorff proved the complementary result that the set of countable order types has cardinality at most that of the continuuum (part (b)). On June 29, 1901, Hausdorff received a copy of Felix Bernstein's Göttingen Inaugural-Dissertation ([Berns 1901]).

Bernstein, a protégé of Cantor's, was one of Hilbert's early doctoral students.[2] His dissertation contained Cantor's example and two proofs of the complementary result,[3] thereby establishing (by the Schröder-Bernstein Equivalence Theorem) what in [H 1901b, 460] Hausdorff calls the Cantor-Bernstein Theorem: The cardinality of the second type class, the class of countable order types, equals that of the continuum.

That the cardinality of the set of countable order types is $2^{\aleph_0}$ was of immediate interest. Trivially contained in this set are the order types of countable, well-ordered sets, i.e., the countable ordinals, the prototype set of cardinality $\aleph_1$. So the countable order types provide a *natural* example showing that $\aleph_1 \leq 2^{\aleph_0}$. Before this, in the absence of any knowledge that the continuum is well-orderable (an aleph), it was not even known whether $\aleph_1$ and $2^{\aleph_0}$ were comparable [Sch 1908, 32n3].[4]

Bernstein himself felt that the significance of this result lay in "the completely parallel definition of both sets [the set $\aleph_1$ and the continuum]" as sets of order types, thereby giving "the numbers of the second number class a concrete geometric interpretation" [Berns 1901, 16; 1905d, 120]. He also stated that "the attainment of such a relationship between the real numbers and the order types, both sets appearing as systems of elements of equal character, is a precondition for the solution of this question [whether the continuum's cardinality is greater than $\aleph_1$ or equal to $\aleph_1$]"[Berns 1901, 34; 1905d, 138].[5]

We do not know how Hausdorff reacted to the news that Bernstein had earlier completed the calculation of the cardinality of the set of countable order types. However, by December 12, 1901, his first paper in set theory, *About a Certain Kind of Ordered Sets* [[*Ueber eine gewisse Art geordneter Mengen*]], was accepted for publication. In it, he singles out a new class of order types and uses them to generalize the Cantor-Bernstein Theorem. These types arise from ordered sets in which no two distinct initial segments are similar. (All well-ordered sets have this property.) Hausdorff calls these sets and their types *gestufte* (which we have translated as *graded*). Significantly for Hausdorff, the countable graded types are an intermediate class wedged between the countable ordinals and the class of countable order types.

Hausdorff's short, dense introductory paragraph begins with acknowledgment of Cantor as the founder of the field of order types and the statement that outside of the ordinal numbers "extremely little is known about general types." Saying that more detailed knowledge [of order types] and questions of type classification belong to the "narrower problem area of set theory," Hausdorff writes:

> [B]ecause, possibly in this way the old question about the cardinality of the continuum can be brought closer to a solution.

In spite of his cautious "brought *closer* to a solution," Hausdorff seems to be joining the hunt for a solution to the *continuum problem*. Lending more credence to this interpretation, he then cites the Cantor-Bernstein Theorem and states:

> [I]t might be important for the comparison of $\aleph$ and $\aleph_1$ to determine possible intermediate stages, i.e., from the totality of all countable [ordered] sets to choose a narrower group that, for its part, contains the totality of all countable well-ordered sets as a component.[6] In particular, for Cantor's conjecture that $\aleph = \aleph_1$, this method promises a facilitation of its proof through the interposition of intermediate stages; while conversely, this way might be more of a detour should it be the case that $\aleph > \aleph_1$.

One can only speculate as to what Hausdorff means in these sentences; he provides no further elaboration. How could the proof of $\aleph = \aleph_1$ be facilitated by the insertion of an intermediate class (say M)? Might it be that with the *right* M proving the equalities $\aleph_1 = \overline{\overline{M}}$ and $\aleph = \overline{\overline{M}}$ would be easy (or at least easier). And why in the case of $\aleph > \aleph_1$ would this approach be more of a detour? Could not the *right* M facilitate the establishment of the inequality $\aleph_1 < \overline{\overline{M}}$ or the inequality $\overline{\overline{M}} < 2^{\aleph_0}$?

In what follows, although Hausdorff does produce an intermediate class, it sheds no light on the continuum problem. The introduction ends with a clear caution to the reader:

> For the time being, it still cannot be determined whether the most appropriate starting point for the exploration of types and for the further elucidation of the cardinality question lies precisely here.

As acknowledged by its author, [H 1901b] is not a contribution to solving the continuum problem. Taken for what it is, [H 1901b] is a small gem of tight, clear exposition, which, with the exception of two speculative sentences, has little to say about the continuum problem.

Hausdorff starts in §1 with important examples of graded types. He uses one such family to establish that the countable graded types have cardinality $2^{\aleph_0}$. The section ends with the observation that the graded sets "belong to the category, considered on occasion by Cantor, of those sets that are similar to themselves in one way." This seems to show Hausdorff's familiarity with the "unpublished" *Acta Mathematica* paper of Cantor where such types are briefly considered [Gr 1970, 89–90].

In §2, an efficiently laid out sequence of theorems establishes that the graded sets (types) are closed under addition and multiplication. The proofs have an almost visual immediacy; an important technique is the tracking of specific elements under (iterated) similarities, which is best comprehended with pencil and paper in hand.

In §3, Hausdorff introduces the subdomain of *forwards and backwards graded* sets, those graded sets whose inverses are also graded. He states a representation theorem for such sets; the infinite ones are dense sums of (suitably chosen) finite sets. (This foreshadows his important result that an ordered set is either scattered or the dense sum of scattered sets [H 1908, 457–458].) Hausdorff shows how to construct forwards and backwards graded sets having the cardinality of the continuum and having the cardinality $\aleph_1$. For the latter, he uses Bernstein's *ultra-rational numbers*, the set of finitely non-zero ω-sequences of countable ordinals ordered by first differences, which provides him with a dense set of cardinality $\aleph_1$ for his construction.[7]

Bernstein stated that his proof [of Cantor-Bernstein] was easily extendible to "higher alephs" [Berns 1901, 40; 1905d, 143]. At the time, the role of the alephs in the world of cardinals was unsettled. Although in his introduction Hausdorff was willing to believe that the continuum's cardinality was an aleph (in anticipation of a promised proof by Cantor), for his final theorem in [H 1901b] he became an agnostic. He offered the following generalization of Cantor-Bernstein:

> If M is a graded set whose cardinal number $\mathfrak{m}$ is equal to its own square, then all order types of cardinality $\mathfrak{m}$ form a set of cardinality $2^{\mathfrak{m}}$. [H 1901b, 473]

The alephs have the arithmetic property in the hypothesis and each initial ordinal ω_α is a graded set of cardinality $\aleph_\alpha$. But if not every cardinal is an aleph, Hausdorff has staked out his claim to a generalization of Cantor-Bernstein, in fact a generalization of Bernstein's own generalization.

Hausdorff precedes the proof of his theorem with a tutorial on $2^{\mathfrak{m}}$. In [Can 1892, 77], Cantor introduced his famous diagonal argument and used it to prove that the set of $\{0, 1\}$-valued functions on the unit interval $[0, 1]$ has higher cardinality than $[0, 1]$ itself. Cantor did not identify these $\{0, 1\}$-valued functions with subsets of the unit interval. Hausdorff is the first to prove that the cardinality of the set of all subsets of a set M is greater than the cardinality of M.[8] He does this by introducing the idea of characteristic function, i.e., he identifies a $\{-1, 1\}$-valued function on the set M with a subset of M, thus interpreting $2^{\mathfrak{m}}$ as the cardinality of the set of all subsets of M. In this context, he essentially repeats Cantor's diagonal argument to prove that $2^{\mathfrak{m}} > \mathfrak{m}$.

Hausdorff goes on to derive the fact that the set S of subsets of M having cardinality $\mathfrak{m}$ has cardinality $2^{\mathfrak{m}}$. His derivation assumes that $\mathfrak{m} + \mathfrak{m} = \mathfrak{m}$ holds for every infinite cardinal. The alephs satisfy this arithmetic condition. However, $\mathfrak{m} + \mathfrak{m} = \mathfrak{m}$ need not hold in the absence of AC.[9]

To show that the set of order types of cardinality $\mathfrak{m}$ has cardinality $\leq 2^{\mathfrak{m}}$, Hausdorff refers to Bernstein's order functions (see Note 3), stating his intention to use them "in exactly the same way." No mention is made of his independent proof in the countable case, as remarked upon in his lecture notes; nor do we know how his proof differed from the two given by Bernstein—if it differed at all.

Hausdorff uses the hypothesis $\mathfrak{m}^2 = \mathfrak{m}$ in calculating the size of the set of order functions. To establish the complementary inequality, Hausdorff shows the existence of a family of $2^{\mathfrak{m}}$ non-isomorphic order types of cardinality $\mathfrak{m}$. For each m in the graded set M, A_m denotes the initial segment of M determined by m. For S a subset of M of cardinality $\mathfrak{m}$, Hausdorff considers the ordered sum over S of the ordered sets $R + A_s, s \in S$, where R is the set of rational numbers. He concludes with a (too?) brief argument that different S give rise to distinct order types.

Although Hausdorff expressed the hope of continuing his "present investigation" on a later occasion ([H 1901b, 473]), the graded types quickly passed from the scene. In [Sch 1908, 41–42], a section is devoted to them in the chapter on "ordered sets," and in [Sch 1913, 208–209], they share the stage (and a section) with Bernstein's "ultracontinuum." In Hausdorff's later papers and in [H 1914a], they are completely absent. The graded types were an intriguing dead end.

Notes

1. The number enrolled appears in the draft of a report written in November 1901 by H. Bruns to the State Minister of Education, requesting that Hausdorff be granted the title of "ausserordentlich Professor" at Leipzig (reprinted in [BeP 1987, 231–234]). Hausdorff was the second person to give such a course. E. Scholz ([Scho 1996, 107]) and G. H. Moore ([Mo 1996, 125–126]) report that Zermelo gave a set theory course at Göttingen in the winter semester of 1900–1901. The cited section in Moore is entitled *Early Courses In Set Theory*. In it, Moore tells of courses by Hurwitz, Borel, and Hilbert given in the 1890s that had substantial set theory in them and of courses given by Zermelo, Hausdorff, and Landau (in that order) in the period 1900-1902 that were the first courses devoted entirely to set theory. Surprisingly, in a teaching career at Halle of over forty years, Cantor never dedicated a course to set theory [BeP 1987, 200].

    The draft report cited above also reveals the impingement of anti-Semitism on Hausdorff's career. The document has an addendum that is remarkable for its candor. In it, the Dean of the Faculty Kirchner explains that in the Faculty's vote of 22 "for" and 7 "against" Hausdorff's appointment the "against" vote is attributable to Hausdorff's "mosaischen Glaubens."

2. Felix Bernstein (1878–1956) was set theory's *wunderkind*. His father was professor of physiology at Halle. As a gymnasium student, Bernstein participated in Cantor's Halle seminar. In winter 1896/1897, while proofreading a paper of Cantor, he found a proof for Cantor's conjectured Equivalence Theorem. Bernstein presented his proof to Cantor's seminar in 1897 [Fre 1981, 84–86]. This result is now known as the Schröder-Bernstein Theorem, even though Schröder's independent 1898 proof was found to have a significant gap (see [Kor 1911]). Dedekind had a proof of the Equivalence Theorem in 1887 that was unpublished until 1932. See [Fr 1968, 77n] or [Ka 2004, §4] for more on the history of proofs of the Equivalence Theorem, and see [Fre 1981] for more about Bernstein's life and work.

    Bernstein defended his Göttingen dissertation, *Untersuchungen aus der Mengenlehre*, in March 1901. It was reprinted in 1905 in the *Mathematische Annalen* with minor additions, deletions, and rearrangements. We will encounter this dissertation again in [H 1904a; 1906b].

3. In [Berns 1901, 7–11; 1905d, 134–138], Bernstein gives Cantor's previously unpublished example: one assigns to each dyadic sequence $(\mu_1, \mu_2, \mu_3, \ldots)$ the countable order type $\mu_1 + (\omega^* + \omega) + \mu_2 + (\omega^* + \omega) + \mu_3 + \cdots$. Distinct sequences produce distinct order types; so the set of countable order types has cardinality $\geq 2^{\aleph_0}$.

    Bernstein's first proof that the set of countable order types has cardinality $\leq 2^{\aleph_0}$ uses what he calls order functions. Following his own dictum that "all concepts of set theory can be reduced to the following three concepts, element, system, mapping (function)," Bernstein views a countable ordered set M as a mapping f from M^2 to the set $\{-1, 0, 1\}$, where $f(a,b) = -1$ for $a > b$, $f(a,b) = 1$ for $a < b$, and $f(a,b) = 0$ for $a = b$ [Berns 1901, 35–39; 1905d, 138–142]. Order functions satisfy additional properties, but it is clear that the cardinality of the set of all functions from pairs of natural numbers to the set $\{-1, 0, 1\}$ is an upper bound for the cardinality of the set of countable order types. Bernstein calculates this upper bound to be $2^{\aleph_0}$, first by imagining the integer lattice points enumerated in a single sequence and then by interpreting the order functions f as triadic representations of real numbers.

Bernstein's second proof relies on Cantor's characterization of the order type η of the rationals as the order type of any countable, densely ordered set without end points [Can 1895, 504–507]. Bernstein *blows up* an arbitrary countable ordered set (order type) μ by placing copies of the rationals to the left and right of it and in every one of its empty intervals. The *blow-up* then has type η by Cantor and the original μ is isomorphically embedded in η. The needed upper bound is then the cardinality of the set of all subsets of the rationals, namely $2^{\aleph_0}$ [Berns 1901, 40–42; 1905d, 143–145].

As far as I know, Bernstein's second argument contains the first published proof of the universality of the rationals with respect to countable, simply ordered sets.

From our vantage point, thanks to the work of W. Sierpiński, we now know that a formal proof of the Cantor-Bernstein Theorem necessarily uses a form of AC. Both of Bernstein's proofs implicitly use AC to calculate the number of countable order types (i.e., to calculate the number of equivalences classes under similarity formed from set of cardinality $2^{\aleph_0}$) [Si 1965, 242–243].

4. Any formal proof that $\aleph_1 \leq 2^{\aleph_0}$ is known to require some sort of choice principle [Si 1965, 374–376].

5. In the introduction to his thesis, Bernstein writes: "At present two problems within set theory are in the forefront of interest. One relates to the continuum, i.e., the set consisting of all real numbers, the other relates to the foundations of set theory." And with respect to the continuum, the problem is: "how many classes [under cardinal equivalence] of this kind can one form from the real numbers?" Bernstein designates this to be "Cantor's continuum problem [das Cantorsche Continuumproblem]" [Berns 1901, 13; 1905d, 118]. According to G. H. Moore, this is the first appearance in print of the phrase *continuum problem* [Mo 1982, 56].

6. For the moment, Hausdorff uses a generic aleph to designate the cardinality of the continuum "since G. Cantor intends to publish a proof soon that each transfinite cardinal number must occur in the 'sequence of alephs'" [H 1901b, 460n2]. This footnote and the cited annotation in his lecture notes, "by oral communication from Cantor," give direct evidence of Hausdorff's mathematical interaction with Cantor.

7. These ultra-rationals appear in [Berns 1901, 52], and they are a subset of the set of all ω-sequences of countable ordinals that Bernstein orders by first differences. This is perhaps the first instance of a *covering set* (in Cantor's terminology) being ordered by the principle of first differences. Ordering covering sets by first differences is taken up by Hausdorff in [H 1904a] and is a major technical device in [H 1906b; 1907a; 1908].

8. In [Fer 1999, 306], José Ferreirós credits Bertrand Russell with being the first mathematician to rework Cantor's 1892 proof to establish that the set of all subsets of a set has cardinality greater than the set itself. He refers to §§346–347 of Russell's 1903 book *Principles of Mathematics*.

9. *For all infinite cardinals* $\mathfrak{m}$, $\mathfrak{m} + \mathfrak{m} = \mathfrak{m}$ is a theorem of ZFC; it cannot be proved in ZF, and it is weaker than AC [Sa 1982]. *For all infinite cardinals* $\mathfrak{m}$, $\mathfrak{m}^2 = \mathfrak{m}$ is equivalent to AC [Ta 1924].

About a Certain Kind of Ordered Sets

F. Hausdorff

In the field of order types that was developed by G. CANTOR, it is really only the special realm of ordinal numbers about which we are somewhat well informed; extremely little is known about general types, the types of non-well-ordered sets. Certainly a more detailed knowledge and a classification of types belongs to the narrower problem area of set theory: because, possibly in this way, the old question about the cardinality of the continuum can be brought closer to a solution. Since by a CANTOR-BERNSTEIN Theorem[1]) the cardinality of the second *type* class (the class of all countable types) is equal to the cardinality $\aleph$ of the continuum[2]), whereas the cardinality of the second *number* class is named by $\aleph_1$, it might be important for the comparison of $\aleph$ and $\aleph_1$ to determine possible intermediate stages, i.e., from the totality of all countable sets to choose a narrower group that, for its part, contains the totality of all countable well-ordered sets as a component. In particular, for CANTOR's conjecture that $\aleph = \aleph_1$, this method promises a facilitation of its proof through the interposition of intermediate stages; while conversely, this way might be more of a detour should it be the case that $\aleph > \aleph_1$. In what follows, one such particular group of order types, in which the ordinal numbers are contained as a special case, is characterized by some main attributes; for the time being, it still cannot be determined whether the most appropriate starting point for the exploration of types and for the further elucidation of the cardinality question lies precisely here.

§1

Comments. In all the following considerations, we shall always speak of *ordered sets* in the sense of their *order types*, i.e., we shall think of their elements, whose individual properties do not matter, as replaced by neutral units that retain their order relations. Since we will not be able to avoid designating individual elements, for example, in order to easily take in the possible exchange of elements for groups of elements, in general we want to retain the ordered sets M themselves instead of their order types $\overline{M}$ and to operate on them according to the rules laid down by G. CANTOR for order

---

[1]) F. BERNSTEIN, Untersuchungen aus der Mengenlehre, Diss. 1901.
[2]) Since G. CANTOR intends to publish a proof soon that each transfinite cardinal number must occur in the "sequence of alephs," there is less reason to avoid this temporary notation (aleph without an index) as prejudicial.

types. Thus if $M = \{m\}$ and $N = \{n\}$ are two ordered sets, the *sum* $M + N$ denotes that ordered set in which the elements m have retained among themselves their ordering in M, the elements n have retained among themselves their ordering in N, while each element m precedes each element n in the ranking ($m \prec n$, $n \succ m$). The addition defined here is associative but not commutative. If, in this sense, $M = P + R$ or $M = P + Q + R$, then the sets P, Q, and R are called *pieces* [[*Stücke*]] of the set M, and clearly P is an initial piece [[Anfangsstück]], Q is a middle piece [[Mittelstück]], and R is an end piece [[Endstück]]. In a partition into two pieces, if the end piece R has a *first element* m, call R the *remainder* [[*Rest*]] determined by m, and call the corresponding initial piece A ($M = A + R$) the *initial segment* [[*Abschnitt*]] of the set M determined by m; thus the initial segment consists of all elements that precede the given element m, and the remainder consists of m itself and all the elements that follow it. In addition, by the *product* $M \cdot N$ of two ordered sets, we understand the ordered set that results when one "substitutes" for each element n of N a set M_n of type of the set M (a set M_n "similar" to the set M, $M_n \simeq M$). This multiplication is associative but not commutative, and in relation to addition, it is one-sidedly distributive; $M(P+Q) = MP + MQ$, but $(P+Q)M \neq PM + QM$. By the "substitution" (which in this case is not limited to only similar sets M_n) of the ordered sets M_n for the elements n of N is understood that ordering by virtue of which each element of M_n has the same order relationship to each element of $M_{n'}$ as n has with respect to n'.

462   *Definition. A set is called graded* [[*gestuft*]] *if none of its initial segments are similar to one another. The order type of a graded set is called a graded order type.*

We also immediately note that no initial segment is similar to the set itself. Of course, were it the case that $M \simeq A$, where A is the initial segment of M belonging to the element m, then the element m would be sent to an earlier element m' by the similarity mapping of M onto A in such a way that the corresponding initial segments Λ and Λ' would be similar. Thus the types $\overline{A}$ are all different from each other and from $\overline{M}$.

It will be desirable first of all to convince ourselves of the *existence* of graded sets by examples. As a rule, in the proof that any defined set is graded, we will first establish that M and A are not similar because usually the non-similarity of different initial segments can be concluded in analogous ways — meanwhile, taken logically, the first part of this proof scheme is not sufficient, and by execution of the second part, it is not even necessary.

Example 1. Each *well-ordered* set is graded. The types $\overline{A}$ of its initial segments are the finite and transfinite ordinal numbers $0, 1, 2, \ldots, \omega$, $\omega + 1, \ldots$, ordered by magnitude, concluding with the ordinal number $\overline{M}$.

Example 2. Consider the set of rational numbers $R = \{r\}$ *ordered by magnitude*; as G. Cantor[1]) has shown, its type η is countable, everywhere

---

[1]) Math. Ann. Bd. 46 (1895), p. 481.

dense [[überall dicht]], without a first and last element, and uniquely determined by these properties. In this set, for each element r fix a *finite* set Q_r of ν elements in such a way that to each r there corresponds one and only one integer ν. This can happen in infinitely many ways; for if one puts R into a simply ordered sequence $r_1\, r_2\, r_3 \cdots r_\nu \cdots$, one need only assign to each $r = r_\nu$ its position index or, as well, assign to each $r = r_\nu$ the integer a_ν, provided that $a_1\, a_2\, a_3 \cdots a_\nu \cdots$ denotes any increasing sequence of positive integers ($1 \leqq a_1 < a_2 < a_3 < \cdots$). A countable graded set $M = \{Q_r\}$ results from the above substitution.

The proof is easy. Let m be any element of M, A the corresponding initial segment. Now m must belong to one of the finite sets Q, say to Q_r. In the set R, take any element s following r ($s \succ r$); surely there is such an element since R has no last element. The group Q_s belonging to s consists of some number of elements, say five. Now clearly groups of 5 elements following one after the other will also occur in A, but none of them stands in the same ordering relations to its neighbors as Q_s stands in respect to its; for example, the first element has no immediate predecessor, and the fifth element has no immediate successor. The set Q_s itself is the only such free-floating group of five elements, and therefore M cannot be similar to the initial segment A. Completely analogous reasoning shows that two different initial segments A and A' cannot be similar.

In this way, a very comprehensive category of countable graded sets is established, sets which arise from the substitution of *different finite* sets in the everywhere dense type η. For each of these sets M, the *inverse* M^* that arises from M by inverting all order relations is also graded.

A generalization of this method through which one arrives at graded sets of higher cardinality consists of inserting *different well-ordered* sets in place of individual elements in an everywhere dense type.

Example 3. Well-ordered sets can also be substituted in the type ω^* of the descending numerical sequence $\ldots, 3, 2, 1$ so that a graded type results. For instance, consider the well-ordered sets whose ordinal numbers are $\omega, \omega^2, \omega^3, \ldots, \omega^\nu, \ldots$. If one adds these from left to right, one obtains $\omega + \omega^2 + \omega^3 + \cdots$ in inf. $= \omega^\omega$, again an ordinal number. However, if one joins them in the sense of descending powers, one obtains the type

$$\cdots + \omega^\nu + \cdots + \omega^3 + \omega^2 + \omega,$$

which clearly is not well ordered; however it is graded.

The proof essentially depends on a property that belongs to the powers of ω and to no other numbers of the second number class, namely, each *remainder* of a set of type ω^ν has the same type ω^ν. Now let Q_ν be a set of type ω^ν, and contrary to hypothesis, let $M = \cdots + Q_3 + Q_2 + Q_1$ be similar to the initial segment A determined by the element m; denote the occurring order relations preserving mapping of M onto A by $A = \varphi(M)$. Under the mapping, m itself goes to $m' = \varphi(m)$, and it is the case that $m' \prec m$; the image of m' is again an element $m'' = \varphi(m') \prec m'$, etc.

So there arises a decreasing sequence of elements $m \succ m' \succ m'' \succ \cdots$. Now, for instance, if m belonged to the well-ordered set Q_μ, since in a well-ordered set any decreasing sequence of elements is finite, there must occur a *last* element n in the sequence $m\,m'\,m''\cdots$ that still belongs to Q_μ, whereas the immediately following element $\varphi(n)$ already belongs to an earlier set Q_ν ($\nu > \mu$). However, this leads to a contradiction; for as the successive mappings show, the *remainder* of M belonging to m is similar to the aggregate of all elements between m' and m (including m', excluding m) or to those between m'' and m' or finally to those between $\varphi(n)$ and n. That remainder was of the type $\omega^\mu + \cdots + \omega^2 + \omega$ and this ordinal number is smaller than ω^ν; the interval between $\varphi(n)$ and n, however, begins with a remainder of Q_ν and thus has the type $\omega^\nu + \omega^{\nu-1} + \cdots > \omega^\nu$. Thus both sets cannot possibly be similar. Hence $M \simeq A$ is refuted. The non-similarity between different initial segments of M is proved in an analogous way.

Here also, a generalization is easily possible. If $\alpha_1\,\alpha_2\,\alpha_3\,\cdots$ denotes an increasing sequence of numbers (of the first or second number classes), then

$$\cdots + \omega^{\alpha_\nu} + \cdots + \omega^{\alpha_3} + \omega^{\alpha_2} + \omega^{\alpha_1}$$

is a graded type of the first infinite cardinality. Moreover, the monotone growth of the α is unnecessary; it suffices that for each α_μ there should exist an $\alpha_\nu > \alpha_\mu$, $\nu > \mu$.

The examples already given allow the question of the *cardinality* of the set of all countable graded types to be answered; this cardinality is equal to $\aleph$, the cardinality of the set of *all* countable types or the cardinality of the continuum. On the one hand, namely by the CANTOR-BERNSTEIN Theorem, the sought-after cardinality is $\leq \aleph$; on the other hand, since one can assign simultaneously to each positive increasing sequence of integers $a_1\,a_2\,a_3\,\cdots$ a graded type (for instance, of the third example class) and a real number $x = (\frac{1}{2})^{a_1} + (\frac{1}{2})^{a_2} + (\frac{1}{2})^{a_3} + \cdots$ of the interval $0, 1$, there exists a subset of countable graded types that possesses the cardinality of the continuum. Consequently, the sought-after cardinality is $\geq \aleph$, and thus $= \aleph$ by the SCHRÖDER-BERNSTEIN Equivalence Theorem.

In addition, I note that the graded sets belong to the category, considered on occasion by CANTOR, of those sets that are similar to themselves in only one way, whereas conversely, a set of this category is not necessarily graded (e.g., the inversion of any well-ordered set).

§2

We now prove several propositions about graded sets.

(*a*) In any graded set, no *initial piece* is similar to any other initial piece or to the set itself.

If it were the case that $P + Q \simeq P$, then under the mapping of $P + Q$ onto P, to any element q of Q there would have to correspond an element p of P, thus $p \prec q$, and the initial segment belonging to p would have to be similar to the initial segment belonging to q, which is contrary to the definition.

(*b*) Each *piece* of a graded set is a graded set.

From (*a*) it immediately follows that this proposition is true for initial pieces. If Q is a middle piece or an end piece, P is the aggregate of all the preceding elements, and B and B' are initial segments of Q, then from $B \simeq B'$ it would follow that $P + B \simeq P + B'$, i.e., the similarity between different initial segments of M would follow.

Note that for well-ordered sets not only each piece, but in general, each subset is again well-ordered.

(*c*) A graded set is not similar to any of its *middle pieces*.

This important proposition can be proved as follows. Let $M = P+Q+R$ where P and R are not empty and $M \simeq Q$; let $Q = \varphi(M)$ denote a possible order relations preserving mapping of M onto Q. Hereby $P, Q,$ and R go to the three pieces P_1, Q_1 and R_1 of Q, and $R_1 = \varphi(R)$ is an end piece of Q that attaches to the left of R and forms an end piece $R_1 + R$ of M. Likewise, $R_2 = \varphi(R_1)$ attaches to the left of R_1, etc. Denote by

$$S = \cdots + R_3 + R_2 + R_1 + R$$

the aggregate of all elements picked up by the iteration of the mapping φ; S is an end piece of M. Consequently, by (*b*) S would have to be a graded set; however, as a result of its formation, $S = \varphi(S) + R$. Thus S is similar to an initial piece $\varphi(S)$, which contradicts proposition (*a*). Therefore it cannot be that $M \simeq Q$.

Thus in a graded set the similarities $M \simeq P, M \simeq Q$, and $P \simeq P'$ are excluded. All others *can* occur ($M \simeq R, P \simeq Q, P \simeq R, Q \simeq R, Q \simeq Q', R \simeq R'$).

(*d*) If M is a graded set and N is an arbitrary set, it is never the case that $M \simeq M + N$.

If $M = \varphi(M + N)$ were a similarity mapping, then $\varphi(M) \simeq M$ would be an initial piece of M.

(*e*) If M is a graded set and N and N' are two arbitrary sets, it follows from $M + N \simeq M + N'$ that $N \simeq N'$.

Let $\varphi(M + N) = M + N'$; the image $\varphi(M)$ of M can neither (*a*) be an initial piece of M nor (*d*) be a set that contains M as an initial piece.[a] Consequently, $\varphi(M) = M$ and $\varphi(N) = N'$. Thus $N \simeq N'$.

(*f*) *If M and N are graded sets, then $M + N$ is also a graded set.*

Let A be an initial segment of M, and let B an initial segment of N. The initial segments of $M + N$ are of the form A or M or $M + B$. By (*d*) it cannot be that $M \simeq M + N$. Since A itself is a graded set, it also cannot be that $A \simeq M + N = A + R + N$. Finally, it cannot be that $M + B \simeq M + N$, since from (*e*) it would follow that $B \simeq N$. The set N, however, is not similar to any of its initial segments. Thus $M + N$ is not similar to any of its initial segments. Since B is also graded, $M + B$ is likewise not similar to any of its initial segments; the same holds for M and for A. Thus none of the initial segments of $M + N$ is similar to any other.

---

[a] Instead of (*d*), the erroneous back-reference (*e*) appears in the original.

Also, the addition of arbitrarily many, *finite* in number, graded sets again leads to a graded set. By comparison, the sum of a *transfinite* set of graded sets will be a graded set only under a specified configuration. In this connection, we note the proposition:

(*g*) A *well-ordered* set of graded sets is also a graded set.

Let $S = \{s\}$ be a well-ordered set; the set $M = \{Q_s\}$ arises by putting the graded set Q_s in place of the element s. We assume that M is similar to its initial segment A determined by the element m; let $A = \varphi(M)$ be the similarity mapping under which the element m goes to $m_1 = \varphi(m) \prec m$, which in turn goes to $m_2 = \varphi(m_1) \prec m_1$, etc. In this way, we obtain an infinite decreasing sequence $m \succ m_1 \succ m_2 \succ \cdots$ of type ω^*. Let the element m belong to the set Q_s, the element m_1 to the set Q_{s_1}, etc. Then $s \succeq s_1 \succeq s_2 \succeq s_3 \cdots$. However, from a certain element on, nothing but equality signs must occur in this sequence since S as a well-ordered set contains no infinite decreasing sequences; i.e., in the sequence $m\, m_1\, m_2 \cdots$ an element n will occur that, together with all following elements

$$n_1 = \varphi(n) \prec n, \quad n_2 = \varphi(n_1) \prec n_1, \quad \ldots$$

belongs to one and the same graded set Q_t. Denote the initial segments of this set belonging to $n, n_1, n_2, \ldots$ by $B, B_1, B_2, \ldots$; furthermore, let $B = B_1 + R_1$, $B_1 = B_2 + R_2$, etc. Thus $R_2 = \varphi(R_1), R_3 = \varphi(R_2), \ldots$, and

$$T = \cdots + R_3 + R_2 + R_1$$

is a piece of Q_t and thus a graded set. However, since $T = \varphi(T) + R_1$, T is similar to its initial segment $\varphi(T)$; this contradiction shows that the assumption $M \simeq A$ is to be rejected. Thus a well-ordered set of graded sets is not similar to any of its initial segments, and since each initial segment is again such a well-ordered set of graded sets, no initial segment is similar to any other initial segment. So M is a graded set.

(*h*) If a finite number of elements of a graded set is replaced by graded sets, a graded set again results.

Consider an element m, and let $M = A + m + R$. Then by (*b*), A and R are graded. If m is replaced by the graded set Q_m, the set $A + Q_m + R$ results, which by (*f*) is likewise graded. — The set remains graded if this process is repeated a finite number of times.

By comparison, the proposition does not hold for a transfinite set of elements. For instance, in Example 2 if one replaces the *last* element of *each* group Q_r by a well-ordered (thus graded) set of type ω, then the entire group takes on the type ω, and the set $M = \{Q_r\}$ takes on the type $\omega \cdot \eta$, which is no longer graded. Hence in contrast to (*g*), and moreover, in contrast to a known property[1]) of well-ordered sets, a graded set of well-ordered or graded sets is not in general a graded set. On the other hand, it will be shown that by the substitution of *one and the same* graded set for *all* the

---

[1]) SCHOENFLIES, Die Entwicklung der Lehre von den Punktmannigfaltigkeiten, p. 41, Theorem VII.

elements of a graded set, i.e., by the *multiplication* of two graded sets, a graded set again results.

(i) Let M be a graded set and P an initial piece of M; furthermore, let S be a set that contains no piece of type ω^*, and let T be an arbitrary set. Then the set MS is not similar to any set of the form $MT + P$, and it is similar to a set of the form MT only when $S \simeq T$.

The proof depends essentially on property (c) that M cannot be similar to any of its middle pieces. From this, it follows that any piece of the set MS (produced by the substitution of $M_s \simeq M$ for each element s of S) can be similar to the set M itself only when it is distributed over *at most two immediately adjacent* sets M_s. If P_s, Q_s, R_s denote an initial piece, a middle piece, and an end piece of M_s and s_1, s ($s_1 \prec s$) are two elements of S that immediately follow each other, then, accordingly, each piece of MS that is similar to M has one of the four forms R_s, M_s, $R_{s_1} + M_s$, $R_{s_1} + P_s$. For the same reason, each piece of MS that is similar to an initial piece P of M has one of the four forms R_s, Q_s, P_s, $R_{s_1} + P_s$.

Now suppose it were the case that $MS \simeq MT + P$; let

$$MT + P = \varphi(MS), \quad MS = \psi(MT + P) = \psi(MT) + \psi(P)$$

denote the similarity mapping of both sets. Then $\psi(P)$ must have one of the last four noted forms, and as a final piece of MS, this can only be R_s; consequently, S must have a last element s, and the corresponding set M_s is partitioned into $P_s + R_s$ with $R_s = \psi(P)$. Write $S = S_1 + s$ where S_1 denotes the aggregate of all the elements of S except the last one. Then

$$MS = MS_1 + M_s = MS_1 + P_s + R_s = \psi(MT) + \psi(P);$$

consequently,

$$MS_1 + P_s = \psi(MT), \quad MT = \varphi(MS_1) + \varphi(P_s).$$

In exactly the same way, it then follows that T must have a last element t and that $M_t = P_t + R_t$, $R_t = \varphi(P_s)$, and that with $T = T_1 + t$

$$MT_1 + P_t = \varphi(MS_1), \quad MS_1 = \psi(MT_1) + \psi(P_t).$$

From this, it follows once again that S_1 must have a last element s_1, etc.; thus this never ending process requires that there should exist in S a sequence $s \succ s_1 \succ s_2 \succ \cdots$ of consecutive elements and that there should exist in T a corresponding sequence $t \succ t_1 \succ t_2 \succ \cdots$ in such a way that

$$P \simeq R_s, \quad P_t \simeq R_{s_1}, \quad P_{t_2} \simeq R_{s_2}, \quad \ldots$$
$$P_s \simeq R_t, \quad P_{s_1} \simeq R_{t_1}, \quad P_{s_2} \simeq R_{t_2}, \quad \ldots.$$

The process could only end if one time no initial piece would remain in these successive overlapping images, thus one time it would be the case that $P \simeq M$ instead of $P \simeq R$; however, this is impossible by (a).

Now since we have assumed that in S no such decreasing sequence of adjacent elements (no piece of type ω^*) should exist, the impossibility of a similarity between MS and $MT + P$ is hereby proved.

In order to prove the second part of statement (i), we assume $MS \simeq MT$. Let $MT = \varphi(MS)$ and $MS = \psi(MT)$. Now it easily follows that the image of a set M_t must again be just a set M_s. Thus the image can neither be distributed over two such sets nor be contained in one. For were $\psi(M_t) = R_{s_1}$ or $= R_{s_1} + P_s$ or $= R_{s_1} + M_s$, then the remaining initial piece P_{s_1} would again call for a remainder R_{t_1} as image, the corresponding P_{t_1} would again call for an R_{s_2}, etc.; so once more the above unbounded sequence $s \succ s_1 \succ s_2 \succ \cdots$ would appear, which is excluded by hypothesis. Therefore $\varphi(M_s) = M_t$ and $\psi(M_t) = M_s$, and thereby the elements s, t themselves are put into a one-one correspondence that leaves order relations unchanged, i.e., $S \simeq T$.

(k) *If M and N are graded sets, then the product $M \cdot N$ is also a graded set.*

If A denotes an initial segment of M and B denotes an initial segment of N, then the initial segments of the product set MN are of the form $MB+A$ or, only in the case where M has a first element, of the form MB. Now note that graded sets have the assumed property of set S in (i) of containing no piece of type ω^*; since any such piece is not graded, because of (b) it certainly cannot occur in a graded set. Consequently, MN is neither similar to any initial segment $MB + A$ nor to any initial segment MB, for $N \simeq B$ would follow from the second case. But also no initial segment of MN is similar to any other initial segment. No similarity between $MB + A$ and MB' can exist, and a similarity between MB and MB' can exist only if $B \simeq B'$. Finally, were $MB + A \simeq MB' + A'$ without A' being the image of A, then A' would have to be the image of an end piece of A (or conversely), and if P denotes the remaining initial piece of A, it would then have to be that $MB + P \simeq MB'$, which again is impossible. Thus it follows from $MB+A \simeq MB'+A'$ that A and A' are corresponding images of one another and, furthermore, that $MB \simeq MB'$, i.e., $A \simeq A'$ and $B \simeq B'$. So both the mentioned initial segments correspond to the same element. With that, our assertion is proved.

§3

By (f) and (k), the graded sets form a "field" [["Körper"]], a closed domain [[Bereich]] in relation to the basic operations (addition and multiplication). In this domain, subtraction and division can only be defined under certain conditions. If M is an initial piece of S (see (e)), N is uniquely determined by the equation $M + N = S$, whereas assuming that N is an end piece of S, the solution for M generally turns out to be multiple-valued. For instance, among well-ordered sets the equation $\mu + \omega = \omega + \omega$ is solved by $\mu = \omega, \omega + 1, \omega + 2, \ldots$. Furthermore, the equation $MN = T$, where generally speaking T is of the required form, is uniquely solvable for N (see (i)), but it is in general multiple-valued for M; e.g., among well-ordered sets the equation $\mu\omega = \omega^2$ is solved by $\mu = \omega$, also however by each transfinite number in the interval $\omega < \mu < \omega^2$.

"Subfields" [["Unterkörper"]], i.e., such subdomains [[Thielbereiche]] of the aggregate of all graded sets that are likewise invariant under addition and multiplication, can be formed in a variety of ways. The well-ordered sets furnish such an example, moreover, so do the graded sets that have first elements, and likewise, the *forwards* [[*vorwärts*]] *and backwards* [[*rückwärts*]] *graded sets*. A graded set M is called forwards and backwards graded if the inverse M^* is also graded. Now since

$$(M+N)^* = N^* + M^*, \quad (MN)^* = M^*N^*,$$

the addition and the multiplication of two forwards and backwards graded sets again lead to such a set.

There are still a few more things to say about such sets. A forwards and backwards graded set is not similar to any of its pieces (not even to end pieces); moreover, no end piece is similar to any other end piece, and no piece is similar to one of its own pieces. They must contain neither pieces of type ω nor pieces of type ω^*. If one collects all the elements that immediately follow one another into a group Q_n, this group is necessarily *finite*; and the set $N = \{n\}$ from which the forwards and backwards graded set $M = \{Q_n\}$ arises by substitution of finite groups no longer contains two consecutive elements n since the two associated groups Q_n would certainly then be combinable into a single group. Thus N is either a single element or *everywhere dense*, i.e., between any two elements of N there always lies at least one more element (thus infinitely many).

(*l*) *A forwards and backwards graded set, in case it is not finite, arises through the substitution of finite sets for the elements of an everywhere dense set.* But of course these finite sets must be suitably chosen. The surest way would be to choose them all to be *different* from one another; but then the everywhere dense set is only countable, thus similar to one of the four types η, $1 + \eta$, $\eta + 1$, $1 + \eta + 1$, where η denotes the unique, countable, everywhere dense type without first and last elements (cf. the note on p. 462).[b] Thus we again arrive at the construction of graded sets in Example 2; bring the everywhere dense set $N = \{n\}$ into the simple sequence form $n_1 \, n_2 \, n_3 \, \cdots$; consider, on the other hand, an increasing sequence of positive integers $a_1 \, a_2 \, a_3 \, \cdots$, and replace the element $n = n_\nu$ in N by the finite group Q_n of a_ν elements.

But one can also construct forwards and backwards graded sets of the cardinality of the *continuum*, whereby thus surely, the finite sets cannot all be different, for at least one must occur *infinitely* (even the cardinality of the continuum) often. Let x be any real number of the interval $0 < x < 1$, the endpoints excluded, and let $X = \{x\}$ be the linear continuum, i.e., the set of all x ordered by magnitude whose order type λ is everywhere dense too. Now the proposition holds:

(*m*) *If the elements x of X are replaced by graded sets Q_x of the first*

---

[b] The incorrect reference "p. 458" appears in the original.

infinite cardinality, no two of which are similar, there arises a graded set $M = \{Q_x\}$.

From all that has gone before, the proof needs only to be indicated. Under any mapping of M onto one of its initial segments, the image of any set Q_x must again be such a set since only the pieces Q_x and their subpieces are of the first infinite cardinality, whereas each piece properly including a Q_x has the cardinality of the continuum. Moreover, since no Q_x is similar to any other one, $M = A + R$ cannot be mapped onto A because any Q_x occurring in R has no image in A. Likewise, a similarity between initial segments A and A' is immediately excluded if their corresponding elements m and m' belong to different sets Q_x; if they belong to the same set Q_x, the similarity $A \simeq A'$ would have as a consequence the similarity of certain initial segments of Q_x and thus be contrary to the assumption that Q_x is a graded set.

If, however, in (m) one chooses for Q_x nothing but forwards and backwards graded sets, then $M = \{Q_x\}$ also becomes such a set. The hypotheses of (m) are fulfilled if, for example, one assigns to each value x, say in virtue of $x = (\frac{1}{2})^{a_1} + (\frac{1}{2})^{a_2} + (\frac{1}{2})^{a_3} + \cdots$, an increasing sequence $a_1 \, a_2 \, a_3 \cdots$ of positive integers and if one understands by Q_x the forwards and backwards graded set formed from the set of rationals R, ordered by magnitude, by the substitution of a_ν elements for $r = r_\nu$; no such Q_x is similar to any other one. The everywhere dense set N from which our graded set M arises by the substitution of *finite* groups is the product RX with order type $\eta\lambda$; both N and M are of the cardinality of the continuum.

Perhaps it is also of interest to know a forwards and backwards graded set of the cardinality $\aleph_1$ of the second number class (the second number class itself is only forwards graded, as in general are all well-ordered sets whose types are numbers of the third number class). To this goal, one has only to replace the linear continuum X of proposition (m) by an everywhere dense type $U = \{u\}$ that itself, as well as each of its pieces, has cardinality $\aleph_1$; then if the elements u are replaced by countable graded sets Q_u that are not similar to each other, a graded set $M = \{Q_u\}$ of the second infinite cardinality results. The "ultra-rationals," i.e., the finite groups of numbers of the first two number classes, given by F. BERNSTEIN (Diss. p. 52), furnish such a set. For arbitrary finite ν, let $\alpha_1 \, \alpha_2 \cdots \alpha_\nu$ be an aggregate of numbers from the first or second number classes (zero included in the first). Write this as an infinite number sequence

$$u = \alpha_1, \alpha_2, \ldots, 0, 0, 0, \ldots$$

and bring such u into the form of a simply ordered set $U = \{u\}$ as follows: let $u \prec u'$ if u and u' agree in the first $\mu - 1$ places for some finite μ and $\alpha_\mu < \alpha'_\mu$. Consider an element u'' whose first μ places agree with u while $\alpha''_{\mu+1} > \alpha_{\mu+1}$, then $u \prec u'' \prec u'$. Consequently, U is everywhere dense, and since $\alpha''_{\mu+1}$, except for an initial segment, can run through the entire second number class, the piece of U contained between u and u' is certainly not

countable. On the other hand, U has cardinality
$$\aleph_1 + \aleph_1^2 + \aleph_1^3 + \cdots = \aleph_0 \cdot \aleph_1 = \aleph_1.$$
Thus U itself and each of its pieces is of the second infinite cardinality $\aleph_1$.

A generalization of the CANTOR-BERNSTEIN Theorem that the set of all types of cardinality $\aleph_0$ (of all countable types) has cardinality $2^{\aleph_0}$ (the cardinality of the continuum) forms the conclusion of the present investigation, which I hope to enter into again on another occasion.

(n) If M is a graded set whose cardinal number $\mathfrak{m}$ is equal to its own square, then all order types of cardinality $\mathfrak{m}$ form a set of cardinality $2^\mathfrak{m}$.

For example, this holds for $\mathfrak{m} = \aleph =$ the cardinality of the continuum and, moreover, for each $\mathfrak{m}$ that occurs in the aleph sequence (BERNSTEIN had stated the theorem for this case without proof, p. 40).

Some observations about $2^\mathfrak{m}$ have to be made in advance of the proof. First, these powers denote the cardinality of all two-valued functions of the elements m (of all "coverings" of M by two values); i.e., if $f(m) = \pm 1$, the totality of all functions $f(m)$ has cardinality $2^\mathfrak{m}$. Each such function uniquely determines a certain subset P of the set M, namely, the set all elements p for which $f(p) = +1$; therefore $2^\mathfrak{m}$ is also the cardinality of the set $\{P\}$ of all subsets P of M. Incidentally, it follows very easily from this that $2^\mathfrak{m} > \mathfrak{m}$.[1]) Since each element m itself is such a P, $\{P\}$ contains a subset of cardinality $\mathfrak{m}$. However, had $\{P\}$ itself the cardinality $\mathfrak{m}$, then it could be brought into the form $\{P_m\} \sim \{m\}$. So one immediately finds a new P different from all the P_m by putting together a P out of all those elements p which do *not* occur in the corresponding P_p. So the sets P and P_m surely do not have the element m in common. — For the following, we still need the observation that the set $\{S\}$ of all those subsets that are of cardinality equal to the whole set M has cardinality $2^\mathfrak{m}$. Since each transfinite cardinal is equal to its double ($\mathfrak{m} = \mathfrak{m} + \mathfrak{m}$), one can split M into two subsets A and B of cardinality $\mathfrak{m}$; the totality of all subsets of M that consist of A and a subset of B has cardinality $2^\mathfrak{m}$ and is a part of $\{S\}$, as $\{S\}$ is again a part of $\{P\}$ of cardinality $2^\mathfrak{m}$; thus by the Equivalence Theorem $\{S\}$ has cardinality $2^\mathfrak{m}$.

The proof of theorem (n) divides itself, likewise according to the Equivalence Theorem, into two parts; one has to show that the cardinality of the type class [$\mathfrak{m}$] (the set of all types of cardinality $\mathfrak{m}$) is $\leq 2^\mathfrak{m}$ and then that it is $\geq 2^\mathfrak{m}$. The first part is settled using "order functions," exactly as was done by F. BERNSTEIN. Consider the two-valued function of element pairs $f(m,n) = \pm 1$, and stipulate that $f(m,n) = +1$ be equivalent to $m \prec n$, $f(m,n) = -1$ equivalent to $m \succ n$. Each type τ of cardinality $\mathfrak{m}$ can be characterized in infinitely many ways by such an order function of element pairs of M. On the other hand, not every function $f(m,n)$ represents such a type τ; rather, to that end, it is necessary that in general $f(n,m) = -f(m,n)$ and

---

[1]) Cf. CANTOR, Berichte d. D. M. V. I, p. 77, also SCHOENFLIES, Die Entwicklung der Lehre von den Punktmannigfaltigkeiten, p. 26, theorem VII.

that with $f(m,n) = f(n,p)$ at the same time $f(m,n) = f(m,p)$. However, if these binary and ternary relations hold, $f(m,n)$ represents a definite type τ. Accordingly, the aggregate of all functions $f(m,n)$, which is of cardinality $2^{\mathfrak{mm}} = 2^{\mathfrak{m}}$, has a subset of cardinality of the type class $[\mathfrak{m}]$; the cardinality of this type class is $\leqq 2^{\mathfrak{m}}$.

For the second part of the proof, we make use of the hypothesis that M is graded; let m be any element of M, A_m the initial segment belonging to m, and $S = \{s\}$ a subset of M equivalent to M. In addition, let R be a non-graded set, each initial piece and final piece of which is also non-graded; in particular, we choose R to be a countable, everywhere dense set of type η. We consider the ordered set $T = \{R + A_s\}$ that results from S by substituting the ordered set $R + A_s$ for each element s. The cardinality of $R + A_s$ is $\leqq \aleph_0 + \mathfrak{m} = \mathfrak{m}$; thus the cardinality of T (since S is of cardinality $\mathfrak{m}$) lies between $\mathfrak{m}$ and $\mathfrak{m}^2 = \mathfrak{m}$, i.e., T is an ordered set of cardinality $\mathfrak{m}$, and its type belongs to the type class $[\mathfrak{m}]$. Now it is to be shown that all these types τ are different from one another, i.e., the sets T and T' cannot be similar if S and S' are different. This depends on the specific structure of T, by virtue of which the pieces A_s do not intermix with their associates but stand out as *graded* pieces, whereas each piece projecting from an A_s is non-graded. Thus if two sets of the form T are similar to each other, the *whole* pieces A_s must correspond to each other as images. However, if S and S' are different, there is at least one element m which belongs to S but not to S' (or conversely); and to the piece $R + A_m$ contained in T, there corresponds nothing similar in T'. Consequently, the relationship between S and T is one to one, and the type class $[\mathfrak{m}]$ contains a set $\{T\} \sim \{\tau\}$ equivalent to $\{S\}$, i.e., a subset of cardinality $2^{\mathfrak{m}}$; the cardinality of the type class $[\mathfrak{m}]$ is $\geqq 2^{\mathfrak{m}}$. Thereby theorem (n) is completely proved.

[Declared ready to print 12/12/1901]

Introduction to "The Concept of Power in Set Theory"

In an announcement dated November 1901, which appeared on page 3 in the *Jahresbericht der Deutschen Mathematiker-Vereinigung* of 1902, the managing board of the **DMV** describe the changes they are about to make to the *Jahresbericht*. Among these, is the inauguration of feature called a *Sprechsaal*:

> A "Sprechsaal" shall serve for the communication of suggestions, comments, additions, and proposals for the improvement of works that have already appeared, and the like; however, self-contained mathematical articles shall be excluded from acceptance.

Hausdorff's *The Concept of Power in Set Theory* [[*Der Potenzbegriff in der Mengenlehre*]] appeared in the second year of the Sprechsaal's existence and was its sole entry in 1904. In keeping with the **DMV** guidelines, [H 1904a] is a research announcement. In it, Hausdorff reports on three items: I and II concern the extension of power formation to bases that are general order types and whose exponents are the inverses of ordinals (I) or ordinals (II); while III is about the already known operation of cardinal power formation and gives a recursion formula for powers of alephs where the base aleph has non-limit index.

The power formed in item I is obtained by ordering Cantor's *covering set* according to the principle of first differences, a principle that is to play a prominent role in [H 1906b].[1] Item I is illustrated by three examples: the powers 2^{ω^*} and ω^{ω^*} are identified without corroborating details[2]; the type ϑ^{Ω^*}, where ϑ is the type of the unbounded reals,[3] is said to be a type in which no fundamental sequence has a limit, a property that Hausdorff notes is also enjoyed by the "infinitary pantachie [of DuBois-Reymond]." Types such as ϑ^{Ω^*} are systematically studied in [H 1906b], and in [H 1907a], DuBois-Reymond's infinitary pantachie is judged to have been a fantasy.

The power formed in item II uses a fixed element in the base and the decomposition that it induces. Hausdorff states that his approach generalizes Cantor's definition of power for ordinals in the first two number classes [Can 1897, 231–235]. Verification of this is carried out in [H 1906b, 115–117]. In [H 1906b, 108], Hausdorff dramatically deems the definitions in items I and II to have "stood side by side, isolated and unheralded." In [H 1906b], both

I and II are subsumed under a more general concept of power formation for arbitrary order types.

Item III of [H 1904a] is the most significant historically. For powers of alephs with non-limit index μ, it ostensibly presents what in [Ta 1925] is first called "Hausdorff's recursion formula":

$$\aleph_\mu^{\aleph_\alpha} = \aleph_\mu \cdot \aleph_{\mu-1}^{\aleph_\alpha}.$$

As we shall see, the story behind this is almost operatic.

Hausdorff's formula has as its genesis a formula *proved* by Felix Bernstein in his dissertation [Berns 1901, 49–50]:

$$\aleph_\mu^{\aleph_\nu} = 2^{\aleph_\nu} \cdot \aleph_\mu.$$

The non-trivial case in Bernstein's proof is for $\mu > \nu$. In a crucial place he argues as follows:

> Let α_μ run through all numbers in the class $\{\alpha_\mu\}$ of cardinality $\aleph_\mu$. If one understands by $\overline{\alpha}_\mu$ the cardinal number belonging to α_μ, then in any case
>
> $$\aleph_\mu^{\aleph_\nu} \leq \sum_{\alpha_\mu} \overline{\alpha}_\mu^{\aleph_\nu}.$$
>
> This is because each subset of cardinality $\aleph_\nu$ formed from the elements of $\aleph_\mu$ can be found, in accordance with equation [*sic*] 7 [the inequality $\aleph_\mu > \aleph_\nu$], in an initial segment of the well-ordered set of numbers $\{\alpha_\mu\}$.[4]

Unfortunately, this last assertion is not true for $\mu = \omega$ and $\nu = 0$. Its untruth was brought to public notice in a spectacular fashion.

At the Third International Congress of Mathematicians held at Heidelberg in August, 1904, the Budapest mathematician Julius König delivered a lecture that shook the set theory community. Using Bernstein's formula for $\nu = 0$ together with own his result that $\mathfrak{s}^{\aleph_0} > \mathfrak{s}$ for any $\mathfrak{s}$ that is the sum of an ω-sequence of increasing cardinal numbers (König's lemma), König proved that the cardinality of the continuum is not an aleph; i.e., the continuum is not well-orderable.[5] The twin dogmas of Cantorian set theory, that every set could be well-ordered and that $2^{\aleph_0} = \aleph_1$, seemed to be overturned at one time.

J. W. Dauben and G. H. Moore report that within less than a day of König's lecture Zermelo had found a gap in his argument [Da 1979, 249], [Mo 1982, 87]. G. Kowalewski's 1950 memoir, *Bestand und Wandel*, seems to be the only source for this story.[6] In fact, it is highly doubtful that the error was discovered so quickly and that Zermelo was the one who found it.

There is no question that soon after König's lecture, Bernstein's formula attracted great attention [Sch 1922, 100]. In his polite but defensive response to [H 1904a], *Zur Mengenlehre*, which appeared in the *Jahresbericht*'s Sprechsaal [Berns 1905c, 199], Bernstein makes four comments.[7] In the first of these, he states:

> After Herr König had lectured on his interesting investigation, unfortunately as yet unpublished, it was very soon noticed by various parties (especially by Herr König himself) that the proof of equation (1) [the formula] is incomplete in the sense correctly identified by Herr Hausdorff.[8]

In his paper in the Proceedings of the 1904 Congress (reprinted in the *Mathematische Annalen*, 60 (1905)), König retracts the theorem of his lecture:

> The assumption that the continuum is equivalent to a well-ordered set would be false if Bernstein's theorem were correct in general. Unfortunately however, its proof has an essential gap since for $\aleph_\omega$ and for each of the "singular" well-ordered sets [the $\aleph_{\mu+\omega}$] considered above, the assumption that each countable subset lies in an initial segment of the entire set is no longer valid.[9]
>
> Above all, I bring this up in order to expressly withdraw the conclusion that I drew in my Congress lecture under the assumption of the truth of Bernstein's theorem.[10] [Ko 1904, 147; 1905, 180]

König cites no other sources for the discovery of Bernstein's error, and Bernstein's Sprechsaal note explicitly mentions König in this regard. As we shall see, Hausdorff in writing [H 1904a] was certainly aware of the contents of König's paper. However, Hausdorff seems to have independently pinpointed the problem with Bernstein's proof.

In the immediate wake of the turbulent meeting at Heidelberg, several of the participants, Hilbert, Hensel, Hausdorff, and Schoenflies found themselves vacationing in Wengen, a resort city in the Swiss Alps. König's lecture was still the focal point of their conversations. Cantor, who was staying in the vicinity, joined in, offering new rebuttals to König's conclusion [Sch 1922, 100–101].

In a letter to Hilbert,[11] written on September 29, 1904, two days after his return to Leipzig, Hausdorff reports:

> After the continuum problem had plagued me at Wengen almost like an obsession, here [in Leipzig] I of course looked first at Bernstein's dissertation. The error is exactly in the suspected place, p. 50 ...

Hausdorff then quotes the (earlier cited) part of Bernstein's proof in which the $\aleph_\nu$-element subsets of $\aleph_\mu$ are asserted to lie in initial segments of $\aleph_\mu$ and he explains the "simple counterexample" of $\mu = \omega$ and $\nu = 0$. He notes that Bernstein's proof also fails for the alephs for which König "urgently needs" the formula. Continuing, he writes:

> On the way home, I had already written to Herr König to this effect, to the extent that I could do it without the aid of Bernstein's work, but I have received no answer, I am thus rather more inclined to regard König's proof as false and König's theorem as the height of improbability. On the other hand, you will, to be sure, hardly have acquired the impression that Herr Cantor should have found in the last weeks what he has been seeking for thirty years to no avail, and so after the Heidelberg Congress, your problem No. 1 appears to stand exactly where you left it at the Paris Congress.
>
> But perhaps, while I write this, one of the parties to the dispute is already in possession of the truth, I am very anxious to see the printed proceedings of the Congress.

That Hausdorff had communicated with König raises the intriguing possibility that Hausdorff was the unacknowledged source for König's discovery of the error in Bernstein's proof. In any case, as far as Schoenflies is concerned, Hausdorff deserves the credit for finding the mistake. He noted this briefly in [Sch 1905, 186n2] and more effusively in his 1922 rembrance of Cantor:

> As is well-known, we are beholden first of all to Hausdorff [in [H 1904a]] for an accurate examination of the domain of validity of Bernstein's aleph relation; his result, which he had already arrived at in Wengen, necessitated the above mentioned restriction [to finite μ] of the Theorem [Bernstein's] and with that, the withdrawal of the conclusion [König's] drawn for the continuum. [Sch 1922, 101]

In [H 1904a] itself, no mention is made of the origins of the recursion formula. From his letter to Hilbert and from Schoenflies's recollections, it seems clear that it arose from Hausdorff's intensive scrutiny of Bernstein's formula at Wengen.

After stating the recursion formula for non-limit μ in [H 1904a], Hausdorff remarks that for the non-trivial case $\alpha < \mu$, the proof relies on the fact that

> the left side is the cardinality of the covering set whose elements are formed from $\aleph_\alpha$-element subsets of a set M of cardinality $\aleph_\mu$ and ... each such subset lies in an initial segment of M ...

This is Bernstein's proof (!), but in his restriction to non-limit μ, Hausdorff has delineated *where* the proof works and in his statement of the recursion formula he has ascertained *what* it proves. No further justification is given for the above fact.[12] (He does point out that it need not hold for limit μ, and in particular it fails for $\mu = \omega$.) The recursion formula never appears again in Hausdorff's published writings, not even in [H 1914a].

The failure of Bernstein's proof to justify his formula did not mean the formula could be discarded as false.[13] Hausdorff ends by stating the formula and noting:

> The formula ... is, therefore, to be considered as unproved for the time being. Its correctness seems rather problematical, since as Herr J. Konig has shown, the paradoxical result that *the cardinality of the continuum is not an aleph and that there are cardinal numbers that are greater than every aleph* [his italics] would follow from it. [H 1904a, 571]

This indicates Hausdorff's familiarity with the content of [Ko 1904]. In his retraction, saying that "he believes the matter [as he has related it] offers a certain interest beside historical accuracy," König speculated on what the consequences would be for the continuum if its cardinality failed to be an aleph. He asserted that it meant that $2^{\aleph_0}$ would be greater than every aleph. From the "greater than every aleph" claim, he immediately derived Bernstein's formula and so asserted its truth was equivalent to the non-well-orderability of the continuum [Ko 1904, 147; 1905, 180].[14]

Hausdorff's choice of the phrase "paradoxical result" at least shows his inherent skepticism towards the propositions that the continuum is not well-orderable and that there are cardinals that are greater than every aleph. (G. H. Moore interprets this choice as reflecting Hausdorff's belief at this time that the reals *can* be well-ordered and that every cardinal *is* an aleph [Mo 1982, 87].)

Notes

1. The *covering set* [[*Belegungsmenge*]] of N by M consists of all coverings of N by M where "[b]y *a covering of the set N by the elements of the set M* ... we understand a rule through which a definite element of M is connected to each element n of N ..." [Can 1895, 486]. In this case, a covering was Cantor's way of introducing an arbitrary function from N into M, and the covering set was the construct he needed to define cardinal exponentiation $\mathfrak{m}^\mathfrak{n}$. As bold as this expansion of the function concept was, Cantor could not yet abandon the idea that a function was determined by an explicit rule. (See [Ka 1996, 8 and 48n19].)

2. In [H 1904a], Hausdorff uses an old notation of Cantor, α_* (instead of α^*), to denote the inverse of the type α. The powers 2^{ω^*} and ω^{ω^*} are discussed in complete detail in [H 1914a, 158] and [H 1914a, 151], respectively. In the appendix ([H 1914a, 456]), it is remarked that 2^{ω^*} is the type of any bounded, perfect, nowhere dense set of reals. Earlier, in [H 1907a, 91] the same is asserted for the type ν^{ω^*} ($\nu'(\omega)$ in the notation of [H 1907a]) for any finite $\nu > 1$.

3. Although Hausdorff uses ϑ to represent the unbounded linear continuum, Cantor introduced the symbol ϑ for the bounded linear continuum in [Can 1895, 510]. Hausdorff reverts to Cantor's usage in [H 1906b, 143].

4. Bernstein finished his argument by using complete induction. Since $\overline{\alpha}_\mu < \aleph_\mu$, by induction $\overline{\alpha}_\mu^{\aleph_\nu} = \overline{\alpha}_\mu \cdot 2^{\aleph_\nu}$. Even with the failure at $\mu = \omega$, the proof still works for finite μ. In [Berns 1905d, 151–152], Bernstein imposed the condition that μ and ν be finite.

5. Julius König (1849–1913), who was trained at Heidelberg, had a reputation for "acuity and complete reliability"[Kow 1950, 201]. This, together with the simplicity of his argument, lent weight to his conclusion. There were only two components to his proof, his own lemma and Bernstein's formula. Ironically the proof of his lemma makes implicit use of the Axiom of Choice, which Zermelo was soon to introduce in his own proof that every set can be well-ordered [Ze 1904]. See Note 13.

6. The accounts of the events at Heidelberg in [Kow 1950] and [Sch 1922] have proved most influential. Using these sources, J. W. Dauben details how Heidelberg affected Cantor [Da 1979, 247–250]. G. H. Moore's version relies on the same sources, but he also gives a complete discussion of the mathematical content of the controversy, the *sub rosa* usage of AC, and how it all played out as the role of AC became clearer [Mo 1982, 86–88]. Later in his book Moore tells of a letter from the French philosopher Couturat to Russell reporting that no one at the Heidelberg Congress had discovered an error in König's proof [Mo 1982, 123]. The source of Couturat's information was H. Fehr, who is listed as attending in the Congress Proceedings.

   Schoenflies was a participant at the Congress; he and Cantor and Hilbert are even named as the discussants for König's lecture. However, Kowalewski is not listed as attending in the Congress Proceedings and the account in his memoir ([Kow 1950, 198–203]) does not claim to be at first hand. In fact, his version occurs in the part of his memoir on his years teaching at Bonn, and it is in a set piece describing two "very important events" that happened in 1904: König's Heidelberg lecture is one, the other is the introduction of the Lebesgue integral.

   After concluding that Kowalewski's story about Zermelo was (charitably) unreliable, I found that Walter Purkert had earlier, using much of the same evidence, declared it "*falsch*" [Pu 2002, 10–11].

7. Bernstein's second, third, and fourth comments can be summarized as follows: for $\nu > \mu$, the formula is true and can be written as $\aleph_\mu^{\aleph_\nu} = 2^{\aleph_\nu}$ (also, he notes, independently proved by Jourdain); the results of his dissertation in no way depend on the general truth of his formula; he only used the case $\mu = 1, \nu = 0$, and his proof is still valid for finite μ and ν; the omission of a limiting condition took place in a auxiliary formula that was not used at any point to draw broader conclusions. In the *Annals* reprint of his dissertation, the finiteness of the formula's indices is added, completely restoring the validity of the work at the point where the original omission occurred.

8. In the last footnote of [Sch 1905], Schoenflies reports that in a soon to appear note ([H 1904]) Hausdorff shows that Bernstein's proof does not extend from finite indices to ω. Tellingly, Schoenflies does not mention Zermelo or König as sharing in this discovery.

9. Denied the full force of Bernstein's formula, König's main result became: the continuum is not an aleph of the form $\aleph_{\mu+\omega}$. In the *Mathematische Annalen* reprint, König added a paragraph after the statement of this theorem: "one easily generalizes the theorem so that the continuum cannot be equivalent to a well-ordered set for which no immediately preceding well-ordered set exists" [Ko 1905, 146]. Shortly thereafter, a correction struck this claim (*Mathematische Annalen* **60** (1905), 462).

10. König's International Congress paper was reprinted in the *Mathematische Annalen* of 1905. Bernstein's 1901 dissertation was about to be reprinted in the same journal. Under advice from David Hilbert, its editor, Bernstein submitted a paper to the *Mathematische Annalen*, (again) entitled *Zur Mengenlehre* [Berns 1905c]. His comments are very much along the lines of those summarized in Note 7, except at the end, where perhaps showing particular annoyance with König's portrayal of events, he writes: "Herr König used the theorem in question as grounds for far-reaching conclusions, ones that lay claim to a basic significance. Therein lies the fundamental difference between his use of the theorem and mine. Moreover, it is unlikely that the Continuum Problem could be solved with the concepts and methods developed so far."

    In the *Jahrbuch über die Fortschritte der Mathematik* review of [Ko 1904], the reviewer G. Hessenberg refers to [Berns 1905c] as "*Bernstein's peevish reply* [[*Bernsteins* gereizte Entgegnung]]."

    There is some irony in this situation, in that Bernstein had intended to speak on his own proof of CH at the 1904 congress but was unable to attend. He published a sketch of his (unsuccessful) argument in an article entitled *Die Theorie der reelen Zahlen* in the Jahresbericht der Deutschen Mathematiker-Vereinigung **14** (1905), 447–449.

11. This handwritten letter is in Hilbert's Nachlaß at the Niedersächsische Staats- und Universitätsbibliothek in Göttingen (Hilbert 136). Excerpts from the letter are reprinted in [Mo 1989, 108-109] (translated) and in [Pu 2002, 11–12].

12. It is hard to see how Hausdorff *could* justify it; essentially the formula works because $\aleph_\mu$ is "regular" for non-limit μ. However, Hausdorff did not isolate the property of regularity until 1907–1908. In [H 1906b, 124], he introduced the relation of cofinality between a set and its subsets, and in [H 1907b, 542; 1908, 442], he used cofinality to partition the initial ordinals (and alephs) into two classes, those that are "regular" and those that are "singular." In [H 1908, 443], he omits the "quite simple" proof that *each initial number whose index is not a limit number is regular*. (A formal proof requires AC, for in ZF one cannot even prove that $\aleph_1$ is regular.)

13. The ZFC formalization of König's derivation of "the continuum is not well-orderable" shows that the unrestricted Bernstein formula is provably false in ZFC. In ZFC the continuum is well-orderable, so $2^{\aleph_0} = \aleph_\beta$ for some $\aleph_\beta$; by König's lemma, which is a theorem of ZFC, $\aleph_{\beta+\omega}^{\aleph_0} > \aleph_{\beta+\omega}$. But $\aleph_{\beta+\omega} \cdot 2^{\aleph_0} = \aleph_{\beta+\omega} \cdot \aleph_\beta = \aleph_{\beta+\omega}$. Thus $\aleph_{\beta+\omega}^{\aleph_0} \neq \aleph_{\beta+\omega} \cdot 2^{\aleph_0}$ and Bernstein's formula fails for $\mu = \beta + \omega$, $\nu = 0$.

14. A theorem of Hartog insures that in ZF no cardinal is greater than every aleph [Har 1915]; so in ZF, if the continuum is not well-orderable, there is some some $\aleph$ with which $2^{\aleph_0}$ is incomparable.

The Concept of Power in Set Theory

By FELIX HAUSDORFF in Leipzig

The following remarks, which I intend to follow up on elsewhere, relate to powers in which

    I. the base is an order type, the exponent is an inverse ordinal number,
    II. the base is an order type, the exponent is an ordinal number,
    III. the base, as well as the exponent, is an aleph.

To begin with, recall that in the domain of order types the concept of power only has to satisfy the following conditions ("rules of powers") [[("Potenzregeln")]]

$$\mu^1 = \mu, \quad \mu^\alpha \cdot \mu^\beta = \mu^{\alpha+\beta}, \quad (\mu^\alpha)^\beta = \mu^{\alpha \cdot \beta},$$

whereas the condition $\mu^\alpha \cdot \nu^\alpha = (\mu \cdot \nu)^\alpha$ cannot be imposed owing to the absence of commutativity of multiplication.

I. When M is an ordered set and A is a well-ordered set, the *covering set* [[*Belegungsmenge*]] (A/M), which was considered by Herr G. CANTOR (Math. Ann. 46, p. 487) and which serves as the definition of the cardinal power $\mathfrak{m}^\mathfrak{a}$, can be easily converted to a simply ordered set. Let μ and α be their order types, and, moreover, let

$$x = (x_0, x_1, x_2, \ldots, x_\omega, x_{\omega+1}, \ldots) = \{x_\beta\}$$

be a well-ordered set of type α with elements x_β ($\beta < \alpha$), each of which runs through a set of type μ. These sequences x, which are elements of the covering set, are to have the order relationship of first differing elements; i.e., let $x < y$ if for a particular β and each $\gamma < \beta$, $x_\beta < y_\beta$ and $x_\gamma = y_\gamma$. The resulting type of the set $\{x\}$ has, according to the notation for multiplication presently adopted by Herr CANTOR, the formal properties of the power $\mu^{\alpha*}$, where α_* is understood to be the type inverse to α. As examples, I mention: $2^{\omega*} = \nu^{\omega*}$ (ν finite), a type which results from the linear continuum ϑ when one inserts *two* consecutive points at all the rational places and adds a boundary point to both the left and the right; $\omega^{\omega*} = 1 + \vartheta$, the left-bounded linear continuum; $\vartheta^{\Omega*}$ (Ω, the first number in the third number class), a type that shares with the "infinitary pantachie" the property that *no fundamental sequences* in it have limits. The product of infinitely many *distinct* order types can be defined analogously to this power notion.

II. Cantor's concept of power (Math. Ann. 49, p. 231) for ordinal numbers of the first two number classes can be generalized to the case of an arbitrary order type as base, whereby indeed, the power is just defined uniquely through a particular decomposition of the base. From the ordered set M of type μ, choose a particular element m, which is kept fixed in all the following operations and which marks the place where the already constructed sets are to be "inserted" in M as middle pieces. Then if
$$M = P + m + R, \quad \mu = \pi + 1 + \varrho,$$
form in turn
$$\mu^2 = \mu(\pi + 1 + \varrho) = \mu\pi + \mu + \mu\varrho,$$
$$\mu^3 = \mu^2(\pi + 1 + \varrho) = \mu^2\pi + \mu\pi + \mu + \mu\varrho + \mu^2\varrho,$$
$$\cdots \cdots \cdots \cdots \cdots \cdots \cdots \cdots \cdots \cdots \cdots$$
$$\mu^\omega = \cdots + \mu^3\pi + \mu^2\pi + \mu\pi + \mu + \mu\varrho + \mu^2\varrho + \mu^3\varrho + \cdots.$$
Thus μ^ω is the limit resulting from $\mu, \mu^2, \mu^3, \ldots$ through expansion to the left and right. Continue like this according to the following definition-formalism, which corresponds to G. Cantor's first and second "generating principles" [["Erzeugungsprinzip"]]:
first,
$$\mu^{\alpha+1} = \mu^\alpha \cdot \mu = \mu^\alpha \pi + \mu^\alpha + \mu^\alpha \varrho;$$
second, if $\alpha\ \alpha_1\ \alpha_2\ \alpha_3 \cdots$ is an increasing sequence of numbers with α_ω their limit and
$$\mu^{\alpha_1} = \pi_1 + \mu^\alpha + \varrho_1, \quad \mu^{\alpha_2} = \pi_2 + \mu^{\alpha_1} + \varrho_2, \quad \mu^{\alpha_3} = \pi_3 + \mu^{\alpha_2} + \varrho_3, \quad \ldots,$$
then μ^{α_ω} would be defined by
$$\mu^{\alpha_\omega} = \cdots + \pi_3 + \pi_2 + \pi_1 + \mu^\alpha + \varrho_1 + \varrho_2 + \varrho_3 + \cdots.$$

All these powers are uniquely defined following the choice of the fixed element m; in order to indicate the dependence on m, if necessary, instead of μ^α a more explicit notation ought to be chosen, say $(\pi + 1 + \varrho)^\alpha$. The validity of the rules of powers, but note carefully, only for powers based on the same m, is proved by transfinite induction (the inference from β to $\beta+1$ and from $\beta\ \beta_1\ \beta_2 \ldots$ to β_ω). Cantor's concept of power is a special case of ours, namely, when M is a well-ordered set and m is its first element. The power μ^ω constitutes an interesting example of the general concept when μ is a number from the second number class.

III. If $\aleph_\mu$ is an aleph that has an immediate predecessor $\aleph_{\mu-1}$, then the recursion formula
$$\aleph_\mu^{\aleph_\alpha} = \aleph_\mu \cdot \aleph_{\mu-1}^{\aleph_\alpha}$$
holds. The proof for the case $\alpha < \mu$ (the formula is also true otherwise, but trivial) relies upon the fact that the left side is the cardinality of the covering set whose elements, which are formed from $\aleph_\alpha$ elements, are *subsets* of a set M of cardinality $\aleph_\mu$ and *the fact that each such subset lies in an initial segment of M*, i.e., it has an element of M as maximum or as upper limit.

This recursion principle, by virtue of which a power of $\aleph_\mu$ can be reduced to equal powers of earlier alephs, *fails however for those $\aleph_\mu$ that have no predecessor* (μ a limit number); for in this case, the subsets in question need not lie in an initial segment of M. So for example, $\aleph_\omega^{\aleph_0}$ cannot be expressed, at least in this way, by powers of $\aleph_\nu^{\aleph_0}$ (ν finite) since the countable subsets of $\aleph_\omega$ can contain numbers from *each* number class of finite order. The formula

$$\aleph_\mu^{\aleph_\alpha} = \aleph_\mu \cdot 2^{\aleph_\alpha}$$

obtained by Herr F. BERNSTEIN through unbounded recursion ("Untersuchungen aus der Mengenlehre," Diss. Halle 1901, p. 50) is, therefore, to be considered as unproved for the time being. Its correctness seems rather problematical since, as Herr J. KÖNIG has shown, the paradoxical result that *the cardinality of the continuum is not an aleph and that there are cardinal numbers that are greater than every aleph* would follow from it.

Introduction to "Investigations into Order Types I, II, III"

With the five articles (three in [H 1906b] and two in [H 1907a]) that appeared under the umbrella title of *Investigations into Order Types* [[*Untersuchungen über Ornungstypen*]], Hausdorff began a series of publications that represented his deepest work on ordered sets. Only some of this material appeared in his seminal 1914 text, *Grundzüge der Mengenlehre*. The first part ([H 1906b]) of *Investigations into Order Types* consists of three articles: *I. The Powers of Order Types; II. The Higher Continua; III. Homogeneous Types of the Second Infinite Cardinality* [[*I. Die Potenzen von Ordnugstypen; II. Die höheren Kontinua; III. Homogene Typen zweiter Mächtigkeit*]].

In his introduction, Hausdorff clearly separates [H 1906b] from what has gone before:

> As far as I know, all the *essential* [his italics] results of these works are new, since the hitherto existing investigations of others refer almost exclusively to subsets of the linear continuum or to well-ordered sets.

Acknowledging that "details from the following investigation may already be found in the existing literature" and further that "individual examples from type categories that arise systematically with me through the algorithm of power formation have [already] been devised for occasional purposes," he states:

> However, the correspondence of my abstract investigations with those of others, whose main purpose is the application to analysis and geometry, probably does not extend beyond such details. [H1906b, 106]

A reader of [H 1906b] would find no hint that the world of set theory had been irrevocably altered in 1904 by Zermelo's proof in [Ze 1904] of the Well-Ordering Theorem and by his explicit statement of the Choice Principle. Hausdorff cites his own starting point for [H 1906b] as the "assets of pure set theory" as laid out in [Can 1895; 1897] and [Sch 1900].

Zermelo's proof, which arrived amid an ongoing battle over the nature of the *set* of ordinals and the use of unrestricted comprehension (the Burali-Forti and Russell paradoxes) drew a host of critics and was the occasion for even more intense questioning of the foundations of set theory (and mathematics).[1] Hausdorff's avoidance of these issues was deliberate. He ends his introduction with a clear affirmation of his noncombatant status in the battles then underway:

> I have not gone into the foundations and paradoxes of set theory that have been discussed in recent years so passionately and, for the most part, fruitlessly. And I have sought to avoid points of contention through suitably qualified formulations, so that the premises of my theorems are always sufficient but might not always be necessary.[2] [H 1906b, 107]

In *The Powers of Order Types*, Hausdorff develops a method of power formation for arbitrary order types, which yields Cantor's powers for ordinals in the first two number classes as a special case and which generalizes the two methods mentioned in [H 1904a].[3] Power formation for arbitrary types is approached via Cantor's *covering sets*. Cantor introduced the "covering set" in order to define cardinal exponentiation [Can 1895, 486-488]. For ordered sets M and A of types μ and α, respectively, a covering of A by M is an A-sequence of elements from M ("A-sequence" is our term, not Hausdorff's); the covering set (of A by M) is the collection of all such A-sequences. Attempting to order the covering set by the principle of first differences generally leads to a partial ordering that is not total. Hausdorff solves this problem as follows: he chooses an element m of M, and he considers only those coverings in which the set of places occupied by elements not equal to m is finite. The resulting collection, which is totally ordered by first differences, is called *the power of the first class with base μ, principal element m, and argument α*, denoted by $\mu_m(\alpha)$.[4] Among these powers, *everywhere dense* types abound, since for $\mu > 1$ and for α without a last element, $\mu_m(\alpha)$ is everywhere dense[5] (that is, it has no pairs of consecutive elements) [H 1906b, 113–114]. Hausdorff easily generalizes this power definition to obtain powers of higher classes. Particularly relevant to the second article of [H 1906b] are powers of the second class, $\mu'_m(\alpha)$; such a power consists of only those coverings where the set of places occupied by elements not equal to m is a well-ordered set of cardinality $\leq \aleph_0$. Significantly, when α is a finite or countable ordinal, the power $\mu'_m(\alpha)$ is independent of m.

In *The Higher Continua*, Hausdorff adopts and enriches Cantor's language for discourse about ordered sets. Cantor developed the theory of ordered sets in [Can 1895, 496–512]. There, using *fundamental sequences* (well-ordered subsets of type ω or ω^*), he transferred to order types some of the terminology of the *theory of point sets* that he employed in analysing the continuum and its subsets: an order type μ is *dense in itself* if every element is the limit of a fundamental sequence; μ is *closed* if every fundamental sequence has a limit; and μ is *perfect* if it is both dense in itself and closed [Can 1895, 508–510]. In [Can 1897], Cantor pursued the important special case of well-ordered sets; he introduced the terms *Abschnitt* and *Rest* for certain segments: *Abschnitt* to refer to all the elements less than some fixed element and *Rest* to refer to all the elements greater than or equal to some fixed element.

Hausdorff uses Cantor's imported *point set* terminology, and he creates a more expressive vocabulary for talking about the subsets and segments of ordered sets.[6] For subsets, he defines the crucial notion: M is *cofinal* (*coinitial*) with its subset A.[7] For segments, he introduces a raft of terms that distinguish initial, middle, and final segments (and pieces). At a higher level, Hausdorff introduces the terms *isotomic, isomeric,* and *homogeneous* to categorize certain order types by their regular segment structure. He calls a type *isotomic* if all its initial segments are similar and all its end segments are similar; a type is *isomeric* if all its middle segments are similar, while a type is *homogeneous* if all its segments (initial, middle, and final) are similar. Except for the trivial cases (the isomeric $\mu = 2$, the homogeneous $\mu = 1$), isomeric and homogeneous types are everywhere dense. There are some simple relationships between these categories: for example, if a bounded type is isomeric, removing its endpoints produces a homogeneous type. Finally, Hausdorff transfers Dedekind's definition of continuity for the reals to order types. A type μ is *continuous* (*in Dedekind's sense*) if for any decomposition $\mu = \alpha + \beta$, either α has a last element or β has a first element—but not both. A continuous type is everywhere dense.

The heart of *The Higher Continua* is a study of the powers $\mu'(\alpha)$ of the second class for α a finite or countable ordinal. In this case, the entire covering set is totally ordered by the principle of first differences and principal elements are superfluous. Hausdorff starts by exhibiting a μ for which all the powers $\mu'(\alpha)$ are identical. Interestingly, this is the occasion for a long footnote justifying "proof by transfinite induction" for the first two number classes; in this footnote, he praises Schoenflies for championing this proof method, and he dismisses its anonymous naysayers [H 1906b, 127–128]. (Later, Schoenflies credits Hausdorff with coining the term *transfinite induction* in this very footnote [Sch 1908, 4n1].) For bounded μ, Hausdorff discovers a useful sufficient condition that guarantees that the powers $\mu'(\alpha)$ are all distinct: every set of "exclusive" intervals (no two have more than one element in common) in μ has type different from μ. In particular, the $\aleph_1$ types $\vartheta'(\alpha)$, α a finite or countable ordinal and ϑ the type of the bounded linear continuum, are all distinct since any set of exclusive intervals of ϑ is at most countable [H 1906b, 129–132].

The article culminates with a series of theorems in which appropriate sufficient conditions on μ and α result in the powers $\mu'(\alpha)$ having Cantor's properties (dense in itself, closed, perfect) or Hausdorff's properties (isotomy, isomery, homogeneity) or Dedekind's property (continuity). Finally, putting all these separate pieces together, Hausdorff exhibits a family of $\aleph_1$ distinct types, the types $\vartheta'(\omega^\alpha)$, each having the cardinality of the continuum and sharing many of the order properties of the bounded linear continuum; namely, they are perfect, isomeric, continuous, and invertible ($\mu = \mu^*$). Because of this, he claims for them the "right" to be called "continua of a higher level." Reflecting on the $\vartheta'(\omega^\alpha)$ and their properties, Hausdorff writes:

> [It is] a result that, contrary to views expressed elsewhere, appears quite remarkable and affords a deep insight into the inexhaustible abundance of higher order types. When one considers that through the one attribute of everywhere denseness (up to possible boundary elements) a countable type is already uniquely determined, it seems somewhat surprising that for types of the cardinality of the continuum the combination of such special properties as perfectness, isomery, and continuity still allows such a wide latitude.[8] [H 1906b, 143]

No specific source for the "views expressed elsewhere" is given, but this may very well be a reference to Felix Bernstein's dissertation and in particular to the final section of its reprint [Berns 1905d, 150–155]. There, Bernstein considers the set X_u of ω-sequences of ordinals $\alpha < \Omega$. X_u is ordered by first differences but with the twist that the α in the first position (third position, etc.) are ordered as in Ω, while those in the second position (fourth position, etc.) are ordered as in Ω^*. (In Hausdorff's notation, X_u is the power $(\Omega^*\Omega)'(\omega)$ of the second class.) Bernstein calls X_u "the ultracontinuum." He proves that X_u has cardinality $2^{\aleph_0}$ and that its finitely non-zero sequences form a dense subset of cardinality $\aleph_1$, which he calls the ultra-rationals.[9] Bernstein also offers a purported proof that X_u is closed; it is not, as Hausdorff points out in [H 1906b, 166n]. As for X_u and its type Θ_u, Bernstein expresses the following judgment:

> [I]t is probably the case that there are only two complete, homogeneous order types of simply ordered, perfect sets that are of cardinality $2^{\aleph_0}$, namely, the continuum and the ultracontinuum. [Berns 1905d, 154]

Certainly, the ultimate result of *The Higher Continua* gives the lie to this.

In the third and final article of [H 1906b], *Homogeneous Types of the Second Infinite Cardinality*, Hausdorff starts with a wonderful set piece in which he develops a continued fraction expansion for pairs of countable ordinals [H 1906b, 144–152].[10] He orders the sequences of quotients that result from these expansions as if they were the continued fraction expansions of ordinary rational numbers. He then extracts three particular "homogeneous" types of cardinality $\aleph_1$ from this ordered set of ω-sequences. In one of them, Ξ_a, no fundamental sequence has a limit. Hausdorff writes of it:

> Apart from P. Dubois-Reymond's "infinitary pantachie," which to this day has not been defined so as to be free of doubt, the type Ξ_a furnishes the first example, as far as I am aware, of an everywhere dense type of smallest possible cardinality *in which no fundamental sequence has a limit* [his italics].[11] [H 1906b, 152]

Unsatisfied with the "fortuitous and incomplete character" of this approach, Hausdorff proposes an *a priori* classification scheme for (non-trivial) homogeneous types, in particular for homogeneous types that only contain sequences (well-ordered subsets) of cardinality $\leq \aleph_1$. Hausdorff states (without proof) that such types are cofinal with ω or Ω and coinitial with ω^* or Ω^*. When he defined the concept of "cofinal" for sets and their subsets in *The Higher Continua*, Hausdorff proved that a type cannot be cofinal with two different initial ordinals ω_ν of finite index ν [H 1906b, 125]. It follows that the types in question can be organized into four groups according to the four possibilities for their *cofinality* and *coinitiality*. The assumption of homogeneity then determines the limit point nature of the elements for the types in each group. For example, if μ is cofinal with Ω and coinitial with ω^*, then each element of μ is the limit of an Ω-sequence and of an ω^*-sequence (Hausdorff calls them $\Omega\omega^*$-elements), while ω-sequences and Ω^*-sequences, if the latter occur, have no limits within μ.

The four groups are further divided into nine "genera" according to whether or not Ω-sequences and Ω^*-sequences actually appear. The final refinement involves the nature of the "gaps" that may occur. A *gap* in μ is a decomposition $\mu = \alpha + \beta$, in which α has no last element and β has no first element. Now by homogeneity, α is either cofinal with ω or Ω, while β is either coinitial with ω^* or Ω^*. So every gap can be assigned one of four "characters." A homogeneous type may contain from zero up to four kinds of gaps. The combined classification scheme produces fifty "species," thirty-two with $\omega\omega^*$-gaps and eighteen without $\omega\omega^*$-gaps. This is a crucial dichotomy.

Once his classification scheme is laid out, Hausdorff tackles the question of which of his fifty species can be represented by actual homogeneous types of cardinality $\aleph_1$. He is immediately confronted with the unsolved continuum problem.

After parenthetically remarking that "after so many other approaches have been tried in vain, a new way to answer the continuum question just might arise here," Hausdorff offers the "simple reflection" that an everywhere dense type contains the the type η (of the rationals) and hence, it is the case that "each everywhere dense type without $\omega\omega^*$-gaps contains [by Dedekind's construction of the reals from the rationals via cuts] the usual continuuum as a subset" [H1906b, 156].[12] As Hausdorff himself puts it so dramatically at the beginning of his next article [H 1907a, 84-85], "the discovery of such a type of the second infinite cardinality would be tantamount to a proof of Cantor's hypothesis ..."

G. H. Moore, who views Hausdorff's "extremely insightful work on order types" (in [H 1906b; 1907a; 1907b]) as "motivated by a desire to solve the Continuum Problem," writes of Hausdorff's "simple reflection":

> In particular, he put forward a new way of posing the problem, in the hope that this would lead to a solution. Namely, CH is true if the following propostion holds: There

is a dense order-type of power $\aleph_1$ having no $\omega\omega^*$-gaps. [Mo 1989, 109]

Whatever unexpressed "hope" Hausdorff may have had for this approach to CH, he used his straightforward observation about everywhere dense types without $\omega\omega^*$-gaps to direct his energies pragmatically toward the tractable portion of his representation problem. He states:

> As long as there is no success, for instance, in constructing an everywhere dense type of the second infinite cardinality without $\omega\omega^*$-gaps and with that in verifying CANTOR's hypothesis $2^{\aleph_0} = \aleph_1$, the question of the existence of homogeneous types of the second infinite cardinality will be restricted to the 32 species with $\omega\omega^*$-gaps. [H 1906b, 156]

One of these thirty two, the species that Hausdorff calls ID1, whose representatives are cofinal with ω and coinitial with ω^*, have $\omega\omega^*$-gaps, and have neither Ω-sequences nor Ω^*-sequences, is represented by both the rationals and the irrationals. The unsolved continuum problem again frustrates Hausdorff's desire to represent ID1 by a type of cardinality $\aleph_1$.[13]

In a tour de force, Hausdorff exhibits eight bases, certain sums of ω, ω^*, Ω, and Ω^*, that in conjunction with the arguments ω, Ω and appropriate choices of principal elements yield powers of the first class that represent the thirty two species with $\omega\omega^*$-gaps. All of these powers, except a countable one (it has type η of the rationals) representing the species ID1, have cardinality $\aleph_1$.

At the end of *Homogeneous Types of the Second Infinite Cardinality*, Hausdorff turns his attention to the invertible types of two species that he deems "closest" to the linear continuum, the already considered (higher) *continua* and what he now designates as the *ultracontinua*. He borrows the undefined term "ultracontinuum" that Bernstein used in his dissertation and gives it the following precise meaning: it is a type that is bounded, invertible, perfect, isomeric, and has both Ω-sequences and Ω^*-sequences. The (higher) continua in his second article have the first four properties, but they have neither Ω-sequences nor Ω^*-sequences. As Hausdorff points out in a footnote (p. 166) acknowledging Bernstein's coinage of "ultracontinuum," neither of the ordered sets that Bernstein named "the ultracontinuum," $\Omega'(\omega)$ in his dissertation and $(\Omega^*\Omega)'(\omega)$ in its reprint, satisfy the formal definition: "the first lacks invertibility and isomery and the second lacks closedness." (Because of the latter, Hausdorff asserts the invalidity of [Berns 1905d, 153 f.].[14])

Hausdorff produces an ultracontinuum Θ for which the $\aleph_1$ types $\Theta'(\omega^\alpha)$ are distinct ultracontinua, each of the cardinality of the continuum. This parallels his earlier result that the $\aleph_1$ types $\vartheta'(\omega^\alpha)$ are distinct continua. Hausdorff obtains Θ from the power $(\Omega + \Omega^*)'(\omega)$ by *collapsing* certain pairs of consecutive elements. (This technique will be used routinely in the first article of [H 1907a].) Hausdorff then shows that no set of "exclusive" intervals in Θ is similar to Θ, so by the criterion established in his second article, the second class powers $\Theta'(\omega^\alpha)$ are distinct ultracontinua.

Notes

1. Gerhard Hessenberg devotes considerable space to foundational issues in his 1906 text, *Grundbegriffe der Mengenlehre*; the word "paradox" appears with some frequency. Hessenberg presents the "Choice Principle," but he assumes a neutral position on it. The turbulence within foundations caused by Zermelo's proof is reflected in [Sch 1908, 5–7] where Schoenflies says that Zermelo's proof "cannot be considered as definitive." The second chapter of G. H. Moore's excellent book, *Zermelo's Axiom of Choice* ([Mo 1982]), provides a thorough account of the attacks launched by Zermelo's critics. In this regard, Zermelo's own reply to his critics [Ze 1908a], which is a masterpiece of polemics, is well worth reading.

2. An example of a "qualified formulation" occurs early in *The Powers of Order Types*. To calculate the cardinalities of his newly defined powers, Hausdorff hypothesizes that one of the relevant cardinals is equal to its own double, while another is equal to its own square. A consequence of Zermelo's Well-Ordering Theorem is that *every infinite cardinal is an aleph*, and both the hypothesized properties hold for alephs. Hausdorff notes that "the stated hypotheses are fulfilled for all cardinal numbers known to us," and he adds parenthetically, "they are only explicitly mentioned in order to leave untouched the question of whether every cardinal number is an aleph, upon which there is not yet universal agreement" [H 1906b, 113].

3. Hessenberg gives a definition of power for arbitrary ordinals (as bases and exponents) that generalizes the one given by Cantor for the finite and countable ordinals [Hes 1906, 597]. He uses his concept of principal number to do this. Hessenberg calls an ordinal number that is similar to each of its remainders a principal number [Hes 1906, 578]. In [Hes 1907], he gives a definition of power for order types, which he acknowledges is a special case of Hausdorff's powers of the first class.

4. Hausdorff justifies calling the $\mu_m(\alpha)$ powers by proving that they satisfy the following laws: (1) $\mu_m(\alpha) \cdot \mu_m(\beta) = \mu_m(\beta + \alpha)$ and (2) for $\nu = \mu_m(\alpha)$ and n the constantly m A-sequence, $\nu_n(\beta) = \mu_m(\alpha\beta)$. The form of the first law is dictated by the interaction between Cantor's definition of multiplication for order types and the ordering of A-sequences by the principle of first differences.

5. In [Can 1895, 504], the term *everywhere dense* [*überalldicht*] was used to describe an intrinsic property of the the rationals: between any two rationals there is always a third rational. Earlier Cantor had defined "a point set P that either partially or completely lies within an interval $(\alpha \ldots \beta)$ and that contains points in each subinterval $(\gamma \ldots \delta)$, no matter how small, of $(\alpha \ldots \beta)$ to be *everywhere-dense* [*überall-dicht*] in $(\alpha \ldots \beta)$" [Can 1879, 2]. (In 1879, Du Bois-Reymond coined the adjective *pantachish* for such sets. See the *Introduction to "Investigations into Order Types IV, V,"* Note 11, p. 110.) Thus the *everywhere dense* rationals are *everywhere dense in* the reals.

According to José Ferreirós, "As regards the notion of dense set, it originated in Dirichlet, and Hankel formulated it clearly (with the words 'filling a segment') well before Cantor or du Bois-Reymond" [Fer 1999, 164].

6. In [H 1901b], Hausdorff had already expanded Cantor's terminology to speak of "pieces" (initial, middle, or final) of ordered sets. A subset A of the ordered set M is a *piece* of M, if whenever $a < b$ are in A, then so is any c with $a < c < b$.

   In spite of his own efforts to provide new terminology, at times Hausdorff felt the gravitational tug of an old Cantorian term. This was especially the case for the concept of *initial segment* for which Hausdorff introduced *Anfangsstrecke* but often returned to Cantor's *Abschnitt*.

7. In today's texts, the order of A and M is reversed and the phrase becomes: [the subset] A *is cofinal with* M. In [H 1908, 442–444; 1914a, 129–132], Hausdorff proves with the aid of a well-ordering that an ordered set is cofinal with a unique "regular" initial ordinal.

   A precursor for the concept of "cofinal" subset can be found in the definition of *zusammengehörig* fundamental sequences in [Can 1895, 508]; such sequences are essentially *mutually cofinal*.

   Hessenberg introduced the idea of a "Kern" of an ordinal μ: a set of ordinals M is a *Kern* of μ, if $A(M) = \{\alpha \mid \alpha \leq \lambda, \text{some } \lambda \in M\} = \{\alpha \mid \alpha < \mu\}$ [Hes 1906, 559]. In Hausdorff's terms, μ is certainly *cofinal* with M. For μ a limit ordinal, Hessenberg shows that among its "Kerns" are ones with a smallest type and that type is an initial ordinal.

8. On the other hand, Cantor *did* characterize the order type of the bounded linear continuum as the unique bounded, perfect order type having a countable subset that is everywhere dense in it [Can 1895, 510–512]. Just before this, he had characterized the order type of the rationals in the open interval $(0,1)$ as the unique unbounded, everywhere dense, countable order type [Can 1895, 504–506].

9. In [Berns 1901, 51–54], Bernstein defined "the ultracontinuum" X_u to be the set of ω-sequences of ordinals $\alpha < \Omega$, ordered by first differences (in Hausdorff's notation, the power $\Omega'(\omega)$). (This may be the first instance of a "covering set" ordered by first differences.) Bernstein stated his (notorious) aleph formula as an aid in calculating the cardinality of X_u. (See the *Introduction to "The Concept of Power in Set Theory,"* p. 24.)

10. Hausdorff introduced a division algorithm for countable ordinals. (Although countability is not really required in its proof.) Cantor used a special case of the division algorithm to produce his normal form for countable ordinals [Can 1897, 235–237]. Hessenberg proved a division algorithm for arbitrary ordinals [Hes 1906, 575] and then used it, together with principal numbers, to prove the full Cantor Normal Form Theorem [Hes 1906, 587–591].

    The continued fraction expansion for pairs of countable ordinals and the resulting ordered set of quotient sequences is recounted in [H 1914a, 185–189] and generalized to pairs of ordinals that are less than some fixed regular initial ordinal.

11. In [H 1904a], the power ϑ^{Ω^*} ($\vartheta'(\Omega)$ in the notation of [H 1906b]), where ϑ is the type of the unbounded reals, is offered as an example of a type in which no fundamental sequence has a limit. This power has cardinality $2^{\aleph_0}$.

12. Let μ be the everywhere dense type without $\omega\omega^*$-gaps in which η is embedded; a gap in η must be *fillable* by elements of μ, otherwise a gap in η would give rise to an $\omega\omega^*$-gap in μ. So the *completion by cuts* of the rationals to the reals works inside μ and serves to embed the linear continuum into μ. (See [H 1914a, 101].) The embedding of η into μ and the realization of the reals makes implicit use of AC.

13. Mahlo removed this exceptional case by constructing a homogeneous subset of the continuum of cardinality $\aleph_1$ having only $\omega\omega^*$-elements and $\omega\omega^*$-gaps [Ma 1909].

    Paul Mahlo (1883–1971), who considered himself a student of Felix Bernstein and who wrote a thesis in geometry, seems to have been an early follower of Hausdorff's work on ordered sets. (Mahlo receives only the briefest mention on p. 454 of the appendix of [H 1914a].) In [Ma 1911; 1912; 1913], Mahlo pursued the regular limit cardinals that Hausdorff in [H 1907b, 1908] called "exorbitant magnitudes," but that he could not fully dismiss. Mahlo also worked on the character sets of dense sets—introduced in [H 1908]. ([Ma 1911]; see also [Sch 1913, 202–205].) Unfortunately, Mahlo's work was not appreciated until much later. His life and difficult career are chronicled in [GoKr 1984].

14. In [Berns 1901, 53-54; 1905d, 154–155], Bernstein sketched a *false* argument (in imitation of Cantor's correct proof of CH for closed sets of reals) for the proposition that a closed subset of X_μ is either countable, has cardinality $\aleph_1$, or has cardinality of the continuum. From this, he went on to argue that the set $\mathcal{A}$ of closed subsets of X_μ has cardinality $2^{\aleph_1}$, and hence by the correspondence between X_μ and the continuum—they both have the same cardinality—one gets a family, having cardinality $2^{\aleph_1}$, of sets (not necessarily closed) of reals where each set is either countable or has cardinality $\aleph_1$ or cardinality $2^{\aleph_0}$. And if $2^{\aleph_0} < 2^{\aleph_1}$, one has extended the classification by cardinality beyond the family of closed subsets of the continuum, a family that has cardinality $2^{\aleph_0}$.

Investigations into Order Types

By
F. Hausdorff

The following three shorter articles form the beginning of a series of contributions to the theory of simply ordered sets, a series that is to be continued later. The first develops the formalism of power formation [[Potenzbildung]] in complete generality—which is indispensable for a deeper study of type theory; the second presents a series of general theorems about certain types formed by raising to powers [[Potenzierung]] and constructs an infinite scale of very remarkable order types that correspond to the ordinary continuum in almost all their properties; the third attempts a first orientation in the area of homogeneous types that contain not just the usual fundamental sequences. As far as I know, all the *essential* results of these works are new since the hitherto existing investigations of others refer almost exclusively to subsets of the linear continuum or to well-ordered sets. In consideration thereof, I have not shied from the creation of a number of new words, especially in the second article, since the available terminology is not in any way sufficient. With that said, it ought not be disputed that details from the following investigations may already be found in the existing literature. Various types have already been constructed according to the very obvious "principle of first differences"; individual examples from type categories that arise systematically with me through the algorithm [[Algorithmus]] of power formation have been devised for occasional purposes. However, the correspondence of my abstract investigations with those of others, whose main purpose is the application to analysis and geometry, probably does not extend beyond such details.

I have not gone into the foundations and paradoxes of set theory that have been discussed in recent years so passionately and, for the most part, fruitlessly. And I have sought to avoid points of contention through suitably qualified formulations, so that the premises of my theorems are always sufficient but might not always be necessary. The following developments go on more or less from the assets of pure set theory that are inventoried in the last two papers by G. Cantor (*Beiträge zur Begründung der transfiniten Mengenlehre*, Math. Ann. 46 (1895) and 49 (1897)) or in the report by A. Schoenflies (*Die Entwicklung der Lehre von den Punktmannigfaltigkeiten*, Leipzig 1900).

I. The Powers of Order Types

In a note "On the Concept of Power in Set Theory" [["über den Potenzbegriff in der Mengenlehre"a]] (Jahresber. D. M. V. 13 (1904), 569-71), I presented two cases in which one can define powers of order types; the base was an arbitrary order type, the exponent, however, had to be an ordinal number or the inverse of one; and both of these special cases—the first of which was an immediate generalization of CANTOR's powers of numbers of the second number class—stood side by side unheralded and isolated. In the following, a far more general concept of power [[Potenzbegriff]] will be developed, which includes all previous cases and brings them into a systematic context, a concept of power in which the *base and exponent are arbitrary order types* of simply ordered sets. Indeed, this procedure doesn't depend on the base and the exponent alone, but rather it depends on certain additional requirements; despite this, the right to consider it as raising to powers [[Potenzierung]] will be justified through the validity of the rules of powers [[Potenzregeln]]. In this respect, one has to confine oneself, aside from the "initial condition" $\mu^1 = \mu$, to the two requirements

$$\mu^\alpha \cdot \mu^\beta = \mu^{\alpha+\beta}, \quad (\mu^\alpha)^\beta = \mu^{\alpha\beta},$$

whereas the third rule of normal powers $\mu^\alpha \cdot \nu^\alpha = (\mu\nu)^\alpha$ cannot be maintained with non-commutative multiplication. As to the first rule, it must be remarked that, as everyone knows, G. CANTOR has already once switched the notation for multiplication out of regard for it and that deviation from the present way of writing it (by which $\alpha\beta$ means: α inserted in β) would only produce confusion. On the other hand, since the most important and interesting types are just those whose exponents are inverses of ordinal numbers, in order to avoid the typographically inconvenient μ^{α^*}, I will define a symbol $\mu(\alpha)$ that has the formal properties of μ^{α^*}, not μ^α, so that the first rule of powers will appear in the form

$$\mu(\alpha) \cdot \mu(\beta) = \mu(\beta + \alpha).$$

Then α^* should be called the *exponent* and α the inverse exponent or, more briefly, the *argument* of the power $\mu(\alpha)$.

The Rank Ordering[b] by First Differences

By this we understand a ranking principle of fundamental significance that allows the following general formulation. Let $A = \{a\}$ be a simply ordered set (of type α and cardinality $\mathfrak{a}$), whose elements a we designate as *places* [[Stellen]] for the sake of distinction from the other elements. Let each place a be assigned a set M_a of type μ_a and cardinality $\mathfrak{m}_a$. Now let each place a be occupied or "covered" by an element x_a taken from the assigned

---

[a] The actual title is "Der Potenzbegriff in der Mengenlehre."
[b] [[Rangordnung]]

set M_a: so a new element arises[1])

$$x = \{x_a\} = \sum_a x_a,$$

a "covering of A by the sets M_a." The totality of the elements x, the covering set [[Belegungsmenge]] of A by the sets M_a, whose cardinal number, incidentally, furnishes a natural definition for the product $\prod_a \mathfrak{m}_a$ of all the cardinal numbers $\mathfrak{m}_a$, arises by each x_a running through its entire set M_a.

Let us consider two distinct elements x, y, where thus not every $x_a = y_a$. The places where they differ, i.e., those where $x_a \neq y_a$, form a subset of A. If this subset has a first element a and for this place $x_a < y_a$ (in M_a), then the ranking [[Rangfolge]] $x < y$ is to be chosen. We name this rank ordering by first differing places or first differences.

One recognizes immediately that the set $\{x\}$ is *not* generally a simply ordered set since among the differing places in two elements there is not necessarily a first. In general, neither of the two relations $x < y$ or $y < x$ exists between two elements. But similarly, one recognizes immediately that the relation $<$, *as long as it is defined*, is "asymmetric and transitive," i.e., if $x < y$, then it is not the case that at the same time $y < x$, and $x < z$ follows from $x < y$, $y < z$.

The Powers of the First Class

Let A and M be arbitrary simply ordered sets of types α, μ and cardinal numbers $\mathfrak{a}$, $\mathfrak{m}$. By identifying all the sets M_a, which to this point have been considered arbitrary, with M, we form an element

$$x = \{x_a\} = \sum_a x_a$$

in which each x_a is an arbitrary element of the set M, thus a covering of A by M; the totality of elements x, the covering set of A by M, has cardinal number $\mathfrak{m}^{\mathfrak{a}}$. The principle of first differing places, applied to the entire covering set, leads no more to a total rank ordering than in the general case.

Hence we allow a restriction to enter. Let a *fixed* element m be chosen from the set M and designated as the *principal element* [[Hauptelement]], the remaining elements designated as *secondary elements* [[Nebenelemente]]. Then stipulate: only those covers x shall be permitted *in which a finite number of secondary elements are used, thus in which all places except a finite number are covered by the principal element m*. These elements x form a subset T_m of the covering set. In this subset, however, the principle of first differences sets each pair x, y of distinct elements into one of the two relations $x < y$ or $y < x$ since the set of differing places is finite and necessarily has a first element. By what was said above, T_m has become a simply ordered set; its order type depends on the *base* μ, the *argument*

---

[1]) The summation sign has "collective" meaning here and does not mean addition; in many cases it is more convenient than { }.

α, and, besides these, on the choice of the *principal element* m; let it be denoted by $\mu_m(\alpha)$.

This type now has the following properties:

(1) For a finite argument
$$\mu_m(\alpha) = \mu^\alpha = \mu \cdot \mu \cdot \mu \cdot \ldots \cdot \mu.$$
In this case, T_m is independent of m and is identical to the entire covering set, which consists of the elements
$$x = (x_0 \, x_1 \, x_2 \, \cdots \, x_{\alpha-1}).$$
If x_0 runs through a set of type μ_0, x_1 through a set of type μ_1, etc., and rank ordering by first differing places is adopted, the type
$$\mu_{\alpha-1} \mu_{\alpha-2} \cdots \mu_1 \mu_0$$
results, thus in our case μ^α.

(2) It is the case that
$$\mu_m(\alpha) \cdot \mu_m(\beta) = \mu_m(\beta + \alpha).$$
Let $x = \{x_a\}$, $y = \{y_b\}$ be the elements of the pair of sets constructed as above with the types
$$\xi = \mu_m(\alpha), \quad \eta = \mu_m(\beta).$$
Then the type $\xi \cdot \eta$ is formed when one orders the element pairs (y, x) by first differences. This rank ordering is identical to that of the elements $\{y_b, x_a\}$ by first differing places if one requires that each place b precede each place a. Hence, here we have a set of type $\beta + \alpha$ whose places are to be covered with elements from μ and in fact, in such a way that the places a and the places b are allowed to separately carry only a finite number of secondary elements. This is the same as the assertion that in $\beta + \alpha$ only a finite set of places, and any finite set of places, is allowed to be covered by secondary elements, i.e., it is the same as the definition of the type $\mu_m(\beta + \alpha)$.

(3) If
$$\nu = \mu_m(\alpha),$$
then
$$\nu_n(\beta) = \mu_m(\alpha\beta),$$
where n denotes the element (the principal element) that arises through the covering of *all* places of α by the principal element m.

The type $\nu = \mu_m(\alpha)$ is formed by the elements
$$x = \{x_a\} = \sum_a x_a$$
and the type $\nu_n(\beta)$ by the elements
$$y = \{y_b\} = \sum_b y_b,$$

in which the x_a are elements of M and the y_b are elements of $\{x\}$. To express the dependence on the place b, we write

$$y_b = \{x_{ab}\} = \sum_a x_{ab},$$

$$y = \sum_b \sum_a x_{ab}.$$

The double indices, summed in this order, indicate the substitution of the type α at all places b, thus a place sequence of type $\gamma = \alpha\beta$, and we now have

$$y = \sum_c x_c = \{x_c\},$$

where $\{c\}$ is a set of type γ and x_c is an element of M. The rank ordering by first differing places for the $\{y_b\}$ matches that for the $\{x_c\}$. Further, it is clear that the secondary elements x_c are found only in the secondary elements y_b (i.e., where y_b differs from $n = \{m\} = \sum_a m$) and there only as secondary elements x_{ab}; thus they form a finite set of finite sets, i.e., they are in a finite set. Conversely, the restriction with respect to the x_c entails in return the restrictions with respect to the y_b and x_a. Hence $y = \{x_c\}$ runs through the type $\mu_m(\gamma)$, which is exactly what we wanted to prove.

With that, the rules of powers are proved; one has only to observe that the principal element must not be arbitrarily changed but rather must be retained with the multiplication of powers, and it must be determined as above with the raising of a power to a power. Under these assumptions, we can summarize our results so far:

(1) The powers with finite exponents are μ, $\mu\mu$, $\mu\mu\mu$, ...

(2) Powers are multiplied by adding exponents (or the arguments in reverse order).

(3) Powers are raised to a power by multiplying the exponents or arguments.

We add the following:

(4) A power is inverted by inverting the base.

I.e., the inverse type of $\mu_m(\alpha)$ is $\mu_m^*(\alpha)$; the argument α must not be inverted! How the retention of m is to be understood is more succinctly illustrated by the schematic figure in the margin than through words. If, e.g., the base μ is finite and its elements are denoted by $0, 1, 2, \ldots, \mu - 1$, then in the sequence of power types

$$\mu_0(\alpha), \; \mu_1(\alpha), \; \ldots, \mu_{\mu-2}(\alpha), \; \mu_{\mu-1}(\alpha)$$

the first and the last are inverses, similarly the second and the second from last, and so on. The types $3_1(\alpha)$, $5_2(\alpha)$, $7_3(\alpha)$, ... are inverses of themselves or *invertible* [[umkehrbar]].

(5) The cardinality of the type $\mu_m(\alpha)$ is independent of m. By choosing in turn the number of secondary elements to be $0, 1, 2, \ldots$, one gets
$$\overline{\mu(\alpha)} = 1 + (\mathfrak{m} - 1)\mathfrak{a}_1 + (\mathfrak{m} - 1)^2 \mathfrak{a}_2 + (\mathfrak{m} - 1)^3 \mathfrak{a}_3 + \cdots,$$
where $\mathfrak{a}_1 = \mathfrak{a}$ is the cardinal number of the type α, $\mathfrak{a}_2$ is the cardinal number of the element pairs of the type α, $\mathfrak{a}_3$ is the cardinal number of the element triples of the type α, etc. If $\mathfrak{a}$ is finite, then $\mathfrak{a}_2 = \frac{\mathfrak{a}(\mathfrak{a}-1)}{2} = \binom{\mathfrak{a}}{2}$, $\mathfrak{a}_3 = \binom{\mathfrak{a}}{3}$, etc. Then $\overline{\mu(\alpha)} = \mathfrak{m}^\mathfrak{a}$. If $\mathfrak{a}$ is transfinite and *equal to its own double*, then as one easily proves, $\mathfrak{a}_2 = \mathfrak{a}^2$, $\mathfrak{a}_3 = \mathfrak{a}^3$, etc., and moreover, $(\mathfrak{m} - 1)\mathfrak{a} = \mathfrak{m}\mathfrak{a}$ for $\mathfrak{m} > 1$; thus
$$\overline{\mu(\alpha)} = \mathfrak{m}\mathfrak{a} + \mathfrak{m}^2 \mathfrak{a}^2 + \mathfrak{m}^3 \mathfrak{a}^3 + \cdots.$$
If $\mathfrak{m}\mathfrak{a}$ is *equal to its own square*, then
$$\overline{\mu(\alpha)} = \mathfrak{m}\mathfrak{a} \cdot \aleph_0 = \mathfrak{m}\mathfrak{a}.$$
Because the stated hypotheses are fulfilled for all cardinal numbers known to us (they are only explicitly mentioned in order to leave untouched the question of whether every cardinal number is an aleph, upon which there is not yet universal agreement), we may summarize by saying that $\overline{\mu(\alpha)}$ is equal to $\mathfrak{m}^\mathfrak{a}$ for finite $\mathfrak{a}$ and that it is equal to $\mathfrak{m}\mathfrak{a}$ for transfinite $\mathfrak{a}$.

A few examples of a more general nature for this concept of power will be desirable.

(A). **If the argument is a type without a last element, then the power is an everywhere dense[c] type.**

Indeed, let x, y be two elements, a their first differing place and $x_a < y_a$, thus $x < y$. In order to emphasize this, we write:
$$x = \sum_p x_p + x_a + \sum_s x_s, \quad y = \sum_p x_p + y_a + \sum_s y_s,$$
where p runs through all places $< a$, and s runs through all places $> a$. By the hypothesis, the set $\{s\}$ is certainly transfinite. Accordingly, we show that x and y cannot be consecutive elements. Were that so, then to begin with x_a and y_a would have to be consecutive elements in M, but in addition, x would have to be the last of the elements beginning with $\sum_p x_p + x_a$, and y would have to be the first of the elements beginning with $\sum_p x_p + y_a$. Then all the elements x_s would have to coincide with the last element in M, and all the elements y_s would have to coincide with the first element of M. But even if M has a first and last element, one of them is certainly a secondary element, and thus one of the required coverings is impossible. (The trivial case $\mu = 1$ is excluded throughout.) Hence x and y are not consecutive, no two elements are. The type $\mu_m(\alpha)$ is everywhere dense for each choice of base and principal element.

---

[c] [überalldichter]

If α is a *countable* type without a last element and μ is likewise countable or finite, then $\mu_m(\alpha)$ is a countable, everywhere dense type; hence by a well-known theorem of CANTOR, $\mu_m(\alpha)$ is equal to the type η of the rational numbers in their natural order, apart from possible boundary elements. In this respect, there exist the following possibilities:

$\mu_m(\alpha) = 1 + \eta$ if m is the first element of M;

$\mu_m(\alpha) = \eta + 1$ if m is the last element of M;

$\mu_m(\alpha) = \eta$ if m is neither the first nor last element of M.

According to this, $2_0(\omega)$, $2_1(\omega)$, $3_1(\omega)$ are the simplest power representations for $1+\eta$, $\eta+1$, and η; the corresponding elements x can be read directly as dyadic, respectively, triadic fractions. On the other hand, if only ordinal numbers are supposed to be used for the base and argument, $1+\eta+1$ is not representable as a power.

Furthermore, if Ω is the least number of the third number class, one gets in

$$2_0(\Omega), \quad 2_1(\Omega), \quad \omega_m(\Omega), \quad \Omega_m(\omega), \quad \text{and the like}$$

the simplest everywhere dense types of cardinality $\aleph_1$. The type $\Omega_0(\omega)$, for example, is the one had by the set of elements

$$x = (x_0 \, x_1 \, \ldots x_{\nu-1} \, 0 \, 0 \, 0 \, \ldots)$$

ordered by first differences, where $x_0 \, x_1 \ldots$ run through the first two number classes

$$0 \, 1 \, 2 \ldots \omega \, \omega+1 \ldots \omega \cdot 2 \ldots \omega^2 \ldots$$

and where from a certain place on nothing but 0s appear. We will make use of such types in Article III to construct "homogeneous" types of the second infinite cardinality.

(B). **Powers whose exponents are ordinal numbers.**

The arguments are then inverse ordinal numbers. For exponents α of the second number class, $\mu_m(\alpha^*)$ is identical to the type in section II of the note cited above that I denoted by μ^α or $(\pi + 1 + \sigma)^\alpha$, where $\mu = \pi + 1 + \sigma$ indicates the decomposition of μ effected by the principal element m. To show the agreement of the definition there with the one in force here, let us consider $\mu_m(\omega^*)$ for instance. Let us understand by p every predecessor of the principal element m in M, by s every successor, and by $x_0, x_1, \ldots$ arbitrary elements in M. All the elements

$$x = (\ldots x_2 \, x_1 \, x_0)$$

can then be distributed in the following pieces, whose ranking from top to bottom is:

$$\begin{pmatrix} \cdots & & & & & \\ \cdots & m & p & x_2 & x_1 & x_0 \\ \cdots & m & m & p & x_1 & x_0 \\ \cdots & m & m & m & p & x_0 \\ \cdots & m & m & m & m & p \\ \cdots & m & m & m & m & m \\ \cdots & m & m & m & m & s \\ \cdots & m & m & m & s & x_0 \\ \cdots & m & m & s & x_1 & x_0 \\ \cdots & m & s & x_2 & x_1 & x_0 \\ \cdots & & & & & \end{pmatrix}$$

and in which rank ordering by first differences governs individual pieces. The type $\mu_m(\omega^*)$ is thus also representable in the form

$$\cdots + \mu^3\pi + \mu^2\pi + \mu\pi + \pi + 1 + \sigma + \mu\sigma + \mu^2\sigma + \mu^3\sigma + \cdots.$$

In entirely the same way, there arises the following method of generating a power with limit number exponents from lesser powers. Let $\alpha, \beta, \gamma, \ldots$ be numbers from the first or second number class differing from 0, and let $\lambda = \alpha + \beta + \gamma + \cdots$ be a limit number, namely, the first number that follows $\alpha, \alpha + \beta, \alpha + \beta + \gamma, \ldots$. For short, the powers already formed $\mu_m(\alpha^*), \mu_m(\beta^*), \mu_m(\gamma^*), \ldots$ may be called $\mu_0\ \mu_1\ \mu_2\ \ldots$; let their principal elements (resulting from a covering by nothing but principal elements m) be $m_o\ m_1\ m_2\ \ldots$, and let the respective decompositions be

$$\mu_0 = \pi_0 + 1 + \sigma_0, \quad \mu_1 = \pi_1 + 1 + \sigma_1, \quad \mu_2 = \pi_2 + 1 + \sigma_2, \quad \ldots.$$

Then $\mu_m(\lambda^*)$ can be represented as the type of the set of coverings

$$x = (\ldots x_{\alpha+\beta+\gamma} \ldots x_{\alpha+\beta} \ldots x_\alpha \ldots x_0) = (\ldots y_2\ y_1\ y_0)$$

ordered by first differences, where y_0 runs through the type μ_0, y_1 runs through the type μ_1, etc., and where only a finite number of secondary elements ($y_\nu \neq m_\nu$) are allowed. Then we get, as above,

$$\mu_m(\lambda^*) = \cdots + \mu_0\mu_1\pi_2 + \mu_0\pi_1 + \pi_0 + 1 + \sigma_0 + \mu_0\sigma_1 + \mu_0\mu_1\sigma_2 + \cdots,$$

the result as it were of the left and right side expansion of the types

$$\mu_0 = \pi_0 + 1 + \sigma_0,$$
$$\mu_0\mu_1 = \mu_0\pi_1 + \mu_0 + \mu_0\sigma_1,$$
$$\mu_0\mu_1\mu_2 = \mu_0\mu_1\pi_2 + \mu_0\mu_1 + \mu_0\mu_1\sigma_2,$$
$$\cdots\cdots\cdots\cdots\cdots$$

As a special case, CANTOR's powers of the second number class (such as ω^ω, ω^α, etc.) belong to those powers whose exponents are ordinal numbers; these are characterized by the fact that the base is also an ordinal number and that m is its least element. Hence CANTOR's power μ^α would be written by us as $\mu_0(\alpha^*)$, and its position in the general system of powers is formulated as follows: *CANTOR's powers are those whose base and exponent are ordinal numbers (of the first and second number classes), and whose principal element is chosen to be the first element of the base.*

In order to give at least one example for the general case, we want to consider the ω-th powers of numbers of the second class, thus powers that produce nothing but countable types. By designating the elements of μ by the types of the corresponding initial segments (the element π is thus the one that effects the decomposition $\mu = \pi + 1 + \sigma$), and by now writing μ_π^ω instead of $\mu_\pi(\omega^*)$, we have as above

$$\mu_\pi^\omega = \cdots + \mu^2 \pi + \mu\pi + \pi + 1 + \sigma + \mu\sigma + \mu^2\sigma + \cdots.$$

For finite μ, we obtain as powers only the three types ω, ω^*, and $\omega^* + \omega$, depending on whether the principal element is the first, the last, or neither of these.

Furthermore, we can make the decomposition $\mu_\pi^\omega = \Pi + 1 + \Sigma$,

$$\Pi = \cdots + \mu^2\pi + \mu\pi + \pi,$$
$$\Sigma = \sigma + \mu\sigma + \mu^2\sigma + \cdots.$$

The component Σ is either 0 if the principal element is the last, or it is equal to CANTOR's power μ^ω; in any case, it need not be considered further. The component Π assumes a particularly simple form if μ is a power of ω, $\mu = \omega^\lambda$. If one writes π $(< \omega^\lambda)$ in "normal form"

$$\pi = \omega^{\pi_\nu} \varrho_\nu + \omega^{\pi_{\nu-1}} \varrho_{\nu-1} + \cdots + \omega^{\pi_2} \varrho_2 + \omega^{\pi_1} \varrho_1,$$

where ν and the ϱ are finite numbers, the π numbers are in the first or second number class, and further $0 \leq \pi_1 < \pi_2 < \cdots < \pi_\nu < \lambda$, then

$$\Pi = \cdots + \omega^{\lambda \cdot 2} \pi + \omega^\lambda \pi + \pi$$

is a type of the form

(a) $$\cdots + \omega^{\pi_3} \varrho_3 + \omega^{\pi_2} \varrho_2 + \omega^{\pi_1} \varrho_1$$

with an infinite number of terms extending to the left, but these, so to speak, are *periodic* inasmuch as $\pi_{\nu+\beta} = \lambda + \pi_\beta$, $\varrho_{\nu+\beta} = \varrho_\beta$.

But also in the general case Π, up to a finite number of final elements, is of this form. The proof is easily furnished by decomposing π or, if π should be finite, μ into the nearest preceding limit ordinal and some finite final component. It is worth noting that the totality of all types (a) has the cardinality of the continuum, while the obtained periodic ones only form a subset of cardinality $\aleph_1$.

The Powers of Higher Classes

The concept of power that has been developed so far is only the first term in a further sequence and accordingly should be designated as a power of the first class. Again, let

$$x = \{x_a\} = \sum_a x_a$$

be a covering of A by M, and fix a principal element m in M. Before, we had restricted ourselves to such coverings where secondary elements ($\neq m$) appear in only a finite number of places a; the set of these special coverings was denoted by T_m, and this subset of the set of coverings, ordered by first differing places, gave the type $\mu_m(\alpha)$. Now we consider the following coverings:

those where a well-ordered set of places of at most the first infinite cardinality is covered by secondary elements: let the set of these coverings be T'_m;

those where a well-ordered set of places of at most the second infinite cardinality is covered by secondary elements: let the set of these coverings be T''_m;

etc., etc.

T_m is contained as a subset in T'_m, which is contained in T''_m, etc.

Each of these subsets of the covering set is transformed into a simply ordered set by the principle of first differences. Then since two elements x and y can differ at most in those places where at least one of them has a secondary element, and thus since these places in A are the union of two well-ordered sets, they themselves form a well-ordered set. Because of this, the set of differing places is a subset of a well-ordered set and therefore has a first element. Hence between any two elements x, y there is the disjunction $x < y$ or $y < x$, and what is more, we know that $x < z$ follows from $x < y$ and $y < z$.

We denote the order types of T'_m, T''_m, ... by $\mu'_m(\alpha)$, $\mu''_m(\alpha)$, ..., and we call these types powers of the second, third, ... class. Thus the general definition is as follows:

After fixing a principal element m in M, we consider those coverings of A by M where the places in A occupied by secondary elements form a well-ordered set of cardinality $\leqq \aleph_\nu$. The set $T_m^{(\nu+1)}$ of these coverings becomes a simply ordered set by the principle of first differences; its order type $\mu_m^{(\nu+1)}(\alpha)$ is called a power of the $(\nu+2)$th class of the base μ with principal element m and with argument α or with exponent α^*.

The proofs of the rules of powers turn out to be pretty much word for word as in the case of powers of the first class; however, they depend essentially on the provision that the cardinality of each well-ordered set of places is supposed to be *at most* $\aleph_\nu$ ($\leqq \aleph_\nu$). If one wanted to remove the inequality sign, respectively, the words "at most," then one would indeed get simply ordered sets again, but they would not have the character of a power. The powers with finite exponents would become empty; moreover, e.g., the provision (in the proof of the second rule) that a countable, well-ordered set of places be chosen out of $\beta + \alpha$ would not coincide with the respective individual provisions for α and β since the former provision can of course be met with a countable subset chosen from α and a finite subset chosen from β. Thus $\mu(\alpha) \cdot \mu(\beta)$ would only be a subset of $\mu(\beta+\alpha)$. Calling attention to this point is enough to make a detailed repetition of the above proofs unnecessary. Hence we can think of the theorems (1) – (4) (p. 112) as having been reproduced unchanged, now of course with the further proviso that we are always dealing with *powers of the same class*.

(5′) With regard to the *cardinality* of the type so obtained, we restrict ourselves to powers of the second class, which are only in consideration for transfinite $\mathfrak{a}$. If we make the same assumptions as in (5), then

$$\overline{\mu'(\alpha)} = \mathfrak{ma} + (\mathfrak{m} - 1)^{\aleph_0}\, \mathfrak{a}_{\aleph_0},$$

where $\mathfrak{a}_{\aleph_0}$ denotes the cardinality of the set of all *well-ordered, countable* subsets of α. This cardinal number depends on the type α, not only on its cardinality $\mathfrak{a}$, because of course, there are transfinite types (the inverses of ordinal numbers) that contain no well-ordered, countable sets at all; hence for them, $\mathfrak{a}_{\aleph_0} = 0$. However, if actual powers of the second class exist, thus $\mathfrak{a}_{\aleph_0} \neq 0$, then

$$2^{\aleph_0} \leq \mathfrak{a}_{\aleph_0} \leq \mathfrak{a}^{\aleph_0};$$

because an ω-sequence exists in α, $2^{\aleph_0}$ infinite subsets also exist, and on the other hand, $\mathfrak{a}^{\aleph_0}$ is the set of *all* countable subsets of A. The power $(\mathfrak{m}-1)^{\aleph_0}$ is replaced by 1 for $\mathfrak{m} = 2$ and by $\mathfrak{m}^{\aleph_0}$ otherwise; so everything summed up, the sought-after cardinality lies between that of the continuum and $(\mathfrak{ma})^{\aleph_0}$. If neither the base nor the argument exceeds the continuum, then a power of the second class has the cardinality of the continuum.

We again give several examples of particular simplicity.

(C). If the order type α contains no well-ordered subsets of cardinality $> \aleph_\nu$ (in particular, if $\mathfrak{a} = \aleph_\nu$), then there are no actual powers of classes $\nu + 3, \nu + 4, \ldots$; i.e, they reduce to powers of the class $\nu + 2$. For instance, there are only powers of the first and second class with the type of the linear continuum as the argument.

If α is an inverse ordinal number (thus the exponent α^* is an ordinal number), then no transfinite, well-ordered subsets exist in α, and all powers of higher classes reduce to the first; this is the case (B) discussed above.

(D). If α is an ordinal number of the $(\nu + 2)$th number class, thus of cardinality $\mathfrak{a} = \aleph_\nu$, then for the concept of powers of class $\nu+2$ the *principal element is irrelevant* since *all* subsets of α are then well-ordered and $\leq \aleph_\nu$. The type $\mu^{(\nu+1)}(\alpha)$ then becomes the type of the *whole covering set* of A by M *ordered by first differences*, whose elements are

$$x = (x_0\, x_1\, x_2\, \ldots\, x_\omega\, \ldots\, x_\beta\, \ldots) = \sum_\beta x_\beta$$

$(0 \leqq \beta < \alpha)$; here each x_β runs through the entire type μ independently of the others. This is the case discussed in section I of my repeatedly cited note; however, it must be mentioned that even here the powers of lower classes exist, but they remain dependent on the principal element. In the following Investigation II, we will deal especially with powers of the second class whose arguments are ordinal numbers of the second number class.

The special position of both cases, where either the argument or the exponent is an ordinal number, can be now recognized. For the former, a concept of powers exists independent of the principal element, and for the latter, only one single class of powers exists.

(E). If every element m in the base μ causes the same decomposition[1]) $\mu = \pi + 1 + \sigma$ (in other words: if all "initial segments" [["Abschnitte"]] of μ have the same type π and all initial segments of μ^* have the same type σ^*), then each of the powers of μ is independent of the principal element. That is, all $\mu_m(\alpha)$ coincide $= \mu(\alpha)$ and all $\mu'_m(\alpha) = \mu'(\alpha)$, etc. Naturally, one must not misunderstand this point, as if (as in the previously discussed case) the restriction on the coverings can simply be lifted; the subsets T_m are one and all similar to, but not perhaps identical with, the whole covering set. Examples of this case are the type η of the rational numbers, the unbounded linear continuum, the type $\omega^* + \omega$, and many others.

(F). **Powers with base 2.** For $\mu = 2$, the only powers that exist are $2_0{}^{(\nu+1)}(\alpha)$ and $2_1{}^{(\nu+1)}(\alpha)$, which are inverses of each other. Here every cover is completely characterized by specifying the subset of A whose places are covered by secondary elements; the places covered by the principal element form the complementary subset. Thus the well-ordered subsets of A of cardinality $\leq \aleph_\nu$ can be viewed as the elements x of our powers. In the case of $2_1{}^{(\nu+1)}(\alpha)$, these subsets are to be ordered thusly: of two subsets, the one that contains the first element a not common to both precedes the other in the ordering.

Transfinite Products of Order Types

Finally, in order to compare the concept of power to the analogous concept of product, we return to our starting point and assign to each place a of the set A an arbitrary set M_a of type μ_a. The definition of the *transfinite sum* of the types μ_a is immediately obvious: by

$$\sum_a \mu_a = \cdots + \mu_a + \cdots + \mu_b + \cdots \qquad (a < b)$$

is to be understood the type of the set obtained by substituting the set M_a in place of the element a. If all $\mu_a = \mu$ are the same, then $\sum_a \mu_a = \mu\alpha$. For a definition of the *transfinite product* of the types μ_a, we once again need to pick out a subset from the covering set of A by the sets M_a that can be ordered by the principle of first differences. To this end, we fix a principal

---

[1]) Such types will be called *isotomic* [[*isotom*]] in the following Article II, §4.

element m_a in every set M_a and call the other elements secondary elements; then let T be the set of those coverings in which only a finite number of secondary elements appear, and let $T^{(\nu+1)}$ be the set of those coverings in which the places a that are occupied by secondary elements form a well-ordered set of cardinality $\leq \aleph_\nu$. These subsets of the whole covering set are simply ordered according to the principle of first differences; their order types, which depend on the choice of all the principal elements m_a or, more briefly, on the choice of the *principal covering* $n = \{m_a\}$, should be defined as a *product of the first class*, respectively, $(\nu+2)$th class of all the order types, where however the factors μ_a that correspond to the sequence of their places are to be read from *right to left*; thus

$$\prod_a \mu_a = \ldots \cdot \mu_b \cdot \ldots \cdot \mu_a \cdot \ldots \qquad (a < b)$$

and

$$\prod_a{}^{(\nu+1)} \mu_a = (\ldots \cdot \mu_b \cdot \ldots \cdot \mu_a \cdot \ldots)^{(\nu+1)} \qquad (a < b).$$

This arrangement is necessary to maintain the agreement with the usual products for a finite α; e.g., for $\alpha = 2$, the product of the set of element pairs (x_0, x_1), ordered by first differences, where x_0, x_1 run through the types μ_0, μ_1, is to be designated by $\mu_1\mu_0$, not by $\mu_0\mu_1$. Of applications of this, up to now, surely most general algorithm [[Algorithmus]] of set theory, the one corresponding to case (D) for powers where the principal covering is irrelevant would be considered first; if α is an ordinal number of the $(\nu+2)$th number class, then the product of the $(\nu+2)$th class is the type of the whole covering set ordered by first differences, in whose elements

$$x = (x_0\, x_1\, x_2\, \ldots\, x_\omega\, \ldots\, x_\beta\, \ldots) = \sum_\beta x_\beta \qquad (0 \leq \beta < \alpha)$$

each x_β runs through the type μ_β, and this product type is to be designated by

$$\prod_\beta{}^{(\nu+1)} \mu_\beta = (\cdots \mu_\beta \cdots \mu_\omega \cdots \mu_2\, \mu_1\, \mu_0)^{(\nu+1)}.$$

If all types μ_a equal μ and the principal element m_a is the same $= m$ in all of them, then the product $\prod_a{}^{(\nu+1)} \mu_a$ turns into the power $\mu_m^{(\nu+1)}(\alpha)$.

II. The Higher Continua

§1
Fundamental Definitions

Let $M = \{m\}$ be a simply ordered set, μ its order type.

The aggregate M^b of all elements (b itself excluded) that *precede* a fixed element b is called an *initial segment* [[Anfangsstrecke]] of M.

The aggregate M_a of all elements (a itself excluded) that *follow* a fixed element a is called an *end segment* [[Endstrecke]] of M.

The aggregate M_a^b of all elements (a and b excluded) that follow a fixed element a and precede a later element b, thus *lying* between a and b, is called a *middle segment* [[Mittelstrecke]] of M.

Initial, middle, and end segments are referred to collectively by the term *segments* [[Strecke]].

Middle segments along with their boundary elements [[Randelementen]] ($a + M_a^b + b$) are called *intervals* [[Intervalle]].

We extend the concept of rank ordering of elements to subsets. If A and B are subsets of M, then:

$m < A$ means m precedes all elements of A;
$m > A$ means m follows all elements of A;
$A < B$ means each element of A precedes each element of B.

The aggregate M^B of all elements that precede a fixed subset B is called an *initial piece* [[Anfangsstück]] of M.

The aggregate M_A of all elements that follow a fixed subset A is called an *end piece* [[Endstück]] of M.

The aggregate M_A^B of all elements that follow a fixed subset A and precede a later B ($B > A$), thus lying *between* A and B, is called a *middle piece* [[Mittelstück]] of M.

Initial, middle, and end pieces are referred to collectively by the term *pieces* [[Stücke]].

A piece can also be defined as a subset of M that together with the two elements a and b also contains every element of the middle segment M_a^b.

If $\mu = \alpha + \beta + \gamma$, then α and $\alpha + \beta$ are initial pieces, γ and $\beta + \gamma$ are end pieces, and β is a middle piece of μ.

Segments are special cases of pieces; indeed, initial segments are special cases of initial pieces, etc. If $\mu = \pi + 1 + \varrho + 1 + \sigma$, then π and $\pi + 1 + \varrho$ are initial segments, σ and $\varrho + 1 + \sigma$ are end segments, and ϱ is a middle segment ($1 + \varrho + 1$ an interval) of M.

M is said to be *cofinal* [[confinal]] with its subset A if no element of M follows A ($M_A = 0$).

M is said to be *coinitial* with A if no element of M precedes A ($M^A = 0$).

M is said to be *coextensive* with A if it is both cofinal and coinitial with A ($M_A = M^A = 0$).

Two separated subsets A and B ($A < B$) are called *contingent* if no element of M lies between them ($M_A^B = 0$).

Theorem A. If M' is a subset of M and M'' is a subset of M', and if M is cofinal (coinitial, coextensive) with M' and M' with M'', then M is also cofinal (coinitial, coextensive) with M''.

Since an element $m > M''$ would either have to belong to M', or if not, precede at least one element m' of M', so that $m' > m > M''$, in both cases M' would not be cofinal with M''. The proofs for the words in parentheses are similar.

If A is a subset without a last element and M_A has a first element b, then b is called *the (upper) limit [[Limes]] of A* and *an (upper) limit element [[Grenzelement]] in M*. In this case, M^b is cofinal with A.

If B is a subset without a first element and M^B has a last element a, then a is called *the (lower) limit of B* and *a (lower) limit element in M*. In this case, M_a is coinitial with B.

The type μ is said to be cofinal (coinitial) with the type α if M contains a subset A of type α with which M is cofinal (coinitial).

If μ is cofinal (coinitial) with α, then the inverse type μ^* is coinitial (cofinal) with α^*.

By $\omega, \omega_1, \omega_2, \ldots$ we understand the smallest ordinal numbers of the 2nd, 3rd, 4th, $\ldots$ number classes; ω_ν is thus the smallest ordinal number of cardinality $\aleph_\nu$. We restrict ourselves to those number classes or alephs *with finite index ν*. Sets of type ω_ν or ω_ν^* are called *sequences*, indeed ω_ν-sequences or ω_ν^*-sequences; ω-sequences and ω^*-sequences are also called *fundamental sequences [[Fundamentalreihen]]*.

The upper limit of an ω_ν-sequence contained in M is called an ω_ν-limit in M; the lower limit of an ω_ν^*-sequence contained in M is called an ω_ν^*-limit in M. The ω-limits and ω^*-limits (limits of fundamental sequences) are also called *fundamental limits [[Fundamentallimites]]*. In reference to these, we retain CANTOR's definitions:

M is called *closed [[abgeschlossen]]* if every fundamental sequence in M has a limit.

M is called *dense in itself [[in sich dicht]]* if every element in M is a fundamental limit.

M is called *perfect* if it is both closed and dense in itself.

Theorem B. A type cannot be cofinal with two different numbers of the sequence $\omega, \omega_1, \omega_2, \ldots$.

Proof. Let A and B be two well-ordered subsets of M of types ω_ν and ω_π, and let M be cofinal with both. The "least multiple" [["kleinste Vielfache"]] $C = \mathfrak{M}(A, B)$, i.e., the aggregate of all elements that belong to at least one of the two sets A or B, is likewise a well-ordered subset of M. Now for instance, let $\omega_\nu < \omega_\pi$, then the type γ of C is certainly $\geq \omega_\pi$, thus $\gamma > \omega_\nu$; consequently, A *lies in a segment [[Abschnitt]]* (= initial segment) of C; thus there are elements in C, so in M, that follow A — in contradiction to the assumption that M is supposed to be cofinal with A. Therefore ω_ν and ω_π cannot be different.

(If one goes beyond the sequence $\omega\,\omega_1\,\omega_2\ldots$, this conclusion no longer holds. For example, $\omega_\omega = \omega + \omega_1 + \omega_2 + \cdots$ is cofinal with ω.)

It also follows immediately from B that μ can not be coinitial with two different inverse ordinal numbers from the sequence $\omega^*\,\omega_1^*\,\omega_2^*\ldots$.

Therefore no element of a set can simultaneously be both an ω_ν-limit and an ω_π-limit ($\nu \neq \pi$) or both an ω_ν^*-limit and an ω_π^*-limit; however it can possibly be both an ω_ν-limit and an ω_π^*-limit.

If μ is cofinal with ω_ν, thus M contains a subset
$$A = (x_0\,x_1\,x_2\,\ldots\,x_\omega\,x_{\omega+1}\,\ldots\,x_\alpha\,\ldots)$$
of type ω_ν after which no element follows, then understand by T_α the aggregate of those elements in M not belonging to A after which x_α but no earlier element of A follows; every element of M then belongs either to A or to one of the pieces T_α, and the decomposition
$$M = T_0 + x_0 + T_1 + x_1 + \cdots + T_\omega + x_\omega + \cdots = \sum_\alpha (T_\alpha + x_\alpha),$$
$$\mu = (\tau_0 + 1) + (\tau_1 + 1) + \cdots + (\tau_\omega + 1) + \cdots = \sum_\alpha (\tau_\alpha + 1)$$
is valid, where the types τ_α in part or all together (for $\mu = \omega_\nu$) can be 0. Thus τ_0 is an initial segment of μ, and the pieces of the form $\tau_{\alpha+1}$ are middle pieces of μ; whereas the pieces τ_α (α a limit number) are certainly not segments since M^{T_α} is cofinal with α and does not have a last element. But τ_α can possibly have a first element, in which case through a change in notation one can get $\tau_\alpha = 0$. For a type cofinal with ω, one has
$$\mu = (\tau_0 + 1) + (\tau_1 + 1) + \cdots = \sum_\nu (\tau_\nu + 1).$$

For example, since the set of the rational numbers ordered by magnitude is cofinal with the set of whole numbers, the type η is cofinal with ω, and because each segment of η itself has the type η,
$$\eta = \eta + 1 + \eta + 1 + \cdots = (\eta + 1)\omega = \eta + (1 + \eta)\omega;$$
an analogous result holds for the type of the unbounded linear continuum. Both types are also coinitial with ω^* and coextensive with $\omega^* + \omega$.

§2

The Powers $\mu(\alpha)$

Let A be a well-ordered, finite or countable set; thus let the type α be a number of the 1st or 2nd number class. We denote the elements of A, which we are going to call "places," by the types of the corresponding initial segments [[Anfangsstrecken]] (initial segments [[Abschnitte]]) in the following way:
$$0\,1\,2\,\ldots\,\omega\,\omega+1\,\ldots\,\omega\cdot 2\,\ldots\,\beta\,\ldots, \quad 0 \leqq \beta < \alpha.$$

By assigning an element x_β taken from an ordered set M of type μ to each place β, we get a *covering* of A by M,

$$x = (x_0\, x_1\, \ldots\, x_\omega\, \ldots\, x_\beta\, \ldots) = \sum_\beta x_\beta.$$

The totality of the elements x is the covering set of A by M and has cardinality $\mathfrak{m}^{\mathfrak{a}}$. We order it by the principle of *first differences*: namely, if β is the *first place* at which the two coverings $x = \sum x_\beta$ and $y = \sum y_\beta$ differ, then $x < y$ holds if $x_\beta < y_\beta$. The type $\xi = \mu(\alpha)$ of the set $\{x\}$, ordered in this way, has the properties of a *power* μ^{α^*}, that is to say,

$$\mu(1) = \mu, \quad \mu(\alpha)\mu(\gamma) = \mu(\gamma + \alpha), \quad \xi(\gamma) = \mu(\alpha \cdot \gamma).$$

Then μ is called the *base* and α the *argument*. In the system of powers that was developed in the previous Note I, $\mu(\alpha)$ was classified as the special case (treated in (D)) of *powers of the second class* that are independent of the choice of a principal element. The notation $\mu'(\alpha)$ assigned there can be replaced by the more convenient $\mu(\alpha)$ since we deal *exclusively* with the powers of the second class in the present investigation.

In this section, a *sufficient* criterion shall be given so that all the powers $\mu(\alpha)$ yield *distinct* order types as α runs through the first and second number classes.

First, we remark that the other extreme, namely, that all the types $\mu(\alpha)$ are identical, can also occur for an appropriately chosen base. Certainly, among the order types known so far no such type is found; the type η of the set of rational numbers is in fact equal to its square and to each of its finite powers, but not to $\eta(\omega)$: since η is countable, but $\eta(\omega)$ has the cardinality $(\aleph_0^{\aleph_0})$ of the continuum. However, the generalized construction of powers from Article I allows such types to be constructed. For if β is a type that satisfies the equation $\beta = \beta\omega$, then one can easily prove *by transfinite induction*[1]) that $\beta = \beta\alpha$ for every number α of the first two

---

[1]) For the case of the first two number classes, proof by transfinite induction consists of applying inferences from α to $\alpha + 1$ and from an ω-sequence to its limit. It can be formulated as follows:

If an assertion $\varphi(\alpha)$ satisfies the following conditions
(1) $\varphi(0)$ is true,
(2) from the truth of $\varphi(\alpha)$ follows that of $\varphi(\alpha + 1)$,
(3) from the truth of $\varphi(\alpha)\ \varphi(\alpha_1)\ \varphi(\alpha_2)\ \ldots$ follows that of $\varphi(\alpha_\omega)$,
then $\varphi(\alpha)$ is true for each α.

In fact, were there numbers for which $\varphi(\alpha)$ were false, then among them there would have to be a first one β; thus $\varphi(\beta)$ would be false, but $\varphi(\alpha)$ would be true for $\alpha < \beta$. And this violates (3) for a limit β, (2) for a number with a predecessor, and (1) for $\beta = 0$.

If $\varphi(\alpha)$ is supposed to be proved for only a *segment* of the second number class, then only correspondingly weakened assumptions are needed; an example of this is the proof of Theorem C that follows in the text.

We have not regarded a formulation of inference by transfinite induction, whose significance as a method was first properly stressed by A. SCHOENFLIES (Bericht p. 45), as superfluous; because even now, the opinion is frequently expressed that transfinite induction requires an unexecutable succession of individual thought acts. As long as the numbers of the second number class were defined only through "generation principles" [["Erzeugungsprinzipien"]], this objection could be considered as justifiable; however, in

128 number classes; e.g., among others, the type η and the left-bounded linear continuum are such types. Now if one forms, with such a β as argument and with an arbitrary base ν and with an arbitrary principal element n, the power of the *second class*

$$\mu = \nu_n(\beta),$$

it then follows from the rules of powers that $\mu(\alpha) = \mu$.

Between the two extremes of equality of all $\mu(\alpha)$ and the distinctness of all $\mu(\alpha)$ lie the cases of periodic recurrence of equal types, which we do not go into further.

Devoting ourselves to a closer examination of the type $\xi = \mu(\alpha)$, we contemplate two elements $x = \sum x_\beta$ and $X = \sum X_\beta$ ($x < X$); the aggregate of all elements between x and X should be called a *middle segment* of ξ; if we add the boundary elements x, X, then we speak of an *interval* $[x, X]$. We

129 say that the interval $[x, X]$ is of *order* β if β is the place of first difference, that is to say, if x_β and X_β ($> x_\beta$) are the first differing elements of x and X; thus one could then write

$$x = (a_0\, a_1\, \ldots \mid x_\beta\, x_{\beta+1}\, \ldots), \quad X = (a_0\, a_1\, \ldots \mid X_\beta\, X_{\beta+1}\, \ldots),$$

where all of the elements to the left of the stroke agree.

The order of an interval of $\mu(\alpha)$ is always $< \alpha$.

Each subinterval of an interval of order β is of order $\geqq \beta$.

We now make an initial hypothesis regarding the base μ:

Hyp. 1. The type μ is bounded, i.e., it has a first and a last element.

Then there are most extensive or maximal intervals of order β, namely, those where $x_\beta\, x_{\beta+1} \ldots$ run through *all* elements of μ, while the preceding elements are fixed. We call such intervals *principal intervals* [[*Hauptintervalle*]] and denote them by

$$(a_0\, a_1\, \ldots \mid_\beta \ldots),$$

also possibly by $(a_0\, a_1\, \ldots\, a_{\beta-1} \mid \ldots)$ if β has an immediate predecessor[1] $\beta - 1$.

All principal intervals of order β are similar and have the type $\mu(-\beta+\alpha)$. Because $\alpha = \beta + (-\beta + \alpha)$, we have $\mu(\alpha) = \mu(-\beta + \alpha) \cdot \mu(\beta)$, and the whole type is formed by inserting the aforementioned principal intervals[2] in the type $\mu(\beta)$.

We call two intervals *exclusive* if they have at most one element in common, and we make a second hypothesis regarding the base μ:

Hyp. 2. No set of exclusive intervals of μ can itself have the type μ.

---

the deduction above, each successive infinity is eliminated just as in the definition of the numbers α as order types of well-ordered, countable sets.

[1]) In the description of subtraction, I allow myself to deviate from G. CANTOR (Math. Ann. 49 (1897), p. 218); if the equation $\alpha + \beta = \gamma$, be it by itself or with the addition of restrictive qualifications, allows a unique solution for α or β, then I denote these by $\alpha = \gamma - \beta$, $\beta = -\alpha + \gamma$.

[2]) For a type μ that does not satisfy Hyp.1, the subsets $(a_0\, a_1\, \ldots \mid_\beta \ldots)$, as defined above, are *middle pieces* of the type $\mu(-\beta + \alpha)$, but not intervals.

Among the types that satisfy both hypotheses are the finite numbers μ (in which there are at most $\mu-1$ exclusive intervals), generally types of well-ordered sets with a last element, and further, the *bounded linear continuum* ϑ, in which each set of exclusive intervals is at most countable, etc.

Let us now consider two sets $\{x\}$ and $\{y\}$ of the types $\mu(\alpha)$ and $\mu(\gamma)$, where
$$x = (x_0\, x_1\, x_2\, \ldots), \quad y = (y_0\, y_1\, y_2\, \ldots),$$
and let us assume that a similarity mapping $y = f(x)$ exists; we call the elements and subsets of x the preimages [[Urbilder]] and the corresponding elements and subsets of y the images [[Abbilder oder Bilder]]. We then prove the lemma:

Theorem C. There exists an element $a = (a_0\, a_1\, a_2\, \ldots)$ in $\{x\}$ with the property that each principal interval $(a_0\, a_1\, \ldots\, |_\beta\, \ldots)$ has an interval of order $\geq \beta$ as its image.

We prove this theorem, which constitutes an assertion for every number β $(0 \leqq \beta < \alpha)$, by transfinite induction.

(C_1). *Theorem C is valid for $\beta = 1$.*

(It is also valid for $\beta = 0$ since the whole sets, which are intervals of order 0, correspond to one another; however, we immediately take $\beta = 1$ in order to make apparent the decisive point, the effect of Hyp. 2.)

The images of the principal intervals of the 1st order $(x_0\,|\ldots)$ cannot all be intervals of the 0th order. For if that were the case, then the image interval of $(x_0|\ldots)$ would be bounded by two elements
$$y = (y_0\, y_1\, \ldots), \quad Y = (Y_0\, Y_1\, \ldots), \quad y_0 < Y_0,$$
and similarly the image of the interval $(x_0'\,|\,\ldots)$ would be bounded by
$$y' = (y_0'\, y_1'\, \ldots), \quad Y' = (Y_0'\, Y_1'\, \ldots), \quad y_0' < Y_0';$$
moreover, for $x_0 < x_0'$, it must be that $Y < y'$, thus that $Y_0 \leqq y_0'$. Hence each element x_0 would determine an interval $[y_0, Y_0]$, and all these intervals of μ would be exclusive; thus in μ there would be a set of exclusive intervals similar to μ, contrary to Hyp. 2. Thus not all principal intervals $(x_0\,|\ldots)$ have images of order 0; i.e., there is at least one element a_0 such that the image of the principal interval $(a_0\,|\ldots)$ is an interval of order ≥ 1.

(C_2). *If Theorem C holds for β, then it also holds for $\beta + 1$ $(< \alpha)$.*

Now let $a_0\, a_1\, \ldots\, a_\gamma\, \ldots$ $(\gamma < \beta)$ be already given so that the principal interval of order β: $(a_0\, a_1\, \ldots\, |_\beta\, \ldots)$ becomes an interval $[y, Y]$ of order $\varrho \geqq \beta$:
$$y = (b_0\, b_1\, \ldots\,|\, y_\varrho\, y_{\varrho+1}\, \ldots), \quad Y = (b_0\, b_1\, \ldots\,|\, Y_\varrho\, Y_{\varrho+1}\, \ldots), \quad y_\varrho < Y_\varrho.$$
We consider the next principal intervals of order $\beta + 1$
$$(a_0\, a_1\, \ldots\, a_\gamma\, \ldots\, x_\beta\,|\, \ldots), \quad x_\gamma = a_\gamma \quad \text{for } \gamma < \beta.$$
Their images are subintervals of an interval of order ϱ, hence of order $\geqq \varrho$. They cannot all be of order ϱ; for again, that would necessitate that the interval $[y_\varrho, Y_\varrho]$ contain a set of exclusive intervals similar to μ. Thus there

is at least one element a_β with the property that the image of the principal interval $(a_0\, a_1\, \ldots\, a_\beta \mid \ldots)$ of order $\beta+1$ is an interval of order $\geq \varrho+1 \geq \beta+1$.

(C$_3$). *If Theorem C holds for an increasing fundamental sequence of numbers $\beta\,\beta_1\,\beta_2\,\ldots$, then it also holds for their limit β_ω ($\beta_\omega < \alpha$).*

Hence let $\beta < \beta_1 < \beta_2 < \cdots$ and $a_0\, a_1 \ldots a_\beta \ldots a_{\beta_1} \ldots a_{\beta_2} \ldots$ be already determined so that the principal intervals

$$(a_0\, a_1\, \ldots\, |_\beta\, \ldots),$$
$$(a_0\, a_1\, \ldots\, a_\beta\, \ldots\, |_{\beta_1}\, \ldots),$$
$$(a_0\, a_1\, \ldots\, a_\beta\, \ldots\, a_{\beta_1}\, \ldots\, |_{\beta_2}\, \ldots),$$

etc., have images of order $\varrho\,\varrho_1\,\varrho_2\,\ldots: \varrho \geq \beta, \varrho_1 \geq \beta_1, \varrho_2 \geq \beta_2, \ldots$. The principal interval of order β_ω

$$(a_0\, a_1\, \ldots\, a_\beta\, \ldots\, a_{\beta_1}\, \ldots\, a_{\beta_2}\, \ldots\, |_{\beta_\omega}\, \ldots)$$

is a common subinterval of all the above, and its image, call its order σ, is thus a common subinterval of the aforementioned image intervals. Consequently,

$$\sigma \geq \varrho_1 \geq \beta_1 > \beta,$$
$$\sigma \geq \varrho_2 \geq \beta_2 > \beta_1,$$
$$\sigma \geq \varrho_3 \geq \beta_3 > \beta_2,$$

etc.; hence σ is greater than all the numbers $\beta\,\beta_1\,\beta_2\,\ldots$ and is therefore at least equal to the limit of this sequence; thus $\sigma \geq \beta_\omega$.

With (C$_1$), (C$_2$), (C$_3$) together, Theorem C is of course proved by means of transfinite induction.

Now the following theorem is immediately obvious:

Theorem D. All powers $\mu(\alpha)$ of a type μ that is bounded on both sides and that contains no set of exclusive intervals similar to itself are distinct.

Indeed, were $\mu(\alpha) = \mu(\gamma)$ for $\alpha > \gamma$, this would give a similarity mapping of both sets and therefore an element $a = (a_0\, a_1\, \ldots)$ in $\mu(\alpha)$ with the property given in C. In particular, $(a_0\, a_1\, \ldots\, |_\gamma\, \ldots)$, the principal interval of order γ, would have to have as its image an interval of order $\geq \gamma$; but such intervals do not even exist in $\mu(\gamma)$, and the aforementioned mapping is thus impossible.

In particular, the $\aleph_1$ powers $\mu(\alpha)$ of a finite or non-limit number base are thus all distinct, as are also the $\aleph_1$ powers $\vartheta(\alpha)$ of the linear continuum ϑ.

§3
Dense in itself, Closed, Perfect Types

In this §, we give theorems that refer to fundamental sequences and their limits in a set of type $\mu(\alpha)$.

Theorem E. If the argument is a limit number or the base is a dense in itself type, then the power is a dense in itself type.

First, let us assume that α is a limit number and that the base μ is arbitrary (naturally $\mu > 1$).

Let $a = (a_0\, a_1\, \ldots\, a_\beta\, \ldots)$ be an element. Then two cases are possible.

1st: *either* there is an ω-sequence $\gamma\, \delta\, \varepsilon\, \ldots$ with upper limit α, so that all the elements $a_\gamma\, a_\delta\, a_\varepsilon\, \ldots$ in μ have *predecessors* $(x_\gamma < a_\gamma\, , x_\delta < a_\delta\, , \ldots)$, predecessors taken in the general sense, not in the sense of "immediate" predecessors. Then a is the upper limit of the following fundamental sequence $A_\gamma < A_\delta < A_\varepsilon < \cdots$,

$$A_\gamma = (a_0\, a_1\, \ldots \mid x_\gamma\, \ldots),$$
$$A_\delta = (a_0\, a_1\, \ldots\, a_\gamma\, \ldots \mid x_\delta\, \ldots),$$
$$A_\varepsilon = (a_0\, a_1\, \ldots\, a_\gamma\, \ldots\, a_\delta\, \ldots \mid x_\varepsilon\, \ldots),$$
$$\text{etc.},$$

where in each of these elements the places to the left of the stroke coincide with the a_β and the relevant predecessor and arbitrary elements follow from then on.

2nd: *or* there is no such ω-sequence. This is only possible if the base μ has a first element 0 and the elements a_β that differ from 0 lie in an *initial segment* [[Abschnitt]](=initial segment [[Anfangsstrecke]]) of α, so that from some particular place γ on $a_\gamma = a_{\gamma+1} = \cdots = 0$. Then, however, there are ω-sequences $\gamma\, \delta\, \varepsilon\, \ldots$ with upper limit α such that all the elements $a_\gamma\, a_\delta\, a_\varepsilon\, \ldots$ have *successors* in μ, and it follows exactly as above that the element a is the limit of an ω^*-sequence. Of course, a can also be an ω-limit, but that depends on the nature of the base. If μ is an unbounded type, then each element in $\mu(\alpha)$ is both an ω-limit and an ω^*-limit.

Second, let μ be dense in itself and α arbitrary; we still need to prove E for the case when α is not a limit number, but rather it has an immediate predecessor $\beta = \alpha - 1$. Since a_β is surely the limit in μ of a fundamental sequence $x_\beta\, x'_\beta\, x''_\beta\, \ldots$, the element $a = (a_0\, a_1\, \ldots\, a_\beta)$ is, for its part, the limit of the fundamental sequence

$$(a_0\, a_1\, \ldots\, x_\beta), \quad (a_0\, a_1\, \ldots\, x'_\beta), \quad (a_0\, a_1\, \ldots\, x''_\beta), \quad \ldots,$$

whose elements agree with those of a up to the last place. With that, E is proved.

Theorem F. If μ is a bounded, closed type, then $\mu(\alpha)$ is likewise a bounded, closed type.

We also count the finite numbers among these types since the definition of closedness is formally satisfied through the absence of fundamental sequences.

For the proof of F, let us consider an ω-sequence of elements $b\, b'\, b''\, \ldots$ in $\mu(\alpha)$. Let the intervals $[b\, b'], [b\, b''], [b\, b'''], \ldots$ be of the orders $\beta'\, \beta''\, \beta'''\, \ldots$ (cf. §2); these numbers are ≥ 0 and $< \alpha$; in addition, $\beta' \geqq \beta'' \geqq \beta''' \geqq \cdots$ since each interval is contained in the following one. Because there are no ω^*-sequences within a well-ordered set, these numbers must reach their minimum β in a finite sequence of steps; we call $[b\, c]$ the first interval of order β; so we have now chosen a subsequence $b\ c\ c'\ c''\ldots$ from the original ω-sequence with the property that all the intervals $[b\ c], [b\ c'], \ldots$ have order β; i.e., all the elements $b\ c\ c'\ldots$ have the same entries $a_0\ a_1 \ldots$ up to but excluding x_β; they belong to the principal interval $(a_0\,a_1 \ldots |_\beta \ldots)$. This analysis is continued for the intervals $[c\ c'], [c\ c''], \ldots$ whose orders $\gamma'\ \gamma''\ldots$ are non-increasing $\geq \beta$ and $< \alpha$; consequently, they have a minimum $\gamma \geq \beta$, let it be attained first in the interval $[c\ d]$, etc. So continuing, we obtain an ω-sequence $b\ c\ d\ e\ldots$, a subsequence of the original one, with the property that all the intervals $[b\ x]$ are of order β, all the intervals $[c\ x]$ are of order γ, all the intervals $[d\ x]$ are of order δ, etc., where

$$0 \leq \beta \leq \gamma \leq \delta \leq \cdots < \alpha.$$

Two cases are possible here.

1st: *either* the numbers $\beta\ \gamma\ \delta\ \ldots$ have a maximum $\kappa < \alpha$.

Then from a certain element k on the intervals $[k\ l], [l\ m], [m\ n], \ldots$ are of order κ. That is, the elements from k on have the form

$$(a_0\,a_1 \ldots |\,x_\kappa \ldots), \quad (a_0\,a_1 \ldots |\,x'_\kappa \ldots), \quad (a_0\,a_1 \ldots |\,x''_\kappa \ldots), \quad \ldots,$$

where $x_\kappa < x'_\kappa < x''_\kappa < \cdots$ is an ω-sequence in μ. By hypothesis, they have a limit a_κ, and also by hypothesis μ has a first element 0; it follows that the ω-sequence above has an upper limit, namely, the element

$$(a_0\,a_1 \ldots a_\kappa\,|\,0\,0\,0\,\ldots),$$

respectively, $(a_0\,a_1 \ldots a_\kappa)$ should κ be the last place of α ($\kappa = \alpha - 1$).

2nd: *or* the numbers $\beta\ \gamma\ \delta \ldots$ have no maximum. Then we can choose from them an ω-sequence and from the elements $b\ c\ d\ldots$ a corresponding ω-sequence (but again, we call these sequences $\beta\ \gamma\ \delta \ldots$ and $b\ c\ d \ldots$) with the property that the intervals $[b\ c], [c\ d], [d\ e], \ldots$ are of increasing orders $\beta < \gamma < \delta < \cdots$. Let κ be the limit of this sequence; $\kappa \leq \alpha$. Then there belongs to:

$b\ c\ d\ e\ f\ \ldots$	a principal interval	$(a_0\,a_1 \ldots |_\beta \ldots),$
$c\ d\ e\ f\ \ldots$	a principal interval	$(a_0\,a_1 \ldots a_\beta \ldots |_\gamma \ldots),$
$d\ e\ f\ \ldots$	a principal interval	$(a_0\,a_1 \ldots a_\beta \ldots a_\gamma \ldots |_\delta \ldots),$

etc. As a consequence, the ω-sequence has an upper limit, namely,

$$(a_0\,a_1 \ldots a_\beta \ldots a_\gamma \ldots a_\delta \ldots |_\kappa 0\,0\,0 \ldots),$$

respectively, $(a_0\,a_1 \ldots a_\beta \ldots a_\gamma \ldots a_\delta \ldots)$ for $\kappa = \alpha$.

Thus because of the hypotheses, each ω-sequence in $\mu(\alpha)$ has a limit. For this only one part of the hypotheses was used, namely, that μ have a first element and that every ω-sequence in μ have a limit. Since μ is supposed to have a last element and every ω^*-sequence in μ is supposed to have a limit, exactly as above, we get that in $\mu(\alpha)$ every ω^*-sequence has a limit. Hence $\mu(\alpha)$ is closed, and with that the proof for F is complete. Combining E and F gives the following:

Theorem G. *If μ is bounded and closed and α is a limit number or if μ is bounded and perfect, then the type $\mu(\alpha)$ is bounded and perfect.*

§4
Isomeric and Homogeneous Types

We define:

A type μ is called *isotomic* [[*isotom*]] if all its initial segments have the same type π and all its end segments have the same type σ.

A type μ is called *isomeric* [[*isomer*]] if all its middle segments have the same type ϱ.

A type μ is called *homogeneous* [[*homogen*]] if all its segments have the same type μ.

As immediate inferences from these definitions, it follows that:

H. $\begin{cases} \textit{Every isotomic type is unbounded.} \\ \textit{Every isomeric type is everywhere dense.} \\ \textit{Every isomeric, unbounded type is also isotomic.} \\ \textit{Every piece of an isomeric type is isomeric.} \\ \textit{Every middle segment of an isomeric type is homogeneous.} \\ \textit{Every homogeneous type is isotomic and isomeric, thus} \\ \quad \textit{unbounded and everywhere dense.} \\ \textit{A type μ is homogeneous if all its initial segments and end} \\ \quad \textit{segments have type μ.} \\ \textit{A type μ is homogeneous if all its middle segments have type μ.} \end{cases}$

The proofs of these statements, excluding a few trivial special cases (sets and pieces of type 1 and the isomeric type 2), are very elementary. An isotomic type cannot have a first element because the initial segment 0 corresponds to it, and thus every element would have to be the first. For the same reason, it cannot have a last element. An isomeric type cannot have a pair of consecutive elements because these enclose the middle segment 0 and then each pair of elements would have to be consecutive. Further, if $b < c$ are two elements of an unbounded isomeric type, choose a third element a ($< b$), so then it is the case that (cf. the terminology in §1)

$$M^b = M^a + a + M_a^b, \quad M^c = M^a + a + M_a^c,$$

and since M_a^b is similar to M_a^c, M^b is also similar to M^c; thus all initial segments of an unbounded isomeric type have the same type π, likewise, all end segments have the same type σ. (The last is through the interpolation of an element $d > c$.) Furthermore, these conclusions also hold for isomeric types bounded on one side or on both sides as long as b is not the first element or c the last element; hence all initial segments (possibly with the first 0 excluded) and all end segments (possibly with the last 0 excluded) of an isomeric type have constant type π, respectively, σ. Put differently, an isomeric type becomes isotomic with the omission of possible boundary elements. Theorems H_4, H_5, H_6 need no commentary, and the proofs of H_7, H_8 are immediately obvious.

Between an isomeric type μ and the types π, ϱ and σ of its initial, middle, and end segments, the relations

(1) $$\begin{cases} \mu = \pi + 1 + \sigma = \pi + 1 + \varrho + 1 + \sigma, \\ \pi = \pi + 1 + \varrho, \quad \varrho = \varrho + 1 + \varrho, \quad \sigma = \varrho + 1 + \sigma \end{cases}$$

exist.

If μ has a first element, then the relations

(2) $$\mu = 1 + \sigma, \quad \pi = 1 + \varrho$$

are added to (1);

if μ has a last element, the relations

(3) $$\mu = \pi + 1, \quad \sigma = \varrho + 1$$

are added; if μ is bounded on both sides, all these relations hold and those following from them

(4) $$\mu = 1 + \varrho + 1, \quad \pi = 1 + \varrho, \quad \sigma = \varrho + 1$$

are valid.

In that case, ϱ is a homogeneous type; thus an isomeric type bounded on both sides becomes homogeneous with the omission of the boundary elements, and a homogeneous type becomes isomeric through the addition of two boundary elements.

I cite as preliminary examples: the type $\omega^* + \omega$ is isotomic; the bounded linear continuum ϑ is isomeric; the unbounded linear continuum, as well as the type η, is homogeneous.

We now turn to investigate our power types $\mu(\alpha)$ with regard to these properties.

Theorem I. If μ is an isotomic type, then so is $\mu(\alpha)$.

If $a = (a_0\, a_1\, a_2\, \ldots)$ and $b = (b_0\, b_1\, \ldots)$ are two elements of $\mu(\alpha)$ and one understands by φ_β a similarity mapping of μ with itself under which the image b_β corresponds to the element a_β, always possible because of the isotomy of μ, then to the element $x = (x_0\, x_1\, x_2\, \ldots)$ let correspond the element

$$y = (y_0\, y_1\, y_2\, \ldots) = (\varphi_0\, x_0,\ \varphi_1\, x_1,\ \varphi_2\, x_2,\ \ldots).$$

This then is a similarity mapping of $\mu(\alpha)$ with itself under which the element b corresponds to the element a; all initial segments of $\mu(\alpha)$ thus have the same type, likewise for all end segments.

Theorem K. If μ is an unbounded, isomeric type, then so is $\mu(\omega^\alpha)$.

Since μ is also isotomic, so too is $\mu' = \mu(\omega^\alpha)$, i.e., all initial segments of μ' have the same type π' and all end segments have the same type σ'. The equality of all middle segments, which is to be proved as well, is based quite fundamentally on the presumed form of the *argument* as a power of ω; namely, these types of the second number class, and only these, are similar to their own end segments. Now let a, b be two elements of μ' and a_β, b_β their first differing elements, thus

$$a = (a_0\, a_1\, \ldots \mid a_\beta\, a_{\beta+1}\, \ldots), \quad b = (a_0\, a_1\, \ldots \mid b_\beta\, b_{\beta+1}\, \ldots), \quad a_\beta < b_\beta.$$

We decompose the middle segment from a to b into the following three pieces:

the aggregate of elements $(a_0\, a_1\, \ldots\, a_\beta\, x_{\beta+1}\, x_{\beta+2}\, \ldots)$,
$(x_{\beta+1}\, x_{\beta+2}\, \ldots) > (a_{\beta+1}\, a_{\beta+2}\, \ldots)$;

the aggregate of elements $(a_0\, a_1\, \ldots\, x_\beta \mid \ldots)$, $a_\beta < x_\beta < b_\beta$;

the aggregate of elements $(a_0\, a_1\, \ldots\, b_\beta\, y_{\beta+1}\, y_{\beta+2}\, \ldots)$,
$(y_{\beta+1}\, y_{\beta+2}\, \ldots) < (b_{\beta+1}\, b_{\beta+2}\, \ldots)$.

Now since each end segment of ω^α again has type ω^α, these three pieces have the following types:

end segment of μ': σ',
μ' inserted in a middle segment of μ: $\mu' \varrho$,
initial segment of μ': π'.

Consequently, $\varrho' = \sigma' + \mu'\varrho + \pi'$ is the type of the middle segment between a and b, thus independent of a and b; whereby, the isomery of μ' is proved.

We now add the following. First of all, there are certain relations between the segment types of μ and μ' that follow from

$$\begin{aligned}\mu' &= \mu(\omega^\alpha) = \mu(1 + \omega^\alpha) = \mu(\omega^\alpha) \cdot \mu = \mu'\mu \\ &= \mu' \cdot (\pi + 1 + \varrho + 1 + \sigma) \\ &= \mu'\pi + \pi' + 1 + \sigma' + \mu'\varrho + \pi' + 1 + \sigma' + \mu'\sigma,\end{aligned}$$

namely,

$$\pi' = \mu'\pi + \pi', \quad \varrho' = \sigma' + \mu'\varrho + \pi', \quad \sigma' = \sigma' + \mu'\sigma.$$

According to Theorem E, μ' is dense in itself, and in fact, each of its elements is both an ω-limit and an ω^*-limit; it follows that π' and ϱ' are cofinal with ω and that σ' and ϱ' are coinitial with ω^*. Thus if we write down an appropriately chosen ω-sequence in ϱ', it is the case that:

$$\varrho' = \varrho' + 1 + \varrho' + 1 + \cdots = (\varrho' + 1)\omega.$$

Furthermore, by choosing one element from each piece $(x_0 \mid \ldots)$, one recognizes that μ' is *coextensive* with μ, i.e., both cofinal and coinitial (§1).

We now seek to make μ' a *homogeneous* type through a special choice of μ.

Then it must be that $\mu' = \pi' = \varrho' = \sigma'$; so μ' is cofinal with ω and coinitial with ω^*. For this, it is also *necessary* that μ be cofinal with ω and coinitial with ω^*, that is, coextensive with $\omega^* + \omega$. For if μ' is to be cofinal with an ω-sequence $a < b < c < \cdots$ contained in it, whose initial elements $a_0 \, b_0 \, c_0 \ldots$ clearly satisfy the condition $a_0 \leq b_0 \leq c_0 \leq \cdots$, then no element of μ can follow all the elements $a_0 \, b_0 \, c_0 \ldots$, and since by hypothesis μ has no last element, $a_0 \, b_0 \, c_0 \ldots$ must contain an ω-sequence with which μ is cofinal. Thus μ must be cofinal with ω, likewise coinitial with ω^*. Indeed, this is also *sufficient*. For if μ is cofinal with ω, μ' is also cofinal with ω (Theorem A), and consequently

$$\mu' = \pi' + 1 + \varrho' + 1 + \varrho' + 1 + \varrho' + \cdots$$
$$= \pi' + 1 + (\varrho' + 1)\omega = \pi' + 1 + \varrho' = \pi';$$

hence $\mu' = \pi'$ and likewise $\varrho' = \sigma'$. On the other hand, if μ' is coinitial with ω^*, then in the same way $\mu' = \sigma'$ and $\varrho' = \pi'$. And if both are the case, μ is certainly coextensive with $\omega^* + \omega$; so $\mu' = \pi' = \varrho' = \sigma'$; hence μ' is homogeneous. The following holds.

Theorem L. If μ is an isomeric type that is coextensive with $\omega^* + \omega$, then $\mu' = \mu(\omega^\alpha)$ is a homogeneous type that is coextensive with $\omega^* + \omega$.

It is very remarkable that in order to strengthen the isomery of μ' to homogeneity, it is not homogeneity, but rather it is coextension with $\omega^* + \omega$ that must be joined as a supplementary condition to the isomery of μ. Of course, this logical connection is a peculiarity of powers of the second class.

The isomeric types *bounded* on both sides behave somewhat differently (we omit those bounded on one side from consideration).

Let $\mu = 1 + \varrho + 1$ be a bounded, isomeric type such that the relations (4) hold. Then in general, $\mu' = \mu(\omega^\alpha)$ is not isomeric, not even isotomic with the omission of boundary elements (as a bounded type, μ' itself cannot be isotomic). If one puts $\mu' = 1 + \varrho' + 1$, then for the mere isotomy of ϱ' the coextension of ϱ' with $\omega^* + \omega$ is needed; to be sure, this condition is sufficient not only for isotomy but also for the homogeneity of ϱ', thus for the isomery of μ'.

In order to establish this in detail, let us first note that, according to Theorem E, the last element of μ' is an ω-limit and the first element is an ω^*-limit. Therefore ϱ' is coextensive with $\omega^* + \omega$.

For any element $a = (a_0 \, a_1 \, \ldots)$ of μ', to find the type of the corresponding initial segment, we must indicate which elements $a_\gamma \, a_\delta \, a_\varepsilon$ of a are not equal to the initial element of μ (let us call it 0); thus

$$a = (0\,0 \ldots a_\gamma \, 0\,0 \ldots a_\delta \, 0\,0 \ldots a_\varepsilon \, 0\,0 \ldots).$$

The sought after initial segment consists of the following principal intervals (§2)

$$(0\,0 \ldots x_\gamma \mid \ldots), \qquad x_\gamma < a_\gamma;$$

$$(0\,0\,\ldots a_\gamma\,0\,0\,\ldots x_\delta\,|\,\ldots), \qquad x_\delta < a_\delta;$$
$$(0\,0\,\ldots a_\gamma\,0\,0\,\ldots a_\delta\,\ldots x_\varepsilon\,|\,\ldots), \qquad x_\varepsilon < a_\varepsilon;$$
etc.;

that is, it consists of a sequence of pieces each of which is formed through the insertion of the type $\mu(\omega^\alpha) = \mu'$ in an initial segment π of μ, thus each of which has the same type $\mu'\pi$; indeed, the set of these pieces has the type of the set of places $\gamma\,\delta\,\varepsilon\,\ldots$, hence a type λ which is $\leqq \omega^\alpha$. Consequently, our initial segment has type $\mu'\pi\lambda$ and depends on λ. In particular, the last initial segment (determined by the final element of μ') has the type $\mu'\pi\omega^\alpha = 1 + \varrho'$ and is cofinal with ω; however, there are even earlier elements whose initial segments are cofinal with ω, e.g., for $\lambda = \omega$.

Hence in any case, for the isotomy of ϱ', it is necessary that for *each* $\lambda < \omega^\alpha$ the type $\mu'\pi\lambda$ be cofinal with ω; in particular, (for $\lambda = 1$) $\mu'\pi$ must be cofinal with ω, from which it follows through reasoning similar to that above (p. 138) that $\pi = 1 + \varrho$ and thus also that ϱ must be cofinal with ω. A completely analogous deduction shows that ϱ must also be coinitial with ω^*.

Indeed, these conditions are sufficient, and certainly not only for the isotomy of ϱ', but also for its homogeneity.

Namely, if ϱ is cofinal with ω, then from (4) it follows that

$$\pi = 1 + \varrho + 1 + \varrho + 1 + \varrho + \cdots = \pi\omega;$$

from this formula in fact, one proves again by transfinite induction that for each λ of the first or second number class $\pi\lambda = \pi$. Consequently, all initial segments of μ' (with the exception of the first 0) have the type $\mu'\pi\lambda = \mu'\pi = 1 + \varrho'$.

In exactly the same way, it follows that if ϱ is coinitial with ω^*, all end segments of μ' (except the last 0) have the type $\mu'\sigma = \varrho' + 1$.

If both are the case, thus ϱ is coextensive with $\omega^* + \omega$, then all initial segments of ϱ' have the same type ϱ', and all final segments likewise have the same type ϱ'; according to Theorem H$_7$, ϱ' is thus homogeneous. With that, we have:

Theorem M. *If $\mu = 1 + \varrho + 1$ is an isomeric type and ϱ is coextensive with $\omega^* + \omega$, then $\mu' = \mu(\omega^\alpha) = 1 + \varrho' + 1$ is also an isomeric type and ϱ' is coextensive with $\omega^* + \omega$.*

§5
Continuous Types

Call a type τ *continuous* [[*stetig*]] (in DEDEKIND's sense) if for each cut [[*Zerschneidung*]] $\tau = \xi + \eta$, either ξ has a last element and η no first element, or ξ has no last element and η has a first element.

A continuous type is everywhere dense.

In conclusion to our reflections upon powers of the second class we prove:

Theorem N. *If μ is bounded and continuous, then $\mu(\alpha)$ is also bounded and continuous.*

Let us imagine a fixed cut $\xi+\eta$ specified in $\mu(\alpha)$. Each interval whose elements belong partly to ξ and partly to η shall be called a *cut* [[*zerschnittenes*]] interval; each interval that is contained completely in ξ or in η shall be called uncut. It is immediately apparent that of two exclusive intervals (§2) at most one can be cut and that an interval that contains a cut subinterval is itself cut.

In particular, let us consider the *principal intervals* (§2) of order β ($<\alpha$)
$$X_\beta = (x_0\, x_1\, x_2\, \ldots \,|_\beta\, \ldots).$$

Since all the principal intervals of order β are exclusive, *for each order β there is at most one cut principal interval*. Moreover, if
$$A_\beta = (a_0\, a_1\, a_2\, \ldots \,|_\beta\, \ldots)$$
is a cut principal interval of order β, then there is certainly one and only one cut principal interval for each order $\gamma < \beta$, namely,
$$A_\gamma = (a_0\, a_1\, a_2\, \ldots \,|_\gamma\, \ldots),$$
because this contains the previous one as a subinterval. The cut principal interval A_β partitions all the remaining principal intervals of order β into two classes $\{X_\beta\}$, $\{Y_\beta\}$ in such a way that $X_\beta < A_\beta < Y_\beta$ and all the elements of X_β belong to ξ, while all the elements of Y_β belong to η. With a somewhat forced but short and understandable choice of words, let us call the number β itself cut or uncut according to whether there is or is not a cut principal interval of order β. So it follows from the above: each predecessor of a cut number is itself a cut number; thus each successor of an uncut number is itself an uncut number. The number 0 is cut since the (only) principal interval of 0th order, the entire type $\mu(\alpha)$, is cut.

Therefore only the following cases are possible:

(1) *Among the numbers β ($<\alpha$) none are uncut.*

Then for each order β there exists one and only one cut principal interval $A_\beta = (a_0\, a_1\, \ldots \,|_\beta\, \ldots)$. The only element common to all of these is $a = (a_0\, a_1\, \ldots)$; then each element $x < a$ belongs to a principal interval $X_\beta < A_\beta$ and is thus an element of ξ; likewise, each element $y > a$ belongs to η. The element a itself can belong to ξ as its last element or to η as its first element.

(2) *There are uncut numbers β ($1 \leqq \beta < \alpha$).*

The set of these numbers has a first element β; then all the numbers $\gamma < \beta$ are cut, and all the numbers $\geqq \beta$ are uncut. Here a further distinction is necessary:

(a) *β is a limit number.* For each $\gamma < \beta$, $A_\gamma = (a_0\, a_1\, \ldots \,|_\gamma\, \ldots)$ is a cut interval. The principal interval of order β
$$A_\beta = (a_0\, a_1\, \ldots\, a_\gamma\, \ldots \,|_\beta\, \ldots)$$
belongs to all intervals A_γ as a common subinterval; as an uncut interval, it belongs entirely to ξ or entirely to η, whereas all the principal intervals $X_\beta < A_\beta$ belong to ξ and all the principal intervals $Y_\beta > A_\beta$ belong to η.

So either ξ has a last element, namely, the last one of A_β, or η has a first element, namely, the first one of A_β.

(b) *β is not a limit number.* Then, setting $\beta = \gamma + 1$ $(0 \leqq \gamma)$, there is a last cut principal interval
$$A_\gamma = (a_0 \, a_1 \, \ldots \, |_\gamma \, \ldots),$$
while the immediately following principal intervals
$$X_{\gamma+1} = X_\beta = (a_0 \, a_1 \, \ldots \, x_\gamma \, | \, \ldots)$$
are all uncut. A part of them, let us call them X_β, belong to ξ, another part
$$Y_{\gamma+1} = Y_\beta = (a_0 \, a_1 \, \ldots \, y_\gamma \, | \, \ldots)$$
belong to η since otherwise A_γ would already be an uncut interval. So $x_\gamma < y_\gamma$. By this means, a cut is produced in the base μ itself, to which, on the basis of the assumed continuity of μ, there corresponds an element a_γ that must be either last among the x_γ or first among the y_γ. An uncut principal interval of order β
$$A_{\gamma+1} = A_\beta = (a_0 \, a_1 \, \ldots \, a_\gamma \, | \, \ldots)$$
is associated to this element a_γ, which being uncut, either belongs entirely to ξ or entirely to η, whereas each $X_\beta < A_\beta$ belongs to ξ, and each $Y_\beta > A_\beta$ belongs to η. Again, either ξ has a last element, the last one of A_β, or η has a first element, the first one of A_β.

Whereby, it is proved that in all cases for a cut $\mu(\alpha) = \xi + \eta$, either ξ has a last element or η has a first element. It is impossible to have both together since μ is everywhere dense, and as is easy to see, it follows immediately that $\mu(\alpha)$ is also everywhere dense. Consequently, the DEDEKIND continuity of $\mu(\alpha)$ is indeed proved.

To conclude this investigation, we specify a type that allows the combination of these separate results: *the bounded linear continuum ϑ.* This type contains no set of exclusive intervals similar to ϑ; it is perfect; it is isomeric since each of its intervals has type ϑ and each of its middle segments has the type of the unbounded linear continuum; with the removal of the boundary elements, it is coextensive with $\omega^* + \omega$; finally, it is continuous in DEDEKIND's sense. Thus through the application of Theorems D, G, M, and N we find:

I. *All powers $\vartheta(\alpha)$ are distinct.*
II. *All powers $\vartheta(\alpha)$ are perfect.*
III. *All powers $\vartheta(\omega^\alpha)$ are isomeric and, with the removal of boundary elements, homogeneous and coextensive with $\omega^* + \omega$.*
IV. *All powers are continuous in DEDEKIND's sense.*

To these we add the easily proved:

V. *All powers $\vartheta(\alpha)$ are invertible* (identical with their inverses).
VI. *All powers $\vartheta(\alpha)$ have the cardinality of the continuum.*

Hence in particular, there are certainly $\aleph_1$ distinct types $\vartheta(\omega^\alpha)$ that have almost all the characteristic properties (perfectness, isomery, continuity, invertibility, cardinality) in common with the continuum, and it is with justification that they may be designated as *continua of a higher level* [[*höherer Stufe*]]: a result that, contrary to views expressed elsewhere, appears quite remarkable and affords a deep insight into the inexhaustible abundance of higher order types. When one considers that through the one attribute of everywhere denseness (up to possible boundary elements) a countable type is already uniquely determined, it seems somewhat surprising that for types of the cardinality of the continuum the combination of such special properties as perfectness, isomery, and continuity still allows such a wide latitude.

III. Homogeneous Types of the Second Infinite Cardinality

§1

In the quest for homogeneous types of the second infinite cardinality, we first pursue an "arithmetic" path, indicated by analogy with the unique countable homogeneous type, by defining for the second number class a kind of *rational number* that everywhere densely fills in the area between the whole numbers. Although this method yields very limited returns, it may indeed be deserving of interest in itself.

In this §, we are always going to understand as "numbers" the types of the *first two number classes with the exception of zero*, thus the elements of the sequence of type Ω

$$1\ 2\ 3\ \ldots\ \omega\ \omega+1\ \ldots\ \alpha\ \ldots.$$

Next we prove a

Lemma: Two numbers α and β ($\alpha > \beta$) uniquely determine a third number ξ so that

$$\beta\xi \leqq \alpha < \beta(\xi+1).$$

For either there is a number ξ such that $\beta\xi = \alpha$, and then there is only one because it is simultaneously the case that $\xi \gtreqless \eta$ and $\beta\xi \gtreqless \beta\eta$. Or there is no such number ξ. Since the numbers $\beta, \beta \cdot 2, \ldots, \beta\xi, \ldots$ must eventually become $> \alpha$, there is a first number η so that $\beta\eta > \alpha$, but $\beta\xi < \alpha$ for $\xi < \eta$. Here η cannot be a limit number because were $\eta = \lim \xi_\nu$, then, as is well known, it would follow that $\beta\eta = \lim \beta\xi_\nu$; but $\beta\xi_\nu < \alpha$; consequently, $\lim \beta\xi_\nu \leqq \alpha$ in contradiction to $\beta\eta > \alpha$. Therefore η is a number of the form $\xi + 1$, and thus there is a uniquely determined number ξ so that $\beta\xi < \alpha < \beta(\xi+1)$.

In a way, the lemma we have just proved constitutes the second number class's "Archimedean Principle."

This having been said, we combine two numbers α and β into a *number pair* $\alpha|\beta$, in which we pay attention to the order of the elements, and we designate the first number α as *numerator* and the second number β as *denominator*.

For now, we restrict ourselves to those number pairs *where the numerator is greater than the denominator*. Thus let $\alpha > \beta = \alpha_1$. The numbers $\alpha\, \alpha_1$ uniquely determine a number ξ_1 so that

$$\alpha_1 \xi_1 \leqq \alpha < \alpha_1(\xi_1 + 1);$$

hence either $\alpha = \alpha_1 \xi_1$ or $\alpha = \alpha_1 \xi_1 + \alpha_2$, $0 < \alpha_2 < \alpha_1$. If the second case happens, $\alpha_1\, \alpha_2$ again determine a number ξ_2 so that

$$\alpha_2 \xi_2 \leqq \alpha_1 < \alpha_2(\xi_2 + 1);$$

hence either $\alpha_1 = \alpha_2 \xi_2$ or $\alpha_1 = \alpha_2 \xi_2 + \alpha_3$, $0 < \alpha_3 < \alpha_2$. So continuing, we recognize that the process must end after a finite number of steps since there are no infinite sequences of decreasing numbers $\alpha > \alpha_1 > \alpha_2 > \cdots$, and thus we obtain a *"continued fraction development"* [[*"Kettenbruchent-*

wicklung"]] *uniquely determined* by the number pair $\alpha|\beta = \alpha|\alpha_1$

(1) $$\left.\begin{array}{rcl}\alpha &=& \alpha_1\xi_1 + \alpha_2 \\ \alpha_1 &=& \alpha_2\xi_2 + \alpha_3 \\ \alpha_2 &=& \alpha_3\xi_3 + \alpha_4 \\ \cdot\ \cdot\ \cdot & & \cdot\ \cdot\ \cdot\ \cdot\ \cdot\ \cdot\ \cdot \\ \alpha_{\nu-2} &=& \alpha_{\nu-1}\xi_{\nu-1} + \alpha_\nu \\ \alpha_{\nu-1} &=& \alpha_\nu\xi_\nu \end{array}\right\} \alpha > \alpha_1 > \alpha_2 > \cdots > \alpha_\nu > 0.$$

The "quotients" $\xi_1\ \xi_2\ \ldots$ are ≥ 1; only the last $\xi_\nu > 1$ since otherwise it would be that $\alpha_{\nu-1} = \alpha_\nu$ and the process would have already ended at the next to last step with $\alpha_{\nu-2} = \alpha_{\nu-1}(\xi_{\nu-1} + 1)$.

On the other hand, the sequence of quotients or the *continued fraction* [[*Kettenbruch*]]

(2) $$\xi = (\xi_1\ \xi_2\ \xi_3\ \ldots\ \xi_{\nu-1}\ \xi_\nu)$$

determines not only one corresponding number pair but infinitely many, since one may choose α_ν arbitrarily; however, the number pair $\alpha|\alpha_1$ is uniquely determined by α_ν and ξ, as one can easily see by reading (1) backwards. From the number pairs belonging to ξ, we select the particular pair $\delta|\delta_1$ that is determined by $\alpha_\nu = 1$ and defined by the equations

$$\begin{array}{rcl}\delta &=& \delta_1\xi_1 + \delta_2 \\ \delta_1 &=& \delta_2\xi_2 + \delta_3 \\ \cdot\ \cdot\ \cdot & & \cdot\ \cdot\ \cdot\ \cdot\ \cdot\ \cdot\ \cdot \\ \delta_{\nu-2} &=& \delta_{\nu-1}\xi_{\nu-1} + 1 \\ \delta_{\nu-1} &=& \xi_\nu. \end{array}$$

Let us call $\delta|\delta_1$ the *reduced* pair belonging to ξ; it follows that every other pair $\alpha|\alpha_1$ belonging to ξ has the form

$$\alpha = \alpha_\nu \cdot \delta, \quad \alpha_1 = \alpha_\nu \cdot \delta_1,$$

and thus that it arises from the reduced pair through left multiplication by an arbitrary number. Therefore the continued fraction development serves to find the "greatest common left divisor" α_ν of the two numbers $\alpha\ \alpha_1$.

If we call two number pairs *equal* when they yield the same continued fraction, it follows that: each number pair is equal to a *reduced* pair and arises from it through left multiplication by an arbitrary number. Two reduced pairs are equal only if their numerators and denominators separately coincide.

In addition, we use the continued fraction representation to carry over the ordering by magnitude of the usual rational numbers to our case. So of two continued fractions, we will designate as smaller the one with the smaller first quotient ($\xi < \eta$ for $\xi_1 < \eta_1$); among the continued fractions beginning with ξ_1, the one-termed (ξ_1) is itself the smallest, and of the rest, the one with a larger second quotient is the smaller, etc. In order to express this ordering in a somewhat simpler way, we recall that ordinary continued fractions can be continued formally with ∞; accordingly, we introduce a symbol Ω which *follows* all the numbers of the first and second number classes, and we agree to also write the continued fraction (2) as the infinite sequence

(3) $$\xi = (\xi_1 \xi_2 \ldots \xi_\nu \, \Omega \, \Omega \, \Omega \, \ldots) = (\xi_1 \xi_2 \xi_3 \ldots).$$

We now have the elements of the form (3) before us, coverings of the sequence ω by elements ξ_λ of the set

$$1 \; 2 \ldots \omega \; \omega+1 \ldots \alpha \ldots \Omega,$$

with the restriction that there should only be a finite number $\nu (\geq 1)$ of elements $\xi_\lambda \neq \Omega$ and that these elements should cover an initial segment [[Anfangsstrecke]] (initial segment [[Abschnitt]]) of ω and that the last of these (ξ_ν) should be > 1.

We order these elements ξ according to the following principle: if ν is the first differing place of the elements ξ and η and $\xi_\nu < \eta_\nu$, then

$$\xi \lessgtr \eta \quad \text{depending on whether } \nu \text{ is } \left. \begin{array}{c} \text{odd} \\ \text{even} \end{array} \right\}.$$

As is easy to see, the set $\{\xi\}$ is hereby transformed into a simply ordered set, whose type we call Ξ. At the same time, we obtain with this a rank ordering for all number pairs $\alpha|\beta$ $(\alpha > \beta)$, in which equal pairs are counted as only one element according to the ranking, and for obvious reasons, we are going to designate this as the *natural rank ordering* of these pairs.

We arrange the number pairs not considered till now, those whose numerators do not exceed their denominators, in a series to the left of Ξ by the stipulations that $\beta'|\alpha' \lesseqgtr \beta|\alpha$ shall hold according as $\alpha'|\beta' \gtreqless \alpha|\beta$ and that all number pairs $\alpha|\alpha$ shall equal the reduced pair $1|1$ and that for $\alpha > \beta$, $\alpha|\beta > 1|1$ is always true. We then get the natural rank ordering of all number pairs and the type

$$\Xi^* + 1 + \Xi;$$

but since, as we shall see, this is not distinct from Ξ, we retain the restriction to number pairs with larger numerators and the type Ξ that we defined above for further investigation.

To abbreviate, we agree that under sum and limit signs α and β $(\beta > 1)$ shall run through the first two number classes, $\lambda \, \mu \, \nu$ shall run through only the first number class, and ξ shall run through the totality (3) of continued fractions; furthermore, for $\xi = (\xi_1 \xi_2 \ldots \xi_\nu)$, we set

$$\alpha + \xi = (\alpha + \xi_1, \xi_2, \ldots, \xi_\nu),$$

(4)
$$(\alpha, \xi) = (\alpha, \xi_1, \xi_2, \ldots, \xi_\nu),$$

This then yields:

$\sum\limits_{\xi}(\alpha,\xi)$, the elements of the middle segment from (α) to $(\alpha+1)$;

$\sum\limits_{\xi}(1,\xi)$, the elements, however, of the initial segment up to (2);

$\sum\limits_{\xi}(\alpha+\xi)$, the elements of the end segment from $(\alpha+1)$ on.

$\sum\limits_{\xi}(\alpha,\xi)$ has the type Ξ^*, and $\sum\limits_{\xi}(\alpha+\xi)$ has the type Ξ. If we decompose the entire type into

$$\sum\limits_{\xi}(1,\xi) + (2) + \sum\limits_{\xi}(1+\xi),$$

then we realize that

$$\Xi = \Xi^* + 1 + \Xi,$$

from which it follows that $\Xi = \Xi^*$ is an *invertible* type and that $\Xi = \Xi + 1 + \Xi$.

Moreover, as is easily seen, Ξ is everywhere dense *but not homogeneous*. Of course, it contains limits of ω-sequences and ω^*-sequences, as well as those of Ω-sequences and Ω^*-sequences, and in fact in the following way. In all circumstances, $\xi = (\xi_1 \xi_2 \ldots \xi_\nu) = (\xi_1 \xi_2 \ldots \xi_\nu \Omega \Omega \ldots)$ is the limit of the sequence

$$\sum\limits_{\beta}(\xi_1 \xi_2 \ldots \xi_\nu \beta),$$

which is an Ω-sequence for *even* ν and an Ω^*-sequence for *odd* ν. Moreover, if ξ_ν is not a limit number, so it has a predecessor $\eta_\nu = \xi_\nu - 1$, then ξ is also the limit of the sequence

$$\sum\limits_{\beta}(\xi_1 \xi_2 \ldots \xi_{\nu-1} \eta_\nu 1 \beta),$$

which is an Ω^*-sequence for *even* ν and an Ω-sequence for *odd* ν. But if ξ_ν is a limit number $= \eta_\omega = \lim\limits_{\mu} \eta_\mu$, then ξ is the limit of the sequence

$$\sum\limits_{\mu}(\xi_1 \xi_2 \ldots \xi_{\nu-1} \eta_\mu),$$

which is an ω^*-sequence for *even* ν and an ω-sequence for *odd* ν. Therefore there are three kinds of elements in Ξ:

(a) $\Omega\Omega^*$-limits like (2) $= \lim\limits_{\beta}(1,1,\beta) = \lim\limits_{\beta}(2,\beta)$,
(b) $\Omega\omega^*$-limits like $(1,\omega)$ $= \lim\limits_{\beta}(1,\omega,\beta) = \lim\limits_{\nu}(1,\nu)$,
(c) $\omega\Omega^*$-limits like (ω) $= \lim\limits_{\nu}(\nu) = \lim\limits_{\beta}(\omega,\beta)$.

Now these three categories of elements, the elements a, b, and c, by virtue of their ordinal distinctness, allow a *separation* in the following way. For each similarity mapping of Ξ to itself, the elements a must again correspond to elements a, the elements b again to elements b, the elements c again to elements c; the same holds for a similarity mapping from Ξ onto one of its pieces and for a similarity mapping of two unbounded (on both sides) pieces of Ξ onto each other.[1]) Thus if we select the set of a-elements from Ξ in the rank ordering that they have in Ξ and we call their type Ξ_a, then in every piece of Ξ that is similar to Ξ the subset of the a-elements again has the type Ξ_a, and the same holds for Ξ_b and Ξ_c. Furthermore, if H is piece of Ξ, unbounded on both sides, and H_a is the type of the a-elements in H, then in each piece of Ξ similar to H, the subset of the a-elements again has the type H_a, and the same holds for H_b and H_c. Supported by this, we will easily be able to prove the homogeneity of the type Ξ_a, respectively, the isomery of the types Ξ_b and Ξ_c. First we just note: Ξ is everywhere dense with respect to all three element categories [[Elementgattungen]], i.e., between two arbitrary elements there always lie elements a, elements b, and elements c. From this it follows: if a certain element is, for example, an ω-limit, then in particular it is the limit of ω-sequences that consist only of a-elements or only of b-elements or only of c-elements; the analogous result holds for all the kinds of limits. Therefore each a-element is an $\Omega\Omega^*$-limit, not only in Ξ but also in Ξ_a; in the same way, every element in Ξ_b is an $\Omega\omega^*$-limit, and every element in Ξ_c is an $\omega\Omega^*$-limit. Furthermore, we note in passing that for the same reasons, not only Ξ, but also Ξ_a, Ξ_b, and Ξ_c are cofinal with Ω and coinitial with Ω^*.

In order to furnish the promised proof that the type Ξ_a is homogeneous and the types Ξ_b and Ξ_c are isomeric, we next investigate the initial segments and end segments of Ξ. The following decomposition of Ξ

$$(1,\xi) \quad (2) \quad (2,\xi) \ \ldots \ (\omega) \quad (\omega,\xi) \ \ldots \ (\alpha) \quad (\alpha,\xi) \ \ldots$$

shows that

$$\Xi = \Xi + (1+\Xi)\Omega,$$

for which we have made use of $\Xi^* = \Xi$. If, for the moment, we set $1+\Xi = \mu$, then $\mu = \mu\Omega$; thus for an arbitrary number α of the 1st or 2nd number class,

$$\mu(\alpha+1) = \mu(\alpha+\Omega) = \mu\Omega = \mu.$$

Since one can always represent a limit number α_ω in the form

$$\alpha_\omega = \alpha + 1 + \beta + 1 + \gamma + 1 + \cdots,$$

it follows that

$$\mu\alpha_\omega = \mu + \mu + \mu + \cdots = \mu\omega.$$

---

[1]) Here, an order-preserving mapping [[gleichsinnige Abbildung]] is presumed. Under an order-reversing mapping [[ungleichsinnige Abbildung]] of Ξ onto Ξ^* the elements a, c, b correspond to the elements a, b, c, from which one draws the conclusion that the types designated as Ξ_b and Ξ_c in the text are inverses of each other and that the type Ξ_a is invertible.

Thus: multiplied on the right by an arbitrary number of the 1st and 2nd number class, μ yields either μ or $\mu\omega$ (these two types are distinct because $\mu = \mu\Omega$ is cofinal with Ω and $\mu\omega$ is cofinal with ω). If we set $\mu\omega = 1 + Z$, so

$$Z = \Xi + (1 + \Xi)\omega,$$

it can now be easily proved that all initial segments of Ξ have either the type Ξ or Z, depending on whether the element in question is an Ω-limit or an ω-limit. Namely:

(1) *every initial segment determined by an element (α) has one of the types Ξ or Z.*

This follows from the decomposition

$$(1,\xi) \quad (2) \quad (2,\xi) \quad \ldots \quad (\beta) \quad (\beta,\xi) \quad \ldots \quad [\beta < \alpha].$$

(2) *every end segment determined by an element (α) has the type Ξ.*

This follows from the decomposition

$$(\alpha,\xi) \quad (\alpha+1) \quad (\alpha+1,\xi) \quad \ldots \quad (\beta) \quad (\beta,\xi) \quad \ldots \quad [\beta > \alpha],$$

or even more simply from the decomposition

$$(\alpha,\xi) \quad (\alpha+1) \quad (\alpha+\xi).$$

(3) *every initial segment determined by an element (α, β) has the type Ξ.*

For $\alpha > 1$, we decompose it into the initial segment up to (α), the element (α), and the middle segment from (α) to (α, β); this last segment has the inverted type of the end segment from the element (β) on, thus $\Xi^* = \Xi$. So our initial segment has the type $\Xi + 1 + \Xi$ or $Z + 1 + \Xi$, and both are equal to Ξ. For $\alpha = 1$, the aforesaid decomposition does not exist, and the similarity of the initial segment up to $(1, \beta)$ with Ξ is immediately evident.

(4) *Every initial segment of Ξ has one of the types Ξ or Z.*

We decompose the initial segment up to $\xi = (\xi_1 \xi_2 \xi_3 \ldots \xi_\nu)$ by interpolating the following elements

$$(\xi_1 \xi_2), \quad (\xi_1 \xi_2 \xi_3 \xi_4 + 1), \quad (\xi_1 \xi_2 \xi_3 \xi_4 \xi_5 \xi_6 + 1), \quad \ldots,$$

which follow one another in the order written and are $< \xi$. The middle segment between the first and second of these interpolated elements obviously has the same type as the initial segment up to $(\xi_3 \xi_4 + 1)$, hence Ξ, and so on; the segment between the last interpolated element and ξ has the type of an initial segment up to (α, β) for each even ν and the type of an initial segment up to (α) for each odd ν, so in any case Ξ or Z. Therefore the type of the entire initial segment is

$$\Xi + 1 + \Xi + 1 + \Xi + \cdots + 1 + \begin{cases} \Xi \\ Z \end{cases}$$

with a finite number of terms; thus $\Xi + 1 + \Xi = \Xi$ or $\Xi + 1 + Z = Z$.

At this point, the additional considerations are very easy. All initial segments of Ξ are of type Ξ or Z; all end segments of Ξ are therefore of type Ξ or $H = Z^*$. Among the four possible combinations, one is eliminated because in Ξ there are no $\omega\omega^*$-limits, and we have, corresponding to the decomposition of Ξ by an element a, b, c:

$$\Xi + 1 + \Xi = \Xi + 1 + H = Z + 1 + \Xi.$$

Each middle segment between two a-elements is an end segment of Ξ and at the same time an initial segment of Ξ, thus of type Ξ.

Each middle segment between two b-elements is an end segment of Ξ and coinitial with ω^*, thus of type H.

Each middle segment between two c-elements is an initial segment of Ξ and cofinal with ω, thus of type Z.

From this it follows:

Each middle segment of Ξ_a has the type Ξ_a: Ξ_a is *homogeneous*.

Each middle segment of Ξ_b has the type H_b: Ξ_b is *isomeric* and H_b is *homogeneous*.

Each middle segment of Ξ_c has the type Z_c: Ξ_c is *isomeric* and Z_c is *homogeneous*.

Let us summarize the results obtained:

The pairs of numbers in their natural rank order form an invertible type of the second infinite cardinality that contains three kinds of elements:

(a) $\Omega\Omega^*$-*limits*; (b) $\Omega\omega^*$-*limits*; (c) $\omega\Omega^*$-*limits*.

These element categories form three separate types Ξ_a, Ξ_b, Ξ_c *with the following properties*:

Ξ_a *is an invertible, homogeneous type of the second infinite cardinality in which each element is an* $\Omega\Omega^*$-*limit*.

Ξ_b *is an isomeric type, and each of its middle segments* H_b *is a homogeneous type of the second infinite cardinality in which each element is an* $\Omega\omega^*$-*limit*.

Ξ_c ($= \Xi_b^*$) *is an isomeric type, and each of its middle segments* Z_c ($= H_b^*$) *is a homogeneous type of the second infinite cardinality in which each element is an* $\omega\Omega^*$-*limit*.

Apart from P. DUBOIS-REYMOND's "infinitary pantachie," which to this day has not been defined so as to be free of doubt, the type Ξ_a furnishes the first example, as far as I am aware, of an everywhere dense type of smallest possible cardinality *in which no fundamental sequence has a limit*.

§2

As natural as the path taken in the previous § may seem, indeed the result found, consisting of three homogeneous types, in itself bears a fortuitous and incomplete character. That is why, for the present, we turn away from actual construction methods for homogeneous types of the second infinite cardinality, and instead we try to classify, a priori, the forms that are at all possible. This systematology, a classification into *four groups, nine genera [[Gattungen]] and fifty species*, holds, by the way, not only for homogeneous types of the second infinite cardinality but generally for homogeneous types *that contain sequences of at most the second infinite cardinality* (thus $\omega, \omega^*, \Omega, \Omega^*$-sequences).

Types that are unbounded on both sides and only contain sequences of at most the second infinite cardinality are cofinal either with ω or Ω and coinitial either with ω^* or Ω^* and therefore can be divided into four groups. In order to now and again save words, we are going to introduce left and right feathered arrows as graphic symbols

whose meaning is:

> one right feather: cofinal with ω.
> one left feather: coinitial with ω^*.
> two right feathers: cofinal with Ω.
> two left feathers: coinitial with Ω^*.

For homogeneous types, which are of course similar to all their segments, this means a classification at the same time with respect to the limit properties of their elements.

The four groups of homogeneous types

I. Types ⟷ cofinal with ω, coinitial with ω^*.
   Each element is an $\omega\omega^*$-limit, and Ω-sequences and Ω^*-sequences do not have limits.

II. Types ⟷↠ cofinal with Ω, coinitial with ω^*.
   Each element is an $\Omega\omega^*$-limit, and ω-sequences and Ω^*-sequences do not have limits.

III. Types ↞⟶ cofinal with ω, coinitial with Ω^*.
   Each element is an $\omega\Omega^*$-limit, and Ω-sequences and ω^*-sequences do not have limits.

IV. Types ↞↠ cofinal with Ω, coinitial with Ω^*.
   Each element is an $\Omega\Omega^*$-limit, and ω-sequences and ω^*-sequences do not have limits.

We make a further distinction according to whether or not Ω-sequences and Ω^*-sequences actually occur.

The nine genera of homogeneous types

IA. Types ⟷ that contain Ω-sequences as well as Ω^*-sequences.
 B. Types ⟷ that contain Ω-sequences, but no Ω^*-sequences.
 C. Types ⟷ that contain Ω^*-sequences, but no Ω-sequences.
 D. Types ⟷ that contain neither Ω-sequences nor Ω^*-sequences.
IIA. Types ⟷↠ that also contain Ω^*-sequences.
  B. Types ⟷↠ that contain no Ω^*-sequences.
IIIA. Types ↞⟶ that also contain Ω-sequences.
  C. Types ↞⟶ that contain no Ω-sequences.
IV. Types ↞↠ .

As the next distinguishing feature, we choose the combinations of *contingent sequences* or, more simply, the *gaps* [[*Lücken*]] of the type in question. We called (II, §1) two subsets of an ordered set contingent if no element of the set lies between them. An ω_ν-sequence that is contingent with an immediately following ω_π^*-sequence determines a decomposition $\mu = \alpha + \beta$ of the set in which α is cofinal with ω_ν and β is coinitial with ω_π^*; we are going to say that this decomposition defines an $\omega_\nu \omega_\pi^*$-gap, in contrast to a decomposition $\alpha + 1 + \beta$ that yielded an $\omega_\nu \omega_\pi^*$-limit. Among the types we have considered, at most four kinds of gaps can occur:

$$\omega\omega^*\text{-gaps}, \quad \omega\Omega^*\text{-gaps}, \quad \Omega\omega^*\text{-gaps}, \quad \Omega\Omega^*\text{-gaps}$$

(more briefly 11-gaps, 12-gaps, 21-gaps, 22-gaps).

In part, only two of these gaps can occur, in part (in ID), only $\omega\omega^*$ gaps can occur. We now make a distinction as to whether a type contains $0, 1, 2, 3$, or 4 kinds of gaps, and we class types with different gaps into different "species." In general, this would give divisions of our genera into 16, 4, and 2 species, respectively; however, some of these are omitted in the homogeneous types because certain sequences have no limits, so corresponding gaps must occur. For example, both Ω-gaps as well as Ω^*-gaps must occur in IA, thus one of the following combinations

$$22; \quad 12, 21; \quad 22, 12; \quad 22, 21; \quad 22, 12, 21,$$

where 11-gaps may be present or absent as you like; so this gives 10 species (instead of 16). One sees that each of the 4 genera IA, IIA, IIIA, IV contributes 10 species and that each of the others contributes 2 species. In the following, we give a synopsis in tabular form of

the 50 species of homogeneous types

by using the numbers 1 through 16 for the possible gap-combinations and by writing beside each the possible genera; then we denote the species by affixing the number after the genus label so that, for example, IIA6 means the species that belongs to the genus IIA and contains 11-gaps, 12-gaps, and 22-gaps. Incidentally, the letters A, B, C, D are now actually unnecessary but shall be retained for more rapid orientation. For reasons that will be apparent later, we are going to separate the species with and without $\omega\omega^*$-gaps.

The 32 Species with $\omega\omega^*$-Gaps

Number	Gaps	Genera
1.	11	ID IIB IIIC IV
2.	11, 21	IB IIB IIIA IV
3.	11, 12	IC IIA IIIC IV
4.	11, 22	IA IIA IIIA IV
5.	11, 21, 22	IA IIA IIIA IV
6.	11, 12, 22	IA IIA IIIA IV
7.	11, 21, 12	IA IIA IIIA IV
8.	11, 21, 12, 22	IA IIA IIIA IV

The 18 Species without $\omega\omega^*$-Gaps

Number	Gaps	Genera
9.	none	ID
10.	21	IB IIIA
11.	12	IC IIA
12.	22	IA
13.	21, 22	IA IIIA
14.	12, 22	IA IIA
15.	21, 12	IA IIA IIIA IV
16.	21, 12, 22	IA IIA IIIA IV

One can easily see which genera and species belong together as *inverses*. Invertible types can be found under the genera IA, ID, IV and the numbers 1, 4, 7, 8, 9, 12, 15, 16. Types without gaps (= continuous types) belong to the species ID9. Types that contain no gaps or only one kind of gap would be appropriately designated as *absolutely homogeneous*; these are contained in 10 species. Invertible, absolutely homogeneous types belong to the four species IA12, ID1, ID9, IV1. The types of the groups I II III, but not group IV, are dense in themselves. Closed types must contain no gaps or only 22-gaps and would thus have to be found under the numbers 9 and 12; however, the types in ID9, IA12 are not yet closed, but rather they become so with the addition of two boundary elements. (With regard to these types, cf. the conclusion of this article.)

The homogeneous types Ξ_a, H_b, Z_c, which were derived in the previous §, belong to the species IV7, IIA6, IIA5. That is to say: it is easy to prove that in the type Ξ every Ω-sequence with which Ξ is not cofinal has a limit, likewise for every Ω^*-sequence with which Ξ is not coinitial; therefore Ξ contains neither Ω-gaps nor Ω^*-gaps, but at most, $\omega\omega^*$-gaps. And these really do occur in Ξ; no element of Ξ lies between the two fundamental sequences

$$(\xi_1\,\xi_2 + 1) < (\xi_1\,\xi_2\,\xi_3\,\xi_4 + 1) < (\xi_1\,\xi_2\,\xi_3\,\xi_4\,\xi_5\,\xi_6 + 1) < \cdots,$$
$$(\xi_1 + 1) > (\xi_1\,\xi_2\,\xi_3 + 1) > (\xi_1\,\xi_2\,\xi_3\,\xi_4\,\xi_5 + 1) > \cdots,$$

where the ω-sequence precedes the ω^*-sequence, because the infinite continued fraction $(\xi_1\,\xi_2\,\xi_3\,\xi_4 \ldots)$ does not belong to Ξ. Therefore Ξ contains only 11-gaps. As a result, Ξ_a contains 11-gaps as well and, in place of the deleted b-elements and c-elements, 21-gaps and 12-gaps but no 22-gaps; in the same way, Ξ_b contains only 11-gaps, 12-gaps, and 22-gaps, and Ξ_c only 11-gaps, 21-gaps, and 22-gaps. With that, the assertion above is proved.

In view of the systematology established, the problem of determining all the homogeneous types of the second infinite cardinality (or say: all the homogeneous types of the cardinality of the continuum) in this generality might seem quite hopeless today. The very problem of constructing in a clear way a class of types as extensive as possible of a single species presumes in certain cases machinery that we developed for types of the species ID9 in the previous article. That is why, for now, we restrict ourselves to the treatment of the fundamental question: *which of our 50 species are really found among types of the second infinite cardinality*. A definitive answer to this question cannot be given at this time as long as the question of the cardinality of the continuum is not settled (on the other hand, after so many other approaches have been tried in vain, a new way to answer the continuum question just might arise here). For simple reflection reveals that each everywhere dense type contains the countable everywhere dense type η; consequently, *each everywhere dense type without $\omega\omega^*$-gaps contains the usual continuum as a subset*. As long as there is no success, for instance, in constructing an everywhere dense type of the second infinite cardinality without $\omega\omega^*$-gaps and with that in verifying CANTOR's hypothesis $2^{\aleph_0} = \aleph_1$, the question of the existence of homogeneous types of the second infinite cardinality will be restricted to the 32 species with $\omega\omega^*$-gaps. Indeed, I have succeeded in constructing a type belonging to each of these, except for the species ID1 in which the continuum question again plays a part. Since this species is at least represented by the countable type η, we can state the theorem:

All 32 species of homogeneous types with $\omega\omega^*$-gaps that have at most the second infinite cardinality exist.

§3

In order to furnish a proof of this theorem, it will be a question of actually constructing some type for each of the 32 required species, and in fact doing it in a clear way that does not lose itself in details and that, for instance, does not necessitate a special argument for each type.

As a construction principle, we use *power formation* as developed in the first note, whose fruitfulness and importance are verified here in an unexpected way; our plan would be very difficult to achieve without this unifying algorithm [[Algorithmus]]. Only *powers of the first class* are suitable for our purposes since true powers of the second class have at least the cardinality of the continuum. We will only need to consider types without a last element as arguments since, owing to

$$\mu_m(\alpha + 1) = \mu \cdot \mu_m(\alpha),$$

in the opposite case the power would contain pieces of type μ of the base and thus would have no properties with regard to homogeneity, gaps, and the like that the base would not already have. Accordingly and in acknowledgment of what was said in §2, here we can only promise types with $\omega\omega^*$-gaps. For if the argument α has no last element, then it surely contains an ω-sequence $a_0\, a_1\, a_2 \ldots$; if we then understand by $(m_0\, m_1\, m_2 \ldots m_\nu)$ that element of $\mu_m(\alpha)$ in which the places $a_0\, a_1 \ldots a_\nu$ are covered by the elements $m_0\, m_1 \ldots m_\nu$, while all other places are covered by the principal element m, and if l is a secondary element $(\neq m)$, then the two fundamental sequences

$$(l) \quad (l\,m\,l) \quad (l\,m\,l\,m\,l) \quad (l\,m\,l\,m\,l\,m\,l) \ldots,$$
$$(l\,l) \quad (l\,m\,l\,l) \quad (l\,m\,l\,m\,l\,l) \quad (l\,m\,l\,m\,l\,m\,l\,l) \ldots,$$

of which one is an ω-sequence and the other an immediately following ω^*-sequence, certainly determine an $\omega\omega^*$-gap since an element between these two would have to carry the secondary element l at infinitely many places, which would violate the definition of powers of the first class.

The arguments ω and Ω are already sufficient for our purposes; we mention a few general formulas and details[1]) beforehand regarding these two powers.

The powers $\mu_m(\omega)$

If we set $\mu_m(\omega) = M$, then first of all, as a result of the power laws

$$\mu_m(\omega) \cdot \mu = \mu_m(1+\omega) = \mu_m(\omega), \text{ thus}$$

(1) $$M\mu = M.$$

We denote the initial segments of μ and M by π and Π, the middle segments by ϱ and P, and the end segments by σ and Σ, possibly using indices. Let $a = (a_0\, a_1\, a_2\, \ldots)$ be an element of M. The initial segment Π determined by it can be decomposed in the following way:

elements	$(x_0 \mid \ldots\ldots)$,	$x_0 < a_0$,
elements	$(a_0\, x_1 \mid \ldots\ldots)$,	$x_1 < a_1$,
elements	$(a_0\, a_1\, x_2 \mid \ldots)$,	$x_2 < a_2$,

etc., where again to the right of the stroke stand arbitrary coverings with a finite number of secondary elements, as in M itself. Consequently, the given pieces of M have the types $M\pi_0, M\pi_1, M\pi_2, \ldots$, where π_ν is the initial segment of μ determined by a_ν; it follows that

(2) $$\Pi = M(\pi_0 + \pi_1 + \pi_2 + \cdots).$$

In exactly the same way, it follows for the corresponding end segments that

(3) $$\Sigma = M(\cdots + \sigma_2 + \sigma_1 + \sigma_0),$$
$$M = \Pi + 1 + \Sigma.$$

In addition, it is to be observed that only a finite number of the π_ν can be distinct from π_m (the initial segment of μ determined by the principal element m) and only a finite number of the σ_ν can be distinct from σ_m. In particular, for the decomposition caused by the principal element $(m\,m\,m\,\ldots)$ of M

(4) $$M = M\pi_m\omega + 1 + M\sigma_m\omega^*$$

holds, from which formula we note the special cases

(4a) $\quad M = 1 + M\sigma_m\omega^* \quad$ if m is the first element of μ,

---

[1]) The conclusions of the previous article regarding powers of the second class could have also been presented in the following more computational form.

(4b) $\quad M = M\pi_m\omega + 1 \quad$ if m is the last element of μ.

If l is a secondary element, then the $\omega\omega^*$-gap determined by the covering $(l\,l\,l\,\ldots)$ allows exactly the same decomposition that aided in the proof of formulas (2) and (3), and we obtain

(5) $\quad\quad\quad\quad\quad M = M\pi_l\omega + M\sigma_l\omega^*,$

a formula that again leads to the special cases

(5a) $\quad M = M\sigma_l\omega^* \quad$ if l is the first element of μ,

(5b) $\quad M = M\pi_l\omega \quad\ $ if l is the last element of μ

(in which the gap comes out of M and in return M becomes coinitial with ω^*, respectively, cofinal with ω).

For every initial segment π and every end segment σ of μ, it now follows from (4) and (5) that

$$M\pi + M = M + M\sigma = M,$$

and for every middle segment ϱ

$$M = M\mu = M(\pi + 1 + \varrho + 1 + \sigma) = M + M\varrho + M;$$

thus

(6) $\quad M(\pi + 1) = M(1 + \varrho + 1) = M(1 + \sigma) = M.$

This important formula is the source of the homogeneity proof, respectively, the isomery proof; it can be expressed as: *M permits right-sided multiplication with each interval of the base* (interval = segment [[Strecke]] together with boundary points [[Randpunkten]]).

From (2) and (3), it follows that for $\pi_m \neq 0$, Π is always cofinal with ω, and that for $\sigma_m \neq 0$, Σ is always coinitial with ω^*. So if m is not the first (last) element of the base, then each element of M is an $\omega(\omega^*)$-limit; if m is neither the first nor last element, then every element in M is an $\omega\omega^*$-limit. These sufficient (not necessary) conditions have the advantage of being independent of the base. In order to construct types of group I, we will certainly use $\mu_m(\omega)$ in which m shall be neither the first nor last element of μ.

The powers $\mu_m(\Omega)$

We again set $M = \mu_m(\Omega)$. If

$$a = (a_0\, a_1\, a_2\, \ldots\, a_\omega\, a_{\omega+1}\, \ldots\, a_\alpha\, \ldots)$$

is an element of M, then the corresponding initial segment and end segment of M are given by

(7) $\quad \Pi = M(\pi_0 + \pi_1 + \pi_2 + \cdots + \pi_\omega + \pi_{\omega+1} + \cdots + \pi_\alpha + \cdots),$

(8) $\quad \Sigma = M(\cdots + \sigma_\alpha + \cdots + \sigma_{\omega+1} + \sigma_\omega + \cdots + \sigma_2 + \sigma_1 + \sigma_0),$

formulas which take the place of (2) and (3) and in which only a finite number of terms are distinct from π_m, respectively, σ_m. Now in place of (4)

(9) $$M = M\pi_m\Omega + 1 + M\sigma_m\Omega^*$$

occurs, together with its special cases

(9a) $\qquad M = 1 + M\sigma_m\Omega^* \quad$ if m is the first element of μ,

(9b) $\qquad M = M\pi_m\Omega + 1 \quad$ if m is the last element of μ.

However, (5), together with its special cases, remains unchanged, and in it, ω cannot be substituted by Ω since the countable sequence ($l\ l\ l\ \ldots$) already defines the gap. Also formulas (1) and (6) still hold. By (7) and (8), every element in M is an $\Omega(\Omega^*)$-limit if m is not the first (last) element of μ, and so we will have powers $\mu_m(\Omega)$ in which m is neither the first nor last element to use for constructing types in group IV.

Here, we need to say something about the Ω-sequences and Ω^*-sequences contained in M. Let

$$x^\beta = (x_0^\beta\ x_1^\beta\ \ldots\ x_\omega^\beta\ \ldots\ x_\alpha^\beta\ \ldots) = \sum_\alpha x_\alpha^\beta$$

be an element of M that depends on the index β, and let $\{x^\beta\} = \sum_\beta x^\beta$ be an Ω-sequence R in M; the summation signs are understood in the collective, not the additive, sense. We consider the set

$$x_\alpha = \sum_\beta x_\alpha^\beta$$

of those elements with which a fixed place α is covered in turn. If x_α contains an *at most countable set of distinct* elements of μ, then we are going to call α a *place of the first kind*; if x_α contains a set of distinct elements of the second infinite cardinality, then call α a *place of the second kind*. If all places are of the first kind, then we call R an *Ω-sequence of the first kind*; if places of the second kind also occur, then call R an *Ω-sequence of the second kind*.

Let R be an Ω-sequence of the first kind. Since

$$x_0^0 \leq x_0^1 \leq x_0^2 \leq \cdots \leq x_0^\omega \leq \cdots \leq x_0^\beta \leq \cdots$$

and x_0 contains only a finite or countable set of distinct elements, the sign $<$ can appear with at most countable frequency; so from a certain index β_0 on, x_0^β is always the same element of μ:

$$x_0^\beta = a_0 \quad \text{for } \beta \geq \beta_0.$$

We call this a_0 the "final element" [["Schlußelement"]] of the place 0. For the elements of R that now follow, all of which begin with a_0, it is the case that

$$x_1^{\beta_0} \leq x_1^{\beta_0+1} \leq \cdots \leq x_1^{\beta_0+\omega} \leq \cdots,$$

and thus again from a certain index $\beta_1(\geqq \beta_0)$ on,
$$x_1^\beta = a_1 \quad \text{for } \beta \geqq \beta_1.$$

Continuing in this way, one recognizes that in an Ω-sequence of the first kind there is a "final element" a_α for every place α, so that from a certain index β_α on
$$x_\alpha^\beta = a_\alpha \quad \text{for } \beta \geqq \beta_\alpha.$$
Among these final elements, only a finite number of secondary elements can be found since otherwise terms of the sequence R would already have to be covered by an infinite number of secondary elements; the totality of final elements
$$a = (a_0\, a_1\, a_2\, \ldots\, a_\omega\, \ldots\, a_\alpha\, \ldots)$$
thus constitutes an element of M that is clearly the limit of the sequence R. We gather from this:

Every Ω-sequence of the first kind in M has a limit.

Ω-sequences of the first kind exist only if m is not the first element of the base.

For an Ω-sequence of the second kind, let α be the first place of the second kind; all the preceding places again have final elements $a_0\, a_1\, \ldots$, and from a certain index γ on, all elements $x^\gamma\, x^{\gamma+1}\, \ldots$ of R begin with these final elements. Then
$$x_\alpha^\gamma \leqq x_\alpha^{\gamma+1} \leqq \cdots \leqq x_\alpha^{\gamma+\omega} \leqq \cdots,$$
but where the sign $<$ now appears with uncountable frequency, i.e., the set x_α contains an Ω-sequence from the base μ. Consequently:

Ω-sequences of the second kind exist only if Ω-sequences occur in the base.

Analogous assertions hold for Ω^*-sequences in M.

Expressed cursorily, one can say: sequences of the first kind arise from the argument Ω, and those of the second kind arise from the base.

Naturally, only $\Omega(\Omega^*)$-sequences of the second kind exist in the types $\mu_m(\omega)$.

Having said this, we pass to the construction of homogeneous

types of the first and fourth groups

by assuming that the *principal element m is neither the first nor last element of the base*. We choose a base of the specific form

(10) $$\mu = \omega + \lambda + \omega^*,$$

where λ is a suitable type that is yet to be determined. Then for the initial segments and end segments of μ:

$$\text{either} \quad \pi = \nu \text{ (finite)} \quad \text{or} \quad \pi = 1 + \pi,$$
$$\text{either} \quad \sigma = \nu \text{ (finite)} \quad \text{or} \quad \sigma = \sigma + 1.$$

By (6)

(11) $$M = M\nu = M \cdot 2 = M \cdot 3 = \cdots,$$

and by (5a) and (5b)

(12) $$\begin{cases} M = M\mu\omega^* = M\omega^*, \\ M = M\mu\omega = M\omega. \end{cases}$$

In (2) and (7), the case $\pi_\alpha = 0$ occurs only a finite number of times; let us think of these as omitted and the numbering of the places adjusted accordingly, so we may now assume that each $\pi_\alpha \neq 0$. But then it is always the case that

(13) $$M(\pi_\alpha + \pi_{\alpha+1}) = M(\pi_\alpha + 1 + \pi_{\alpha+1}),$$

since either $\pi_{\alpha+1} = 1 + \pi_\alpha$ or for $\pi_{\alpha+1} = \nu > 0$,

$$M(\pi_\alpha + \nu) = M\pi_\alpha + M = M\pi_\alpha + M(\nu + 1)$$
$$= M(\pi_\alpha + 1 + \nu).$$

Hence considering the immediately following elements in (2) and in (7), each π_α may be replaced by $\pi_\alpha + 1$. Therefore for $M = \mu_m(\omega)$, by (6)

$$\Pi = M(\pi_0 + 1 + \pi_1 + 1 + \pi_2 + 1 + \cdots) = M + M + M + \cdots = M\omega,$$

and analogously for $M = \mu_m(\Omega)$, $\Pi = M\Omega$. The same holds for end segments, which therefore obtain the types $M\omega^*$ and $M\Omega^*$. Thus considering (12), we find:

In $M = \mu_m(\omega)$, every initial segment has type M and every end segment has type M; consequently, every middle segment (= initial segment of an end segment, or vice versa) also has the type M. *M is a homogeneous type of group I.*

In $M = \mu_m(\Omega)$, every initial segment has type $M\Omega$ and every end segment has type $M\Omega^*$; consequently, the type of every middle segment is $M(\Omega^* + \Omega)$. *M is an isomeric type of group IV, and its middle segment is a homogeneous type of group IV.*

In order to produce the different species, we choose the base so that the $\Omega(\Omega^*)$-sequences contained in it do not have limits, but rather determine gaps; these same gaps are then carried into the power M by the $\Omega(\Omega^*)$-sequences of the second kind, whereas the $\Omega(\Omega^*)$-sequences of the first kind either are missing or have limits and so do not generate new gaps. We reach this goal by keeping the form of (10) through simple combinations of the numbers ω and Ω and their inverses. We set

(14) $$\begin{cases} \omega_{11} = \omega + \omega^*, & \omega_{21} = \Omega + \omega^*, \\ \omega_{12} = \omega + \Omega^*, & \omega_{22} = \Omega + \Omega^*, \end{cases}$$

and then let us say

(15) $$\begin{cases} \mu_1 = \omega_{11}, \quad \mu_2 = \omega_{21}, \quad \mu_3 = \omega_{12}, \quad \mu_4 = \omega_{22}, \\ \mu_5 = \omega_{21} + \omega_{22}, \quad \mu_6 = \omega_{12} + \omega_{22}, \\ \mu_7 = \omega_{21} + \omega_{12}, \quad \mu_8 = \omega_{21} + \omega_{22} + \omega_{12}. \end{cases}$$

Through the presence of ω_{ik}, M will be furnished with ik-gaps, and besides these gaps implanted in it by the base, it can (and must) only contain 11-gaps. Through the raising to powers of the types μ_1 μ_2 ... μ_8 with the argument ω, we indeed obtain homogeneous types of the species I1, I2,..., I8; whereas with the argument Ω, we obtain isomeric types and from their middle segments, homogeneous types of the species IV1, IV2,... IV8. The principal element is arbitrary, with the exclusion of the first and last elements of the base. All types are of the second infinite cardinality except for $\mu_1(\omega)$ of species ID1, which is identical to the countable type η.

Types of the second and third groups

Again, we choose a base of the form (10) and its *first element 0 as the principal element*. The formulas (11) and (13) hold good, only the second formula remains from (12)

(16) $$M = M\mu\omega = M\omega.$$

Also the similarity of all end segments of M holds; i.e.,

all end segments of $M = \mu_0(\omega) = 1 + M\omega^*$ have the type $M\omega^*$,
all end segments of $M = \mu_0(\Omega) = 1 + M\Omega^*$ have the type $M\Omega^*$.

It is a different matter for initial segments. Now in (2) or (7) only a finite number of the π_α are different from $\pi_m = 0$; so that after a renumbering, we can write

$$\Pi = M(\pi_0 + \pi_1 + \cdots + \pi_\nu)$$
$$= M(\pi_0 + 1 + \pi_1 + 1 + \cdots + \pi_{\nu-1} + 1 + \pi_\nu)$$
$$= M(\nu + \pi_\nu) = M\pi_\nu.$$

Thus in each case we have

(17) $$\Pi = M\pi,$$

including the case $\pi = \Pi = 0$.

Here, the homogeneity, respectively, isomery of M cannot yet be proved as in the previous case; however, if we now consider bases of the form (15), since ω_{ik} occurs in μ and is an interval of μ because $\omega_{ik} = 1 + \omega_{ik} + 1$, then according to (6) $M = M\omega_{ik}$; furthermore, each initial segment of ω_{ik} is either ω_{ik} itself or a number α of the 1st or 2nd number class. With that, it immediately follows that in each case

$$M\pi = M\alpha;$$

it indeed follows from (16) by transfinite induction that $M\alpha = M$ for $\alpha > 0$. Thus either $\Pi = M$ or $\Pi = 0$. We have the result:

Each initial segment (except 0, the first) in $M = \mu_0(\omega)$ has the type $M = M\omega$, each end segment has the type $M\omega^*$, and each middle segment has the type $M\omega^*$; M is an isomeric type, and after the omission of the initial element, it is a homogeneous type of group I. Thus this gives nothing new.

Each initial segment (except 0, the first) in $M = \mu_0(\Omega)$ has the type $M = M\omega$, each end segment has the type $M\Omega^*$, and each middle segment has the type $M\Omega^*$; M *is an isomeric type, and after the omission of the initial element, it is a homogeneous type of of group III.*

Regarding gaps, it is as above; the gaps in the base will be transferred to M by $\Omega(\Omega^*)$-sequences of the second kind; M contains no other gaps besides these and $\omega\omega^*$-gaps since Ω-sequences of the first kind do not exist and Ω^*-sequences of the first kind do not yield gaps, but rather they yield limits. Thus the types $\mu_1\, \mu_2\, \ldots\, \mu_8$ raised to powers with the argument Ω give homogeneous types of species III 1, III 2, ..., III 8.

One obviously obtains the types of group II from those just found by *inversion* or also by again choosing the bases (15) and the argument Ω, *but now by picking the last element of the base as principal element.* The case, say, of $\mu_5 = \omega_{21} + \omega_{22}$, which gives the species IIIA5 when its first element is principal, may demonstrate that these amount to the same thing. Through inversion, the base turns into $\mu_5^* = \omega_{22} + \omega_{12}$, the principal element into the last element, and the species IIIA5 into IIA6; with regard to the claims made, the base μ_5^* achieves the same thing as the base $\mu_6 = \omega_{12} + \omega_{22}$. This is an appropriate opportunity to point out that the entire construction of the present § is only one among endlessly many and that with regard to basis, argument, and principal element quite a vast latitude of equally valid possibilities are at one's disposal; in individual cases, with the abandonment of the strived for uniform consideration of all groups and species, one could get by with even simpler bases than (15), e.g., $\mu = 3$, $\mu = \Omega$, especially if one does not immediately aim for the construction of isomeric types, but one also brings in mixed types in the manner of Ξ (§1) and produces isomeric types only through separation of distinct element categories. Reserving all implied questions (e.g., whether there are infinitely many types of each species) for a later investigation in this series, we may consider our temporary project as finished, and we may substitute the following precise statement for the concluding theorem of the previous §:

As bases of powers of the first class, the 8 types $\mu_1\, \mu_2\, \ldots\, \mu_8$ of formula (15) yield all 32 species of homogeneous types of the first and second infinite cardinalities with $\omega\omega^*$-gaps. In fact, one gets types of

group I if the argument is ω and the principal element is any element of the base;

group II if the argument is Ω and the principal element is the last element of the base;

group III if the argument is Ω and the principal element is the first element of the base;

group IV if the argument is Ω and the principal element is a middle element of the base.

The *homogeneous types of the cardinality of the continuum* shall likewise form the subject of a future investigation in which all 50 species with or without $\omega\omega^*$-gaps are to be considered. Though even here, the unsolved continuum question forces the elimination of certain (five) species whose types provably have cardinality greater than the second infinite cardinality; the remaining 45 really do exist, as we will show with the help of *powers of the first and second class*. We make just one more closing remark at this time about those types that are closest to the linear continuum: they are the *invertible* types of species ID9 and IA12. These types, as we expressed on occasion (p. 155), are *absolutely homogeneous, dense in themselves*, and contain no gaps that arise from fundamental sequences; so after the addition of two boundary elements, they are *closed*, that is, *perfect*, and *isomeric*. We are going to designate the types of both these species, with boundary elements affixed, *continua* and *ultracontinua*[1]): thus a continuum is an invertible, isomeric, bounded, perfect type without Ω-sequences and Ω^*-sequences, and an ultracontinuum is the same kind of type with Ω-sequences and Ω^*-sequences. In the previous article, we presented infinitely many ($\aleph_1$) distinct continua, namely, the powers $\vartheta'(\omega^\alpha)$ of the usual linear continuum; we show here that *there are likewise infinitely many* (at least $\aleph_1$) *distinct ultracontinua*. First of all, to construct one, we will form the power of the second class $M = \mu'(\omega)$ with base

$$\mu = \omega_{22} = \Omega + \Omega^*,$$

whose elements we are going to denote by

$$0\ 1\ 2\ \ldots \omega\ \omega+1 \ldots \alpha \ldots \mid \ldots \overline{\alpha} \ldots \overline{\omega+1}\ \overline{\omega} \ldots \overline{2}\ \overline{1}\ \overline{0}.$$

Since the base is bounded and closed and the argument is a limit ordinal, M is a perfect type by Theorem G (II, §3). However, it is not everywhere dense, but rather it contains pairs of consecutive elements such as

$$(\alpha\,\overline{0}\,\overline{0}\,\overline{0}\,\ldots)\quad(\alpha+1\,0\,0\,0\,\ldots)$$

or

$$(\overline{\alpha+1}\,0\,0\,0\,\ldots)\quad(\overline{\alpha}\,0\,0\,0\,\ldots),$$

and its initial segments (except 0, the first) have two kinds of types, namely, $M\omega$ and $M = M\omega + 1$. Now, if one lets each pair of consecutive elements collapse into one element, e.g., it could happen that one deletes all elements of the form

$$(x_0\,x_1\,\ldots\alpha\,\overline{0}\,\overline{0}\,\overline{0}\,\ldots)\quad\text{and}\quad(x_0\,x_1\,\ldots\overline{\alpha}\,0\,0\,0\,\ldots),$$

as easy reflection shows, the type Θ that arises is still perfect, invertible, unbounded, but it is now isomeric as well. Consequently, since it contains

---

[1]) We adopt this term from F. BERNSTEIN: "Untersuchungen aus der Mengenlehre" (Diss., Halle 1901, p. 51 ff.), reprinted in Math. Ann. 61 (1905), p. 152 ff. There "the" ultracontinuum is defined by attributing characteristic features, which shows the author roughly had in mind the above definition. Certainly, neither of the types given by him (in our notation the powers of the second class $\Omega'(\omega)$ and $\Phi'(\omega)$ where $\Phi = \Omega^*\Omega$) are ultracontinua; the first lacks invertibility and isomery, and the second lacks closedness, by which the remarks in the Annals reprinting p. 153 f. become invalid.

Ω-sequences and Ω^*-sequences, it is an ultracontinuum. By Theorems G and M of the previous article, all the powers of the second class $\Theta'(\omega^\alpha)$ are also ultracontinua. In order to show that these are all distinct, by theorem D it suffices to prove *that Θ contains no set of exclusive intervals that is similar to Θ*. Now Θ contains a certain set of type H of the second infinite cardinality, exactly as the linear continuum ϑ contains the countable type η, namely, in such a way that elements of H fall into every interval of Θ; e.g., one such subset H is given by the power of the first class $\mu_m(\omega)$ whose principal element is any middle element of $\mu = \Omega + \Omega^*$. So if Θ contained any set of exclusive intervals similar to Θ, H would have to contain a subset of type Θ. This is impossible by reason of the theorem:

If μ does not contain an everywhere dense subset without $\omega\omega^*$-gaps, then the power of the first class $\mu_m(\omega)$ does not contain any such subset either.

In order to furnish a proof of this assertion, we assume that N is an everywhere dense subset of $M = \mu_m(\omega)$ without $\omega\omega^*$-gaps, and we show that this assumption leads to a contradiction. Since each everywhere dense *countable* set, except for boundary elements, corresponds to η and thus contains $\omega\omega^*$-gaps, so N, thus also M, and thus also μ must be *uncountable* (= of cardinality greater than the first infinite cardinality); also each proper piece (> 1) of N must be everywhere dense and uncountable. Let $x = (x_0\, x_1\, x_2\, \ldots)$ be an element of N. We denote by

$$X_\nu = (x_0\, x_1\, x_2\, \ldots\, x_\nu \mid \ldots)$$

the totality of all elements of N that begin with $x_0\, x_1\, \ldots\, x_\nu$; X_ν is either a single element (= 1) or a proper piece of N, thus everywhere dense and uncountable. The set $\{x_\nu\}$ of the elements of μ that cover the place ν in all the elements of N cannot be finite or countable for every ν; otherwise N would be a subset of $\lambda_m(\omega)$ where λ is an at most countable subset of μ, and N itself would thus be at most countable. If ν is the first index for which the set $\{x_\nu\}$ is uncountable, then among the (at most countable) set of pieces $X_{\nu-1}$ that belong to N at least one

$$A_{\nu-1} = (a_0\, a_1\, \ldots\, a_{\nu-1} \mid \ldots)$$

must occur that decomposes into an uncountable set of pieces

$$X_\nu = (a_0\, a_1\, \ldots\, a_{\nu-1}\, x_\nu \mid \ldots),$$

where x_ν thus runs through an uncountable subset of μ. (For $\nu = 0$, N decomposes directly into an uncountable set of pieces $X_0 = (x_0 \mid \ldots)$.) Among these pieces X_ν, there must occur a transfinite set of proper pieces; for were it only a finite set, one could separate from N a piece that is formed exclusively of improper pieces (single elements) X_ν. It thus would have the type of a subset of μ, so by hypothesis could not be an everywhere dense set without $\omega\omega^*$-gaps. Hence there exists a transfinite number of proper pieces X_ν, and it is possible therefore to select from them three pieces $X_\nu < A_\nu < Y_\nu$,

$$\left.\begin{array}{l} X_\nu = (a_0\, a_1\, \ldots\, a_{\nu-1}\, x_\nu\, |\, \ldots) \\ A_\nu = (a_0\, a_1\, \ldots\, a_{\nu-1}\, a_\nu\, |\, \ldots) \\ Y_\nu = (a_0\, a_1\, \ldots\, a_{\nu-1}\, y_\nu\, |\, \ldots) \end{array}\right\} \quad (x_\nu < a_\nu < y_\nu),$$

so that a_ν is a secondary element of μ ($a_\nu \neq m$). Now we apply exactly the same line of reasoning to the everywhere dense, uncountable piece A_ν of N, and we see that for an index π ($> \nu$) it contains at least three proper pieces $X_\pi < A_\pi < Y_\pi$,

$$\left.\begin{array}{l} X_\pi = (a_0\, a_1\, \ldots\, a_\nu\, \ldots\, a_{\pi-1}\, x_\pi\, |\, \ldots) \\ A_\pi = (a_0\, a_1\, \ldots\, a_\nu\, \ldots\, a_{\pi-1}\, a_\pi\, |\, \ldots) \\ Y_\pi = (a_0\, a_1\, \ldots\, a_\nu\, \ldots\, a_{\pi-1}\, y_\pi\, |\, \ldots) \end{array}\right\} \quad (x_\pi < a_\pi < y_\pi),$$

in which a_π is a secondary element; A_π, for its part, again contains three such pieces $X_\varrho < A_\varrho < Y_\varrho$ ($\varrho > \pi$) in which a_ϱ is a secondary element, etc. At this point, we see that despite all the assumptions made thus far N 169 nevertheless has the $\omega\omega^*$-gap

$$(a_0\, a_1\, \ldots\, a_\nu\, \ldots\, a_\pi\, \ldots\, a_\varrho\, \ldots);$$

because this covering contains infinitely many secondary elements, it does not belong to M and it does not belong to N, and one can select an ω-sequence and a later ω^*-sequence that allow no element of N between them from the sequences of pieces

$$X_\nu\, X_\pi\, X_\varrho \ldots, \quad Y_\nu\, Y_\pi\, Y_\varrho \ldots.$$

With that, our last stated theorem is proved. Its hypotheses apply to the base $\mu = \Omega + \Omega^*$; hence the type $\mu_m(\omega)$ cannot contain an everywhere dense subset without $\omega\omega^*$-gaps; in particular, it contains no subset of type Θ. Therefore no set of exclusive intervals of Θ is similar to Θ itself, all powers $\Theta'(\alpha)$ are distinct, and we can add a supplement to the result of the previous article: *the species of continua as well as the species of ultracontinua are represented by an uncountable set of truly distinct types.*

[Declared ready to print 5/20/1906.]

Introduction to "Investigations into Order Types IV, V"

The second part of the omnibus *Investigations into Order Types* [[*Untersuchungen über Ordnungstypen*]] contains Hausdorff's fourth and fifth articles: *IV. Homogeneous Types of the Cardinality of the Continuum; V. On Pantachie Types* [[*IV. Homogene Typen von der Mächtigkeit des Kontinuums; V. Über Pantachietypen*]]. In *Homogeneous Types of the Cardinality of the Continuum*, Hausdorff seems to be operating (as far as we can tell) with the views expressed in his introduction to [H 1906b], but in the final article, *On Pantachie Types*, it is clear that something has changed. Without much fanfare, Zermelo's 1904 proof of the Well-Ordering Theorem is cited in a footnote ([H 1907a, 117n]) and *external* well-orderings start to appear as needed in constructions. Choice functions, however, are never mentioned and most uses of AC remain implicit.

The fourth article, *Homogeneous Types of the Cardinality of the Continuum*, is a natural follow-up to its immediate predecessor, *Homogeneous Types of the Second Infinite Cardinality* [H 1906b, 144–169]. In it, Hausdorff carries on the program to obtain representatives of his 50 species of homogeneous types whose sequences (well-ordered subsets) are of cardinality $\leq \aleph_1$. Now, however, he seeks representatives having the cardinality of the continuum, which he designates by a generic $\aleph$ (perhaps an unremarked acknowledgement of Zermelo's Well-Ordering Theorem).[1]

In this case, as in the previous one, the unsolved continuum problem "again manifests itself" and imposes a restricted goal. Hausdorff immediately proves by an easy inductive construction (implicitly using AC) that an everywhere dense type μ without $\omega\omega^*$-limits and without $\omega\omega^*$-gaps has both Ω-sequences and Ω^*-sequences and hence has cardinality $\geq \aleph_1$. Under the assumption that μ has cardinality $\aleph_1$, he cleverly uses induction to construct an $\Omega\Omega^*$-gap in μ. The upshot is that *an everywhere dense type without $\omega\omega^*$-limits, without $\omega\omega^*$-gaps, and without $\Omega\Omega^*$-gaps has cardinality $> \aleph_1$*. (In the next article, *On Pantachie Types*, he proves it has cardinality $\geq 2^{\aleph_1}$ [H 1907a, 135].)

Five species meet these criteria and Hausdorff pragmatically excludes them from consideration. He explains:

> [T]he discovery of a pertinent type [of one of the five excluded species] of the cardinality of the continuum would

thus settle the continuum problem in the sense that $\aleph >$
$\aleph_1$. [H 1907a, 85]

Four of the five species have no $\Omega\Omega^*$-elements and thus must contain ω_2-sequences and ω_2^*-sequences because a continuation of Hausdorff's first inductive construction embeds $\omega_2 + \omega_2^*$ into μ. Since these four violate the restriction that sequences have cardinality $\leq \aleph_1$, they are excluded *for cause*. In dismissing them, Hausdorff reveals his view on the nullifying effect of the continuum problem:

> Apart from the currently still imperative precautionary measure of leaving aside problems equivalent to the continuum question because of their suspected inaccessibility, four of the five above mentioned species fall beyond the scope of our investigation ... [H 1907a, 85]

After the removal of these five species, forty-five remain. Proceeding with great efficiency, Hausdorff employs an ingenious bootstrap approach to produce representatives of cardinality $2^{\aleph_0}$ for each of them. He first applies second class powers with argument Ω and bases that are finite or certain sums of $\omega, \omega^*, \Omega, \Omega^*$, together with the *collapsing of consecutive elements*, a technique first employed in the construction of the ultracontinuum Θ in [H 1906b, 166–167], to produce representatives for the sixteen species with $\omega\omega^*$-elements. Then, using representatives of these sixteen species as bases for powers of the first and second class with argument Ω, Hausdorff obtains representatives for the remaining twenty nine species.

The first sixteen species are of more than utilitarian interest. The representatives of nine of them, when used as bases for second class powers with arguments $\alpha < \Omega$, yield $\aleph_1$ distinct types of the same species. So these nine species each have uncountably many distinct homogeneous representatives. Hausdorff, commenting on this abundance, observes:

> [W]e also get a surprising insight into the boundless domain of types of higher cardinality and into the limited power of determination of the apparently significant concept of homogeneity, which we, under the spell of spatial and temporal intuitions, are inclined to take almost as a privilege of the usual continuum. [H 1907a, 98]

For the remaining seven of the sixteen species, the question of uncountably many distinct second class powers of their representatives goes unanswered. Hausdorff notes that should these powers (for $\alpha < \Omega$) fail to be distinct then "the continuum question would be settled in Cantor's sense" [H 1907a, 93].

The final section of *Homogeneous Types of the Cardinality of the Continuum* has the nature of an appendix. Hausdorff seeks to classify *everywhere dense, unbounded* types with sequences of cardinality $\leq \aleph_1$ without reference to homogeneity or isomery. In an approach that is to be generalized in [H 1908], species are formed according to the the kinds of (limit) elements and gaps that occur: $\omega\omega^*, \Omega\omega^*, \omega\Omega^*$, or $\Omega\Omega^*$. Of the 256 formal possibilities,

Hausdorff quickly deduces that 210 are viable. Restricted by cardinality, there are 120 possibilities for types of cardinality $\aleph_1$ (those with $\omega\omega^*$-gaps) and 201 possibilities for types of cardinality $2^{\aleph_0}$.

Actual existence questions are not pursued here, but as a coda Hausdorff defines the *complement* of an everywhere dense, unbounded type μ. He considers decompositions $\mu = \alpha + \beta$, where α has no last element.[2] The set of such α ordered by $\subset$ is shown to be a continuous ordered set.[3] Hausdorff designates it by $[\mu]$ and calls it the "filling" of μ. The α that are the initial segments of elements in μ form an ordered set of type μ; the α that arise from gaps in μ form an ordered set of type $\overline{\mu}$, which he calls the "completion" or "complement" of μ. Inspired by the mutually complementary rationals and irrationals, which have the reals as their common filling, Hausdorff calls an everywhere dense, unbounded type "everywhere discontinuous" if none of its middle segments are continuous. Such a type contains a gap between any two of its elements. For μ everywhere dense, unbounded, and everywhere discontinuous, $\overline{\mu}$ has exactly the same properties, and in fact, μ is the complement of $\overline{\mu}$ and both have the same "filling" $[\mu]$. (Completion and filling are thoroughly discussed in [H 1908, 448–449].)

The final article of [H 1907a], *On Pantachie Types*, is twice the length of the four other articles in *Investigations into Order Types*. In it, Hausdorff very confidently marshals set theory, in particular the theory of ordered sets (which he himself is enlarging), to sort out a controversy among analysts over Paul Du Bois-Reymond's "infinitary rank ordering" of the real functions that monotonically go to $+\infty$ with increasing x.[4] In doing so, he takes on Du Bois-Reymond's all encompassing ordering for the "infinities" of such real functions, the "infinitary pantachie," and Du Bois-Reymond's companion concept of the "ideal boundary between convergence and divergence," two ideas that were particularly vexing to some of Hausdorff's contemporaries.[5] In his application of set theory to a current problem in analysis, Hausdorff is following in the footsteps of Cantor, the theory's creator and his acknowledged master.

Hausdorff's mathemathical creations in the course of this investigation have an interest beyond the analysis problem that inspired them. For example, he gives us the η_ν-sets, which served to inspire the study of homogeneous universal structures and saturated structures in the 1950s and 1960s,[6] the *back-and-forth* construction, the first glimmers of a *maximal principle*, and the first exploration of the order and algebraic properties of a concrete *reduced power*. We also have the first instance where CH is eliminated from a proof—the proof of the existence of a homogeneous pantachie. (There is some overlap between *On Pantachie Types* and *Graduation by Final Behavior* [H 1909a]. It is worth reading the introductory sections of both articles in tandem.)

In the first section of *On Pantachie Types*, Hausdorff turns to the infinitary ordering of the real monotonic functions going to $+\infty$. The "infinitary relations" $<, \sim, >$ among such functions f and g are defined in terms of the behavior of the quotient f/g as x goes to $+\infty$. If this quotient's limit is 0, then $f < g$; if its limit is finite and $\neq 0$, then $f \sim g$; if its limit is $+\infty$, then $f > g$. Of course, tellingly for Hausdorff, none of these conditions need obtain, in which case f and g are "incomparable" ($f \parallel g$ in Hausdorff's notation). Du Bois-Reymond thought of each f as representing an "infinity" (an *Unendlich* in his terminology) and f and g with $f \sim g$ as representing the same infinity. In his infinitary pantachie all these infinities were imagined to form a "universal, coherent scale in which each functional infinity takes a determined place like a point on a line" [H 1909a, 299]. Du Bois-Reymond's "ideal function" that serves as a boundary between convergence and divergence arose from this point of view. He partitioned the infinities according to the convergence or divergence of their representatives' improper integrals $\int_0^\infty f(x)dx$. Those infinities for which the integral is convergent come before all those for which the integral is divergent. Although Du Bois-Reymond himself had proved that this "cut" could not be determined by an actual function, by analogy with cuts in the rationals that represent irrationals, he insisted that there was an "irrational infinity" created by this decomposition—an ideal function to serve as the boundary between convergence and divergence. He did not hesitate to have his ideal function appear under integral signs.[7]

Needless to say, both the infinitary pantachie and the ideal function stirred strong feelings. Cantor was made apoplectic by the appearance of infinitesimal quantities in Du Bois-Reymond's pantachie and by his seemingly cavalier attitude towards the existence of incomparable elements in the infinitary rank ordering.[8] As for the infinitary pantache, after a lawyerly examination of the evidence, Hausdorff declares that because of the existence of incomparable elements its relationship "to a simple ordering by magnitude is completely destroyed," and in fact, "the *infinitary pantachie* in the sense of Du Bois-Reymond *does not exist*" [his italics] [H 1907a, 107].[9] However, *On Pantachie Types* is not just a negative screed intended to oppose the fuzzy ideas of Du Bois-Reymond, some of which Cantor had found distasteful, even abhorrent. As he dispatches the infinitary pantachie, Hausdorff writes in a footnote, "[t]here is no reason to reject the entire theory because of the possibility of incomparable functions as G. Cantor has done ..." [H 1907a, 107n1].

In a rhetorical flourish, Hausdorff states:

> [I]f we designate it as our task to connect the infinitary rank ordering as a whole with Cantor's theory of order types, then nothing remains but to investigate *the sets of pairwise comparable functions that are as comprehensive as possible* [his italics]... [H 1907a, 110]

Obviously, he does designate making this connection as his task, because he proceeds to undertake it. With the understanding that infinitarily equivalent functions are to be identified, the infinitary relation $<$ is a partial ordering.

Hausdorff defines an *ordered domain* to be any set of inequivalent functions that is linearly ordered by the infinitary partial ordering.[10] Then "retaining the term of Du Bois, but abandoning the unsuccessful concept," he calls ordered domains that are as "comprehensive as possible," i.e., maximally linearly ordered, *pantachies*.[11]

Immediately upon the introduction of the term pantachie and before establishing the existence of pantachies, Hausdorff proposes to investigate the properties of their order types without any further assumptions—"since the attempt to actually legitimately construct a pantachie seems completely hopeless." This statement of futility seems puzzling, but upon reflection one realizes that Hausdorff is still referring to Du Bois-Reymond's concept of pantachie and not his own adaptation of it.

His first observation is that since the set of monotonic functions has cardinality of the continuum, any pantachie type has this cardinality as an upper bound; and, in fact, his work in §3 shows it has precisely this cardinality.[12] Hausdorff then proceeds to transform the setting for the study of pantachies from the infinitary ordering of the monotonic real functions to the infinitary ordering of monontonic sequences of positive reals (based on the limit of a_n/b_n as n goes to $+\infty$). He does this by a procedure that we would call *sampling*: f is mapped to the "numerical sequence" $(f(1), f(2), f(3), \ldots)$. As he observes in theorem (A), an ordered domain or pantachie of functions goes to a similar ordered domain of sequences, and if the ordered domain of sequences happens to be a pantachie, then the original ordered domain of functions is a pantachie too [H 1907a, 112].

Ordered domains of numerical sequences can in turn be mapped similarly to ordered domains of continuous functions by linear interpolation; again, if the ordered domain of sequences happens to be a pantachie, then one obtains a pantachie consisting of just continuous functions. These relative results are announced in theorem (B). This theorem, however, begins with the absolute assertion that "[t]here are pantachies consisting of only continuous functions," which seems a mysterious statement when existence proofs have not yet been given.[13]

At this point, Hausdorff drops the monotonicity and infinite growth conditions on both functions and sequences and considers pantachies in the domains of positive, real-valued functions and sequences under their respective infinitary orderings. He remarks that theorems (A) and (B) hold in this new setting.

There is one last important transformation of the overall problem to come. It arises from the wide spread criticism of the somewhat arbitrary choice of the quotient by Du Bois-Reymond in defining the infinitary ordering [H 1907a, 114–115]. Hausdorff repeats the observation of Pincherle that for a suitable Φ the difference $\Phi(f(x)) - \Phi(g(x))$ could serve as well for the definition of an ordering. Hausdorff chooses the sign of the difference $f - g$, a condition already considered by Du Bois-Reymond ([Du 1870, 344]), to define the "final rank ordering." For functions with positive real values, this ordering is defined as follows: $f < g$ ($f \sim g$ or $f > g$) if for some x_0, $f(x) < g(x)$ ($f(x) = g(x)$ or $f(x) > g(x)$) for all $x_0 \geq x$. When specialized to numerical sequences A and B, this is: $A < B$ ($A \sim B$ or $A > B$) if for some n_0, $a_n < b_n$ ($a_n = b_n$ or $a_n > b_n$) for all $n \geq n_0$.[14] For Hausdorff, this is the "essence of Du Bois's idea, namely the graduation of functions and sequences according to final behavior" [H 1907a, 116].

Once finally equivalent functions or sequences are identified, the final relation $<$ is a partial ordering, and the concepts of ordered domain and pantachie are immediately transferrable to the final orderings of functions and numerical sequences. (Of course, these concepts make sense in any partially ordered set, but Hausdorff does not operate at this level of abstraction; here, he is always dealing with concrete examples.) Hausdorff claims once more that theorems (A) and (B) hold in this setting [H 1907a, 116]. The pantachies of the final ordering of numerical sequences are the major focus of [H 1907a], and they are taken up again in [H 1909a]. Hausdorff observes that the infinitary and final orderings are related [H 1907a, 116n]. However, an important distinction between the infinitary and final orderings (for sequences of reals) is the latter's freedom from the topological concept of limit and its relative abundance of similarity transformations. Hausdorff also notes that the concept of final ordering can be extended to arbitrary covering sets. The pantachies for the resulting partial ordering offer another means for constructing types related to μ and α, the types of the base and of the argument of the covering set [H 1907a, 134-135].

As Hausdorff confronts the task of proving that there exist pantachies in the set of sequences of positive reals under final rank ordering, he cites Zermelo's 1904 proof of the Well-Ordering Theorem for the first time in his writings [H 1907a, 117n]. And he proceeds to tackle the problem of the existence of pantachies with his new ability to impose well-orderings.

Hausdorff offers two existence proofs for pantachies; each uses AC in a different way. First, he takes a top-down approach; after well-ordering the $2^{\aleph_0}$ numerical sequences of reals, Hausdorff inductively winnows the well-ordered set of sequences to obtain a pantachie. As he notes, this argument can be extended to produce a pantachie that contains a given ordered domain $\mathfrak{B}_0$. His second argument for showing that $\mathfrak{B}_0$ can be extended to a pantachie really has a different flavor: it works from the bottom up. Hausdorff presents it as an argument by contradiction. Starting with a given ordered domain $\mathfrak{B}_0$ and *assuming that it is not contained in any pantachie*, he inductively defines a well-ordered sequence $\{\mathfrak{B}_\gamma\}$ of ordered domains in which $\mathfrak{B}_{\alpha+1}$ is always a proper extension of $\mathfrak{B}_\alpha$. The following somewhat vague remark justifies the continuation of the sequence $\{\mathfrak{B}_\gamma\}$ at limit indices:

> [F]or each well-ordered sequence (countable or uncountable) of ordered domains, each of which contains all its predecessors, one could give an ordered domain that contains all of them. [H 1907a, 118]

Under the stated assumption, this sequence $\{\mathfrak{B}_\gamma\}$, which contains ever larger ordered domains, is continuable up to any arbitrary ordinal. Considering cardinalities, he concludes that the cardinality of the set of numerical sequences, which is that of the continuum, is greater than any aleph—an impossibility.[15]

Hausdorff does not draw any wider conclusions from these existence proofs; in particular, he does not see a more general context where maximal linearly ordered subsets are proved to exist in arbitrary partially ordered sets.[16] (As I remarked earlier, Hausdorff is not operating at this level of generality.) It is only in [H 1909a, 300–301] that he concludes that his second argument "falls to abstract set theory." And it is in [H 1909a] that this argument's set theoretic reformulation becomes a recognizable *maximal principle*.

After the existence of pantachies is established, Hausdorff studies the common properties of the pantachies for the final ordering of numerical sequences. He proves a series of *interpolation* (not his term) theorems that together can be summarized as follows:

> If $\mathfrak{A}$ is an at most countable set of numerical sequences, there always exists a sequence $X > \mathfrak{A}$ and a sequence $Y < \mathfrak{A}$. If $\mathfrak{A}$ and $\mathfrak{B}$ are two at most countable sets of numerical sequences and $\mathfrak{A} < \mathfrak{B}$, there always exists a sequence X such that $\mathfrak{A} < X < \mathfrak{B}$.[17]

(In [H 1909a, 304–305], Hausdorff presents this as a proposition and calls it the *Fundamental Theorem*.) Hausdorff indicates how to extend these results to sequences under the infinitary ordering and to functions under the final ordering. The former is done more elegantly and the latter in more detail in [H 1909a, 307–308].

From his interpolation theorems, Hausdorff concludes that each pantachie type is everywhere dense, unbounded and neither cofinal with ω nor coinitial with ω^*; its fundamental sequences have no limits, nor does it contain $\omega\omega^*$-gaps. He calls an order type with these properties an H-type. (The designation is not a sign of vanity on his part; H is capital η in the Greek alphabet and is deliberately chosen by the analogy: among the everywhere dense, unbounded types, η—the unique countable such type—is to H as ω is to Ω among well-ordered types.) All the pantachie types and their segments (in the infinitary and in the final orderings) considered up to this point are H-types of cardinality $\leq 2^{\aleph_0}$.

Summing the entries of a given numerical sequence of positive reals leads to the sequence's classification as *convergent* or *divergent*. In a brief digression, Hausdorff seeks to extend his interpolation results to countable sets of convergent and divergent series. Hausdorff proves several interpolation theorems ((L)–(P)) that can be summarized as follows:

> If $\mathfrak{A}$ is an at most countable set of convergent sequences, there always exists a convergent sequence $X > \mathfrak{A}$, and if

> P is any sequence $>\mathfrak{A}$, the convergent X can be made to satisfy $\mathfrak{A} < X < P$; if $\mathfrak{B}$ is an at most countable ordered domain of divergent sequences, there always exists a divergent sequence $Y < \mathfrak{B}$, and if Q is any sequence $< \mathfrak{B}$, the divergent Y can be made to satisfy $Q < Y < \mathfrak{B}$; if in addition $\mathfrak{A} < \mathfrak{B}$, there exist (infinitely many) convergent X and divergent Y such that $\mathfrak{A} < X, Y < \mathfrak{B}$.

(In [H 1909a, 328], Hausdorff presents a streamlined form of this proposition and calls it "a strengthening of the Fundamental Theorem.")

Hausdorff makes no claims for the originality of these results; he cites both [Pr 1890] and [Ha 1894] in [H 1907a, 122n1] and characterizes his own contributions as offering "simplicity and generality." In fact, his first two interpolation theorems (L) and (M) do not go much beyond what appears in [Ha 1894, 326, 328]. Hadamard essentially proved that any finally ordered ω-sequence of convergent sequences can be surpassed by some convergent sequence. Here, Hausdorff has removed the restriction that the countable set be finally ordered. Hadamard also proved that any finally ordered ω^*-sequence of divergent sequences can be preceded by some divergent sequence. Hausdorff's proof for a countable ordered domain of divergent sequences relies on a reduction to Hadamard's special case, which he then reproves. To explain the need for some restriction beyond countability, Hausdorff notes that it is possible to have two divergent sequences P and Q for which any $X < P, Q$ must be convergent. (Hadamard made precisely the same observation in [Ha 1894, 325].)

Although it easily follows from its predecessors, the last interpolation theorem ((P), p. 125), in which both convergent and divergent sequences are shown to exist between a countable set of convergent sequences and a following countable, finally ordered set of divergent sequences, is Hausdorff's best result in this area.[18] Previously such behavior was known for the Bonnet logarithmic scale. (See [Du 1873, 88–91], [Pr 1890, 351–356], and [H 1907a, 109].) Hausdorff uses this result to once again criticize the ideal elements of Du Bois-Reymond; only in [H 1909a, 330] does he remark on the limitations that (P) places on the "usual 'scales' of convergence criteria and divergence criteria," noting that there must be instances where they fail "since they are based on the comparison of an unknown series with a set of known convergent or divergent series where the set is either countable or, with regard to its effect, is replaceable by a countable set."

In §3 Hausdorff takes up H-types again, and in what seems to be a mundane abstraction of the *Fundamental Theorem*, he recharacterizes the H-types as those everywhere dense, unbounded types in which the individual inequalities $x < A, A < x, A < x < B$ $(A < B)$ always have solutions for x when A and B are countable subsets. However, this point of view leads him to introduce descending families of everywhere dense types that he calls η_ν-types; they are defined by allowing the A and B in the previous inequalities to have cardinality $< \aleph_\nu$. H-types are η_1-types. The type of the rationals would be an η_0-type, and in fact it would be the unique countable η_0-type; however, Hausdorff does not use the term "η_0-type."

By directly embedding the second class power $(\omega^* + \omega)'(\Omega)$ into an arbitrary H-type, Hausdorff shows such types have cardinality $\geq 2^{\aleph_0}$; thus pantachie types have cardinality $2^{\aleph_0}$. Assuming that an H-type has no $\Omega\Omega^*$-gaps, Hausdorff shows that it actually embeds the third class power $(\omega^* + \omega)''(\Omega)$ and thus has cardinality $\geq 2^{\aleph_1}$. Since pantachie types are H-types of cardinality $2^{\aleph_0}$, under CH (or under $2^{\aleph_0} < 2^{\aleph_1}$, which we now know is weaker than CH) pantachie types have $\Omega\Omega^*$-gaps.[19] (His proof that there are pantachies with $\Omega\Omega^*$-gaps, which uses only AC, is a major achievement of [H 1909a].) These embedded powers are not homogeneous, but an H-type always embeds a particular *homogeneous* H-type, namely, that of the second class power $3'_0(\Omega)$, where $3 = \{-1, 0, 1\}$.

H-types have other remarkable properties. In analogy with the rationals, the H-types are shown to be *universal* for ordered sets of cardinality $\leq \aleph_1$ (this uses AC). If there is an H-type of cardinality $\aleph_1$, then there is exactly one such type (this does not use AC). This uniqueness proof is the occasion for Hausdorff's invention of the *back-and-forth* construction of an isomorphism.[20] Hausdorff generalizes all these results to η_ν-types, ν finite. He observes that under the assumption of the extended continuum hypothesis: $2^{\aleph_\nu} = \aleph_{\nu+1}$, for ν finite, the η_ν-types of cardinality $\aleph_\nu$ are unique. He is the first to state such an extension of Cantor's continuum hypothesis.

Hausdorff uses transformations that preserve the final ordering to investigate the structure of pantachie types. For this purpose, Hausdorff employs certain "projective" transformations, which require rational operations on numerical sequences of positive reals. He defines the addition, multiplication, and division of these sequences by the corresponding operations on their individual terms. By means of projective (i.e., rational) transformations, Hausdorff shows that the inverse of a pantachie type is a pantachie type and that any segment of a pantachie type is again a pantachie type [H 1907a, 137–140]. The abundance of similarity transformations for the final ordering, as opposed to their scarcity for the infinitary ordering, and the resulting information that these transformations yield make Hausdorff's introduction of the final ordering a particular significant problem simplification.

What more can be said about pantachie types? The continuum hypothesis has a profound effect on H-types and so on pantachie types. Under CH, all pantachie types are similar, homogeneous and equal to $3'_0(\Omega)$. The last significant result of [H 1907a] is the construction of a homogeneous pantachie without resort to CH.[21] To accomplish this, Hausdorff utilizes the algebraic and order structure of numerical sequences in an interesting way.

Hausdorff first expands numerical sequences to allow all reals as entries, not just positive reals. Again, finally equivalent sequences are to be identified, and again, instead of working with equivalence classes of numerical sequences under the relation of final equality, Hausdorff chooses to work with their representatives. This means that equations are not statements of identity, they are assertions of final equality. Now the previously defined addition and multiplication operations on numerical sequences give a commutative ring that is not an integral domain. And for an arbitrary rational function $f(X_1, X_2, \ldots, X_n)$ with real or rational coefficients and any numerical sequences $A_1, A_2, \ldots, A_n$, the expression $f(A_1, A_2, \ldots, A_n)$ is defined in this ring as long as its divisions are legal; division by numerical sequences with infinitely many 0s, i.e., zero-divisors, is forbidden. (Hausdorff does not refer to the set of numerical sequences under addition and multiplication as a *ring*.)

Starting with the sequence $A_1 = (1, 2, 3, 4, \ldots)$, Hausdorff inductively selects (implicitly using AC) a sequence of integer valued numerical sequences, $\mathfrak{A} = \{A_\gamma\}$ of type $\geq \Omega$, such that for $\alpha < \beta$ and $p = 1, 2, 3, \ldots$, $A_\alpha^p < A_\beta$ and such that there is no numerical sequence $X > \mathfrak{A}$. Because of this last property, Hausdorff calls such a sequence "transcendent." Hausdorff calls an ordered domain $\mathfrak{B}$ of numerical sequences a "semifield domain" if $\mathfrak{B}$ is closed under sums, differences, and under multiplication by elements of $\mathfrak{H}_0$, where $\mathfrak{H}_0$ is the set of rational functions in the A_α ($A_\alpha \in \mathfrak{A}$) with rational coefficients. This last closure condition in the definition differs from the one actually appearing in [H 1907a, 143]. I have altered it in accordance with a correction made in [H 1909a, 310n]. In [H 1907a], it was only required that $\mathfrak{B}$ be closed under multiplication and division by each A_α.

The proof that a semifield pantachie exists runs very much along the lines of Hausdorff's second proof that a pantachie exists—but the extra structure requires more work. He first shows that there is a set of numerical sequences that is an ordered domain and has the closure properties required to be a semifield domain; namely, he shows $\mathfrak{H}_0$ is a semifield domain. The hard part is establishing that $\mathfrak{H}_0$ is an ordered domain, and here the growth properties of the sequences in $\mathfrak{A}$ are key. The second step is to show that any semifield domain $\mathfrak{H}$ that is not a pantachie can be extended to a larger semifield domain. For X an element not in $\mathfrak{H}$, but comparable to the elements of $\mathfrak{H}$, $\{AX + B \mid A \in \mathfrak{H}_0, B \in \mathfrak{H}\}$ is a semifield domain properly extending $\mathfrak{H}$. Again, this is amended as specified in the correction in [H 1909a, 310n]. In [H 1907a, 144], the proper extension is taken to be $\{AX+B \mid A, B \in \mathfrak{H}\}$. The third and final step is that the union of any well-ordered, increasing sequence of semifield domains is itself a semifield domain. After listing these steps, Hausdorff echoes his earlier existence proof:

> [F]rom [steps] 2 and 3, it follows that if one [a semifield pantachie] were not to exist, then the construction of semifield domains could be continued up to each transfinite index, and the cardinality of the continuum would have to be greater than any aleph. [H 1907a, 145]

The algebraic structure on the numerical sequences provides enough similarity mappings (namely, the projective transformations) for a proof that any semifield pantachie is isomeric, that is, all its middle segments are similar. Thus each middle segment of a semifield pantachie is a homogeneous type and also a pantachie type. Further projective transformations produce a homogeneous pantachie consisting of numerical sequences of positive reals.

Hausdorff introduces two other kinds of transformations that preserve the final ordering and have nothing to do with the algebraic structure. They might be called combinatorial in nature; one is passing to subsequences ("separation"), the other starts with two similar ordered domains and involves the shuffling of pairs of sequences that correspond under the given similarity ("mixture"). Using this latter kind of transformation, Hausdorff finds a context—pantachies of numerical sequences of positive reals—where Du Bois-Reymond's boundary between convergence and divergence can be realized. Summing the numerical sequences in such a pantachie $\mathfrak{P}$ leads to the decomposition $\mathfrak{P} = \mathfrak{P}_c + \mathfrak{P}_d$, where those that are convergent are in $\mathfrak{P}_c$, while those that are divergent are in $\mathfrak{P}_d$. It is possible for $\mathfrak{P}_c$ to be empty, but $\mathfrak{P}_d$ is never empty; it is impossible for a pantachie to consist solely of convergent sequences [H 1907a, 150]. Using *mixture*, Hausdorff shows that there are pantachies $\mathfrak{P}$ where $\mathfrak{P}_c$ has a last element and pantachies $\mathfrak{P}$ where $\mathfrak{P}_d$ has a first element—having both together is excluded by density. In fact, any gap that exists within a pantachie can be realized as a $\mathfrak{P}_c + \mathfrak{P}_d$ [H 1907a, 148–150]. (Hausdorff returns to the topic of the gaps realized by $\mathfrak{P}_c + \mathfrak{P}_d$ in [H 1909a, 330–334].)

In the final section of *On Pantachie Types*, Hausdorff considers the problem of more precisely specifying the place of the pantachie types within the class of H-types. His belief, however, is that "this problem poses incomparably greater difficulties, and for the time being, it is as likely to find a full solution as is the continuum problem to which it is intimately related."

Undeterred, Hausdorff examines the consequences of what he considers the "most likely conjecture": in a pantachie type, all sequences have cardinality $\leq \aleph_1$. He calls this conjecture the "analogue of Cantor's Continuum Hypothesis in the domain of order types." Assuming the conjecture's truth, he is able to locate pantachie types within the species described in his classification of everywhere dense, unbounded types as described in §4 of *Homogeneous Types of the Cardinality of the Continuum*. There are two possibilities, depending on whether or not $\Omega\Omega^*$-gaps are present.[22] The absence of such gaps means that $2^{\aleph_0} = 2^{\aleph_1} > \aleph_1$. As Hausdorff earlier noted, the truth of CH implies there is exactly one (homogeneous) pantachie type. So a "yes" answer to either of the questions, "Is there a pantachie without $\Omega\Omega^*$-gaps?" or "Does there exist a non-homogeneous pantachie type?" would falsify CH.

A further question that Hausdorff poses concerns the cofinality of the homogeneous pantachies that he constructed via the notion of semifield domains. In that construction, he started by defining an increasing sequence of numerical sequences $\mathfrak{A} = \{A_\alpha\}$ of type $\geq \Omega$ for which there is no numerical sequence X such that $X > \mathfrak{A}$. He called such an $\mathfrak{A}$ "transcendent."

Now, if a transcendent sequence $\mathfrak{A}$ of length ω_ν has the property that for any numerical sequence X there is some A_α in $\mathfrak{A}$ with $X < A_\alpha$, Hausdorff calls $\mathfrak{A}$ an ω_ν-scale. He calls the question of the existence of such a sequence "the Scale Problem." A positive solution to the scale problem would also settle the cofinality question. Hausdorff shows that the existence of a single ω_ν-scale means that every transcendent sequence has length ω_ν and each pantachie is cofinal with ω_ν.

Hausdorff derives the existence of an Ω-scale from CH. He deems that the construction of an Ω-scale without using CH would be a significant step forward:

> Certainly, proving the existence of an Ω-scale, independently of the Continuum Hypothesis, would thus be substantial progress, as it would establish the first close connection between the continuum and the second number class ... [H 1907a, 155]

But he has no illusions that this task is routine. He writes, "the difficulty of the matter is in keeping with its significance."

Restricting himself to numerical sequences with positive integer entries, Hausdorff considers schemes for constructing increasing Ω-sequences of numerical sequences, such as the one described by G. H. Hardy in *A theorem concerning the infinite cardinal numbers*, Quarterly Journal **35** (1903), 87-84, where $A^{\alpha+1} = A^\alpha + 1$ and A^λ, λ a limit ordinal, is obtained from its countable set of predecessors by diagonalization.[23] He explains why such schemes are found wanting for the creation of an Ω-scale.[24]

Hausdorff ends with an attempt (not really successful) to concisely state "the difficulties of the pantachie problem":

> [O]ne is carried forth beyond the countable, but one does not know how far, since the starting point still lies in the countable and no immediate relation with the next higher level, the second infinite cardinality, reveals itself. [H 1907a, 158]

He points out that partial results for both the pantachie problem and scale problem can be obtained for pantachies whose very definition has a relationship to the set of countable ordinals already built in. But even then, the existence of sequences of cardinality $> \aleph_1$ "remains undecided."

Notes

1. In [H 1901b, 460n2], Hausdorff used a generic $\aleph$ to denote the cardinality of the continuum only with the justification that "Cantor intends to publish a proof soon that each transfinite cardinal number must occur in the 'sequence of alephs'."

2. In the language of [H 1901b], α is an "initial piece"(of μ) without a last element, and as such, it is a version of Dedekind's notion of cut.

3. Bertrand Russell described such a construction in §280 of *The Principles of Mathematics* (London: Allen & Unwin), 1903 (referenced by Hausdorff in [H 1907a, 102n]).

   In 1905, Hausdorff wrote a review of Russell's *Principles* (Vierteljahresschrift für wissenschaftliche Philosophie und Sociologie 29 (1905), 119-124). In [Pu 2002, 28], Purkert reports that there are fifty pages of notes, some of them highly critical, on *The Principles* in Hausdorff's *Nachlaß* [Kapsel 49, Fasz. 1068].

4. It seems likely that Hausdorff learned about Du Bois-Reymond's work from [Sch 1898; 1900]; he cites [Sch 1900] as one source for the "latest" on the pantachie question. Hausdorff was a careful reader; in [H 1907a, 115] he provides a counterexample to an erroneous assertion in [Sch 1900, 53–56] that originated with Du Bois-Reymond. Both claimed that passing to function inverses reverses the infinitary ordering. In Schoenflies's discussion of Du Bois-Reymond's "so called infinitary pantachie" ([Sch 1908, 64–66]), he thanks Hausdorff and gives his counterexample. He then goes on to provide a lengthy overview of *On Pantachie Types* in a section entitled *Hausdorff's Pantachie Types* [Sch 1908, 66–71].

5. Gordon Fisher ([Fi 1981]) has written a very informative survey in which he assimilates a prodigious amount of material. In it, "[t]he mixed fortunes of Paul Du Bois-Reymond's infinitary calculus and ideal boundary between convergence and divergence are traced from 1870 to 1914." Fisher sees Hausdorff as occupying a middle position between the critics of Du Bois-Reymond's ideas (Cantor, Dedekind, Peano, Russell, Pringsheim, et al.) and his supporters (Stolz, Borel, Hardy).

6. The model theorists B. Jónsson, M. Morley, and R. Vaught were important contributors to this area. In response to an e-mail inquiry (April, 2003) as to the influence of η-sets upon his work, Michael Morely (quoted with his permission) replied:

   > About 1954, as a graduate student looking for a thesis topic, I deliberately set out to generalize the eta set construction to other algebraic structures. (I think the idea was suggested to me by another graduate student.) It became apparent that amalgamation and unions of chains were the critical conditions. Model theory and the compactness theorem made both of these simpler. These ideas occurred to others. Jónsson in his paper, Homogeneous Universal Relational Systems, [Math. Scand. 8, 1960], explicitly mentions the eta sets.

   In [Fe 2002], Ulrich Felgner details the uses of η_α-sets in algebra, topology, set theory, and model theory.

7. In [1907a, 125n], Hausdorff expresses his disdain for the genesis of the ideal boundary between convergence and divergence and for its operational use:

   > The analogy with the irrational numbers is a logical misconception; one can insert a new thing corresponding to the relation $x^2 = 2$ between the elements of the two classes defined by $x^2 < 2$ and $x^2 > 2$, but between convergence and divergence of positive series there is no third alternative.

As for ideal elements, which "are no longer numerical sequences or functions":

> to let such an element appear, for instance, as a function under the integral sign is to dispense with any meaning.

However, so eminent a mathematician as Èmile Borel found Du Bois-Reymond's approach persuasive and felt no qualms in accepting an ideal element (not an ordinary function) as the boundary between the domains of convergence and divergence. (See [Fi 1981, §6] for an overview of Borel's own work in this area and its relationship to that of Du Bois-Reymond.)

8. Cantor's war against the "infinitely small" is discussed in [Da 1979] and in a posthumous paper of Detlef Laugwitz ([La 2002]) where the sometimes rancorous debate over infinitesimals is viewed (positively) as making clearer the distinction between Cantor's transfinite numbers and the theory of ordered algebraic structures.

9. This is the culmination of Hausdorff's relationship to the infinitary pantachie that began benignly in [H 1904a] with an example of an order type of cardinality of the continuum "that shares with 'the infinitary pantachie' the property that no *fundamental sequences* [his italics] in it have limits," and that later turned sour, when after producing a similar example of cardinality $\aleph_1$ in [H 1906b, 152], he referred to "the 'infinitary pantachie' of P. Du Bois-Reymond, which to this day has not been defined to be free of doubt."

   Later ([H 1907b, 543]), Hausdorff characterizes his work in *On Pantachie Types* as an attempt "to salvage a failed speculation of P. Dubois-Reymond [sic]."

10. The German words *Gebiet* and *Bereich* can both be translated as *domain*. Throughout [H 1907a], Hausdorff only uses *Bereich* to refer to a domain (of functions, of sequences) that is linearly ordered. We always translate *Bereich* as *ordered domain*.

11. Du Bois-Reymond coined the adjective *pantachish* to apply to point distributions in intervals that had elements in every subinterval, no matter how small [Du 1879, 287]. In his book, [Du 1882, 182–183], he explained its derivation from the Greek words $\pi\alpha\nu\tau\alpha\chi\tilde{\eta}$, $\pi\alpha\nu\tau\alpha\chi o\tilde{v}$ (*everywhere*), and he called the point distribution itself a *pantachie*.

12. Hausdorff seems to have been the first to compute the cardinality of the set of real monotonic functions ([H 1907a, 111n]). See [Sch 1908, 19n2].

13. From the context, this is a claim about the infinitary ordering of monotonic functions that go to $+\infty$ with increasing x. Once pantachies of sequences under the final ordering are shown to exist (p. 118), one immediately gets the existence of pantachies consisting of just continuous functions under *the final ordering* by the claimed extension of theorem (B). Here, Hausdorff seems to be lacking a theorem about the existence of pantachies in the positive real sequences under the infinitary ordering that would transfer to the existence of pantachies of continuous functions under the infinitary ordering. His existence proof could be easily changed to do this, but he makes no comment to that effect.

14. The set of distinct equivalence classes formed from the set of numerical sequences of positive reals under the equivalence relation of final equality and the partial ordering induced by < on them is model-theoretically the reduced power of $\langle \mathbb{R}^+, < \rangle$ over the index set $\mathbb{N}$, modulo the filter $Cof(\mathbb{N})$ of cofinite subsets of $\mathbb{N}$.

   Of course, Hausdorff did not have this perspective. Reduced products, as generalizations of ultraproducts, were invented by disciples of Tarski in the late 1950s. *Reduced direct products* by T. E. Frayne, A. C. Morel, and D. S. Scott, Fund. Math. **51** (1962), 195–228, is the standard initial reference. It would be interesting to know if any of the developers of reduced products were aware of Hausdorff's work on this concrete instance of such a structure.

15. In [H 1904a], that "there are cardinal numbers that are greater than every aleph" was considered a "paradoxical result." (See the *Introduction to "The Concept of Power in Set Theory,"* p. 27.)

16. Hausdorff first considers partially ordered sets in general in Chapter VI, §1 of [H 1914a]. There he proves that an arbitrary partially ordered set has a "greatest linearly ordered subset" [eine größte geordnete Teilmenge], and he uses a *choice function* in the construction of such a subset. As his first application, he partially orders the intervals of an arbitrary ordered set by reverse set inclusion and from the existence of a maximal linearly ordered set of intervals concludes that the original ordered set either has a pair of adjacent elements or a symmetric limit or a symmetric gap—and just the latter two possibilities for a densely ordered set. Hausdorff announced such a result for densely ordered sets in [H 1907b, 542] and proved it using a well-ordering in [H 1908, 445–446].

17. Some interpolation results were known for the infinitary ordering of monotonic real functions. In [Du 1875, 365n], we essentially have the following: for any infinitary ω^*-sequence $\mathfrak{A}$ there is an X such that $X < \mathfrak{A}$. In [Ha 1894, 334], Hadamard states an interpolation result for sets of sequences that can be read as follows: for an infinitary ω-sequence $\mathfrak{A}$ and an infinitary ω^*-sequence $\mathfrak{B}$ with $\mathfrak{A} < \mathfrak{B}$, there is an X such that $\mathfrak{A} < X < \mathfrak{B}$.

18. It seems fair to say that as a contribution to the theory of infinite series Hausdorff's interpolation theorems are not well-known; a quick glance at [Kn 1947] turned up references to the work of Du Bois-Reymond and Hadamard—but no mention of Hausdorff's results.

19. That CH implies pantachie types have $\Omega\Omega^*$-gaps also follows from the earlier result that an everywhere dense type without $\omega\omega^*$-limits, without $\omega\omega^*$-gaps, and without $\Omega\Omega^*$-gaps has cardinality $> \aleph_1$ [H 1907a, 85-86].

20. Cantor is often erroneously credited with the invention of the *back-and-forth* construction of mappings. Both Hausdorff in [H 1907a] and E. V. Huntington in [Hu 1905] had the idea of *alternating* between domain and range in such constructions. Huntington's execution of alternation in proving Cantor's characterization of the rationals as an ordered set was flawed. Hausdorff's 1914 text made *back-and-forth* part of the mathematical mainstream. See [Pl 1993] for more on this history.

21. This is the first known instance where CH is eliminated from a proof. Felgner ([Fe 2002, 651]) claims this honor for Hausdorff's non-CH proof that there are pantachies with $\Omega\Omega^*$-gaps in [H 1909a, 320-323]. Hausdorff's proof without CH of the existence of a homogeneous pantachie implicitly uses AC.

22. In [H 1907b, 543], Hausdorff expresses the opinion that for the infinitary and final pantachies it is "probably" the case that all elements are $\Omega\Omega^*$-elements and that there are $\omega\Omega^*$-gaps, $\Omega\omega^*$-gaps, and $\Omega\Omega^*$-gaps.

23. Hardy's construction had nothing to do with the final ordering, but rather was meant to define a subset of the continuum of cardinality $\aleph_1$. His construction was severely criticized by E. W. Hobson who, Hausdorff states, "has wrongly objected to this method" [H 1907a, 155n2].

24. From the post-Cohen era, we now know that the scale problem cannot be solved in ZFC. See [He 1974].

Investigations into Order Types[1])

By
FELIX HAUSDORFF

IV. Homogeneous Types of Cardinality of the Continuum

§1
The formulation of the problem

In immediate reference to the preceding note III, *Homogeneous types of the second infinite cardinality*, we consider the question of which of the 50 species of homogeneous types with sequences of first and second infinite cardinality put forward there (pp. 154–155) are represented by types of cardinality ($\aleph$) of the continuum. The methods for constructing such types are of course also based on the principle of power formation [[Potenzbildung]]; they are in fact considerably different from those that were successful for homogeneous types of the second infinite cardinality: a state of affairs in which the unsolved continuum question again manifests itself. Even the results in both cases are different. Of the homogeneous types of the second infinite cardinality, 32 species were successfully constructed (including one species that was represented by the countable type η), namely, all those with $\omega\omega^*$-gaps [[*Lücken*]]; from the start, we had to exclude the 18 species without $\omega\omega^*$-gaps *since an everywhere dense* [[*überall dichter*]] *type without $\omega\omega^*$-gaps has cardinality at least that of the continuum* and the discovery of such a type of the second infinite cardinality would be tantamount to a proof of CANTOR's hypothesis that $\aleph = \aleph_1$. On the other hand, as we shall show, of the homogeneous types of cardinality $\aleph$, 45 species are constructible and the remaining 5 are excluded because they require a cardinality *higher than the second infinite cardinality*; the discovery of a pertinent type of the cardinality of the continuum would thus settle the continuum question in the sense that $\aleph > \aleph_1$. Apart from the currently still imperative precautionary measure of leaving aside problems equivalent to the continuum question because of their suspected inaccessibility, four of the five species mentioned

---

[1]) The first articles in this series have appeared in these Reports **58** (1906), pp. 106–169, with the titles
    I. The Powers of Order Types
    II. The Higher Continua
    III. Homogeneous Types of the Second Infinite Cardinality
and are cited in what follows by I. i. O. T.

above fall beyond the scope of our investigation since they necessarily contain sequences of the third infinite cardinality.

Namely, the following theorem holds:

A. *An everywhere dense type without $\omega\omega^*$-limits and $\omega\omega^*$-gaps certainly contains Ω-sequences and Ω^*-sequences; if it also has no $\Omega\Omega^*$-gaps, then its cardinality is $> \aleph_1$.*

Let μ be an everywhere dense type as in the hypothesis; then certainly between any two elements a_0, b_0 ($a_0 < b_0$) lie two more a_1, b_1 ($a_1 < b_1$), and once again between these lie a_2, b_2 ($a_2 < b_2$) and so on. Since μ does not contain any $\omega\omega^*$-gaps and $\omega\omega^*$-limits, between the two resulting fundamental sequences

$$a_0\, a_1\, a_2 \cdots \quad \cdots b_2\, b_1\, b_0$$

there again lie at least two more elements $a_\omega < b_\omega$, and between these lie two more $a_{\omega+1} < b_{\omega+1}$, etc. One sees that for each index of the second number class this procedure yields an element pair a_α, b_α; therefore μ contains a subset of type $\Omega + \Omega^*$ and has cardinality $\geqq \aleph_1$. We prove the second part of the statement as follows: we show that if μ is of cardinality $\aleph_1$, then it certainly contains an $\Omega\Omega^*$-gap. Imagine that we have put the elements of μ into the form of a well–ordered set

$$(0)\,(1)\,(2)\,\cdots\,(\omega)\,\cdots\,(\alpha)\,\cdots.$$

We denote the element pair $(0)\,(1)$ by $a_0\, b_0$ (thus either $(0) = a_0$, $(1) = b_0$ or $(0) = b_0$, $(1) = a_0$, according to whether $(0) < (1)$ or $(1) < (0)$ in μ). Let $(\alpha_2)\,(\alpha_3)$ be the first two elements falling between $a_0\, b_0$, and let them be denoted by $a_1\, b_1$; denote the first two elements $(\alpha_4)\,(\alpha_5)$ falling between $a_1\, b_1$ by $a_2\, b_2$, etc. Consequently, we obtain a sequence of elements of the second infinite cardinality

$$(0)\,(1)\,(\alpha_2)\,(\alpha_3)\,\cdots\,(\alpha_\omega)\,(\alpha_{\omega+1})\,\cdots\,(\alpha_{2\xi})\,(\alpha_{2\xi+1})\,\cdots$$

that forms a subset in μ of type $\Omega + \Omega^*$; the element pair $(\alpha_{2\xi})\,(\alpha_{2\xi+1})$ is identical with the just now mentioned $a_\xi\, b_\xi$. But this subset represents an $\Omega\Omega^*$-gap of μ. For if (β) is not an element of this set and if $(\alpha_{2\xi})\,(\alpha_{2\xi+1})$ is the first element pair following (β), so $\beta < \alpha_{2\xi}$, then (β) cannot lie between $a_\xi\, b_\xi$ since otherwise, according to the given instructions, $(\beta)\,(\alpha_{2\xi})$ ought to have been chosen as the pair $a_\xi\, b_\xi$ instead of $(\alpha_{2\xi})\,(\alpha_{2\xi+1})$. Thus (β) lies outside the middle segment $a_\xi\, b_\xi$ and all the ones which follow, and there exists no element which would lie within all the middle segments $a_\xi\, b_\xi$.

Theorem A, now proved, allows the following immediate generalization, where by $\omega\, \omega_1\, \omega_2\, \cdots\, \omega_\nu$ are understood the smallest numbers of the 2nd, 3rd, 4th, ..., $(\nu + 2)$th number classes:

B. *An everywhere dense type that contains neither limits nor gaps of the form $\omega\omega^*$, $\omega_1\omega_1^*$, $\cdots$, $\omega_{\nu-1}\omega_{\nu-1}^*$ certainly contains ω_ν-sequences and ω_ν^*-sequences; if it also does not contain any $\omega_\nu\omega_\nu^*$-gaps, then it has cardinality*

$> \aleph_\nu.$[1])

In the tableau of our fifty species, there are five species without $\omega\omega^*$-limits that are free of $\omega\omega^*$-gaps and $\Omega\Omega^*$-gaps; four of those also contain no $\Omega\Omega^*$-limits. Thus we have the result:

C. *The five species*

$$IIA11,\ IIA15,\ IIIA10,\ IIIA15,\ IV15$$

can only be represented by types of cardinality higher than the second infinite cardinality; the first four can only be represented by types which contain sequences of the third infinite cardinality.

§2

Types of the first group

Now we try to furnish an existence proof for the remaining 45 species. Here, we also have a wide range of methods at our disposal. We give a mixed procedure of raising to powers [[Potenzierungen]] of the first and second class with the arguments ω and Ω, which to us seems to be the clearest.

First, representatives for the 16 species of group I (homogeneous types with $\omega\omega^*$-limits) shall be constructed; for this, we utilize powers of the second class with argument ω. One such power is

$$M = \mu'(\omega),$$

the type of the *entire* covering set of ω by μ (so there is no preferred principal element) ordered according to first differences, i.e., of the set of elements

$$x = (x_0\, x_1\, x_2\, \cdots)$$

in which each x_ν runs through the type μ independently of the others.

We denote the initial segments [[Anfangsstecke]], middle segments [[Mittelstrecke]], and end segments [[Endstrecke]] of the base μ by π, ϱ, σ and those of the power M by Π, P, Σ. For the decomposition $M = \Pi + 1 + \Sigma$ produced by x, we have

(1) $\quad \begin{cases} \Pi = M(\pi_0 + \pi_1 + \pi_2 + \cdots), \\ \Sigma = M(\cdots + \sigma_2 + \sigma_1 + \sigma_0), \end{cases}$

where $\mu = \pi_\nu + 1 + \sigma_\nu$ is the decomposition of the base produced by x_ν. In particular, a covering with nothing but equal elements $(m\, m\, m \cdots)$ yields $(\mu = \pi + 1 + \sigma)$

(2) $\qquad\qquad M = M\pi\omega + 1 + M\sigma\omega^*.$

From this it follows that

(3) $\qquad\qquad M = M(\pi + 1) = M(1 + \sigma).$

On the other hand,

(4) $\qquad\qquad M = \mu'(1 + \omega) = M\mu;$

---

[1]) Theorems A and B still permit a further strengthening: in the mentioned cases, the cardinality is not only $> \aleph_1(\aleph_\nu)$, but it is $\geqq 2^{\aleph_1}(2^{\aleph_\nu})$. Cf. the upcoming Note V, §3, p. 135.

and for $\mu = \pi + 1 + \varrho + 1 + \sigma$,
$$M = M(\pi+1) + M\rho + M(1+\sigma) = M + M\rho + M,$$
(5) $$M = M(1 + \varrho + 1).$$

So M remains unchanged under right multiplication by the base or any of its intervals (interval = segment with endpoints [[Strecke mit Randpunkten]]).

Just as in I. i. O. T. III, we now choose a base composed additively from a finite number of the following four summands

(6) $$\begin{cases} \omega_{11} = \omega + \omega^*, & \omega_{21} = \Omega + \omega^*, \\ \omega_{12} = \omega + \Omega^*, & \omega_{22} = \Omega + \Omega^*. \end{cases}$$

According to (2) and (4), the decompositions
$$\pi, \sigma = 0, \mu \quad \text{or} \quad 1, \mu \quad \text{or} \quad \mu, 0$$
then give
(7) $$M = 1 + M\omega^* = M\omega + 1 + M\omega^* = M\omega + 1;$$
hence
$$M = M(\omega + 1) = M(1 + \omega^*).$$

And from this it follows by transfinite induction that for each α of the first two number classes

(8) $$\begin{cases} M\alpha = 0 & \text{if } \alpha = 0; \\ M\alpha = M & \text{if } \alpha \text{ is not a limit number;} \\ M\alpha = M\omega & \text{if } \alpha \text{ is a limit number.} \end{cases}$$

Moreover, it follows that each initial segment π_ν of the base, if it is not 0, comes from an earlier initial segment π by adding a 1 or a limit number, thus
$$\pi_\nu = 0 \quad \text{or} \quad = \pi + 1 \quad \text{or} \quad = \pi + \alpha_\omega;$$
hence
$$M\pi_\nu = 0 \quad \text{or} \quad = M \quad \text{or} \quad = M\omega.$$

And if one substitutes this into (1), it then follows that Π can also have only one of these 3 types. Thus
$$\Pi = 0 \quad \text{or} \quad = M\omega \quad \text{or} \quad = M = M\omega + 1,$$
similarly
$$\Sigma = 0 \quad \text{or} \quad = M\omega^* \quad \text{or} \quad = M = 1 + M\omega^*.$$

Consequently, M is not everywhere dense, but rather M contains pairs of consecutive elements; the elements whose initial segment is M have immediate predecessors, and those with final segment M have immediate successors. These elements and the boundary elements of M are one-sided ω^*-limits, respectively, ω-limits, and all the remaining ones are two-sided $\omega\omega^*$-limits.

We now think of all the elements of M with immediate successors as deleted, whereby M is transformed into a new type $M_1 = 1 + P_1 + 1$. Easy considerations along the lines of those we employed in III, §1 for the separation of the elements of the mixed type Ξ now show that all the initial segments of M_1 have the type $\Pi_1 = 1 + P_1$ and that all end segments have the type $\Sigma_1 = P_1 + 1$. All the initial segments and end segments of P_1 have the type P_1; so P_1 is a homogeneous type and M_1 is an isomeric type. Since all elements of P_1 are $\omega\omega^*$-limits, it belongs to group I, and, moreover, it has the cardinality of the continuum.

The question regarding gaps still remains to be dealt with; we mention beforehand the following remarks, which will find application in the next section. From coverings of a well-ordered argument α by an arbitrary base μ (these coverings ordered according to the principle of first differences), one can form ω_ν-sequences of the following kind for $\alpha \geqq \omega_\nu$: let x be a fixed covering and let x^β be one that is $< x$ and which differs from x for the first time in place β; thus

$$x = (x_0\, x_1 \cdots \mid x_\beta\, x_{\beta+1} \cdots),$$
$$x^\beta = (x_0\, x_1 \cdots \mid a_\beta\, a_{\beta+1} \cdots), \qquad a_\beta < x_\beta.$$

Then if β runs through an ω_ν-sequence $\beta_0\, \beta_1 \ldots$ contained in the argument, x^β runs through an ω_ν-sequence of coverings $x^{\beta_0}\, x^{\beta_1} \cdots$; we call such a sequence an ω_ν-sequence *of the first kind* or an ω_ν-sequence *arising from the argument*. On the other hand, if the base contains an ω_ν-sequence, then set up a sequence of coverings whose elements belonging to place β run through an ω_ν-sequence from the base while the preceding elements remain unchanged, i.e., for fixed β let the a_β in x^β run through an ω_ν-sequence; call the ω_ν-sequence of coverings produced an ω_ν-sequence *of the second kind* or an ω_ν-sequence *arising from the base*. Based on the assumption that the argument is an ordinal number, one easily proves[1]) that each ω_ν-sequence of coverings is cofinal [konfinal] with a sequence of either the first or the second kind. Indeed, now let $x^0\, x^1 \cdots x^\beta \cdots$ be an ω_ν-sequence of coverings; let (β, γ) denote the place of the first difference between x^β and x^γ; so that (β, γ) is thus a place in the argument or an ordinal number $< \alpha$. One sees immediately that for $\beta < \gamma < \delta$ the number (β, δ) is the smaller of the two numbers (β, γ) and (γ, δ), thus

$$(\beta, \delta) \leqq (\beta, \gamma), \quad (\beta, \delta) \leqq (\gamma, \delta).$$

Hence the numbers $(0,1)\,(0,2) \cdots (0,\beta) \cdots$ never increase; since there are no infinite descending sequences in α, from a certain index β_1 on they must attain a minimum value κ_0, so that $(0, \beta) = \kappa_0$ for $\beta \geqq \beta_1$. One deals with the numbers $(\beta_1, \beta_1 + 1)$, $(\beta_1, \beta_1 + 2), \ldots$ accordingly and finds an index $\beta_2 > \beta_1$ for which $(\beta_1, \beta) = \kappa_1$ for $\beta \geqq \beta_2$. Continuing like this, one selects from the given sequence a subset, likewise of type ω_ν, which we again wish to denote by $x^0\, x^1 \cdots x^\beta \cdots$ and in which $(\beta, \gamma) = \kappa_\beta$ for $\gamma > \beta$; thus it has a value depending only on β but not on γ; moreover, according to the above

---

[1]) Similar considerations are found in I. i. O. T. II, §3, p. 133; cf. III, §3, p. 160 as well.

remarks,
$$\kappa_0 \leqq \kappa_1 \leqq \cdots \leqq \kappa_\beta \leqq \cdots.$$
Here, either the inequality sign occurs $\aleph_\nu$ times and our ω_ν-sequence is then cofinal with an ω_ν-sequence of the first kind; or it occurs fewer than $\aleph_\nu$ times, in which case there exists a maximum κ for the numbers κ_β and the place κ is eventually covered by steadily increasing elements and we have an ω_ν-sequence of the second kind. — We have considered here the totality of all coverings of α by μ; within a power of a specific class with base μ and argument α, which possibly contains only a part of all the coverings, there can of course be no other ω_ν-sequences but those that are cofinal with ω_ν-sequences of the first or second kind, except such sequences, which exist in the total covering set, can be omitted from a restricted subset of coverings. What is to be understood by ω_ν^*-sequences of the first or second kind needs no discussion; the first kind arise from ω_ν-sequences of the argument, the second kind from ω_ν^*-sequences of the base.

Let us now return to our types of the first group. M_1 contains the gaps of M and only these since by the omission of each one-sided ω-limit with an immediate successor no new gaps have arisen. For its part, M contains the gaps of the base and only these because sequences of the first kind in M are fundamental sequences and have limits. Either a sequence of the second kind corresponds to a base sequence [[Basisreihe]] with a limit, and then it has a limit itself because μ possesses boundary elements, or it corresponds to a base sequence which precedes or follows a gap and M exhibits the same gap. (A third case, yet to be considered, in which the base sequence in μ defines neither a limit nor a gap, but rather μ is cofinal or coinitial with it, is excluded due to the boundedness of μ.) Accordingly, one obtains homogeneous types P_1 of all 16 species of group I from the choice, for instance, of the following bases[1]):

$ID1$	ω_{11}	$ID9$	ν
$IB2$	$\omega_{11} + \omega_{21}$	$IB10$	ω_{21}
$IC3$	$\omega_{11} + \omega_{12}$	$IC11$	ω_{12}
$IA4$	$\omega_{11} + \omega_{22}$	$IA12$	ω_{22}
$IA5$	$\omega_{11} + \omega_{21} + \omega_{22}$	$IA13$	$\omega_{21} + \omega_{22}$
$IA6$	$\omega_{11} + \omega_{12} + \omega_{22}$	$IA14$	$\omega_{12} + \omega_{22}$
$IA7$	$\omega_{11} + \omega_{21} + \omega_{12}$	$IA15$	$\omega_{21} + \omega_{12}$
$IA8$	$\omega_{11} + \omega_{21} + \omega_{12} + \omega_{22}$	$IA16$	$\omega_{21} + \omega_{12} + \omega_{22}$

Only species $ID9$, whose base ν (any finite number > 1) is not of the form in (6), might need a specific comment, unless the simple remark that the power $\nu'(\omega)$ can be read as the collection of ν-adic fractions would suffice;

---

[1]) Instead of $\omega_{21} + \omega_{12}$, for example, one can of course choose either $\omega_{12} + \omega_{21}$ or $\omega_{12} + \omega_{21} + \omega_{12}$ or any finite sum of ω_{21}, ω_{12} as base. The difference in comparison with powers of the first class (III, p. 163) should be pointed out; besides the base-gaps, these always contain $\omega\omega^*$-gaps, and the bases on the left as well as those on the right in the following tableau only give types of species 1–8 in this case.

these represent the interval (0,1) of real numbers in which two such fractions are assigned to a rational number with a power of ν in the denominator. The continuum of real numbers between 0, 1, endpoints included, arises through the removal of these double representations (i.e., precisely through the transition from M to M_1); thus M_1 is identical with CANTOR's linear continuum ϑ, whereas $M = \nu'(\omega)$, as is easy to see, represents the order type of a linear, nowhere dense, perfect point set[1]).

The M_1-type of species $IA12$ is the simplest "ultracontinuum" Θ (see I. i. O. T. III, §3).

The homogeneous P_1-type of species $ID1$ is identical with the order type of the irrational numbers in their natural ordering. One recognizes this roughly as follows: the $\omega\omega^*$-gaps of P_1 or M_1 or M are bounded by fundamental sequences of the second kind; so they correspond to gaps of the base $\omega + \omega^*$. To mark the position of the gap in M, we insert an element l into the base that fills its only gap $(\omega + l + \omega^*)$; then all the gaps of M are represented by the coverings

$$(l), \quad (x_0\, l), \quad (x_0\, x_1\, l), \quad (x_0\, x_1\, x_2\, l), \quad \ldots,$$

where $x_0\, x_1 \ldots$ run through the base μ. The order type of all these gaps is obviously unbounded, everywhere dense, countable and so equal to the type η of the rational numbers. According to the discussion in §4, η is the "complement" of P_1, but also conversely, P_1 is the complement of η; i.e., P_1 is the type of the gaps of η or the type of all the irrational numbers ordered according to size. Of course, this type is capable of many other representations; for example, if one thinks of the rational numbers > 1 expanded as continued fractions, then one finds that their type is also given by the power $\psi'(\omega)$ of the second class where $\psi = \omega^*\omega$.

Each of the 16 constructed isomeric types M_1 raised to a power of the second class with argument ω^α again yields an isomeric type[2]) of the same species (I. i. O. T. II, Theorem M, p. 140).

If M_1 does not contain a set of exclusive intervals similar to itself, then all these powers are distinct (ibid. Theorem D, p. 131). That this case actually occurs can be proved for species 9–16 in exactly the same way as for the ultracontinua (ibid. p. 167). Thus each of the species $I\,9$–16 is certainly represented by an uncountable set of distinct homogeneous types. Since M_1 contains a subset $N = \mu_m(\omega)$, i.e., a power of the first class with some principal element, in such a way that its elements fall into each interval of M_1, if M_1 were to contain a set of exclusive intervals similar to M_1, the set N would have to contain a subset of the same type as M_1; and this is impossible if M_1 is free of $\omega\omega^*$-gaps. The same holds for the species $ID1$

---

[1]) [H 1907a, 91n2] Cf. A. SCHOENFLIES, Die Entwicklung der Lehre von den Punktmannigfaltigkeiten, (Leipzig 1900), Part 2, Chapter 3.

[2]) [H 1907a, 92n1] Through the same raising to powers, P_1 would also give homogeneous types (ibid. Theorem L, p. 138) but not necessarily of the same species, since new $\omega\omega^*$-gaps would appear because of the unboundedness of the base.

since in this case the set N is countable and does not contain any subset of the cardinality of the continuum. I have not been able to settle the question for the remaining species 2–8; in this case, should the powers $M_1'(\alpha)$ not all be distinct, thus should N contain a subset of type M_1, then the continuum question would be settled in CANTOR's sense since the set N mentioned above is only of the second infinite cardinality.

§3
Types of the remaining groups

We now utilize the isomeric types of species I 1–16 found in §2 as bases for new powers, and we denote them by $\mu = 1 + \varrho + 1$, their initial segments different from 0 by $\pi = 1 + \varrho$, and their end segments different from 0 by $\sigma = \varrho + 1$. Since ϱ is homogeneous and cofinal with ω, it follows that

$$\varrho = \varrho + 1 + \varrho + 1 + \cdots = \sigma\omega,$$

$$\pi = \pi + \pi + \pi + \cdots = \pi\omega,$$

and similarly $\varrho = \pi\omega^*$ and $\sigma = \sigma\omega^*$, from which it still follows that for each number α (> 0) of the first two number classes $\pi\alpha = \pi$. We take Ω as argument and the last, the first, or a middle element of the base as principal element, depending on whether we want to represent a type of group $II, III,$ or IV; finally, powers of the first or second class are to be chosen according to whether we want to construct types with or without $\omega\omega^*$-gaps.

Types with $\omega\omega^*$-gaps

For the powers

$$M = \mu_m(\Omega)$$

of the first class, the following fundamental formulas hold (I. i. O. T. III, §3, p. 159):

(9)
$$\begin{cases} M = M\pi_m\Omega + 1 + M\sigma_m\Omega^*, \\ M = M\pi_l\omega + M\sigma_l\omega^*, \\ \Pi = M(\pi_0 + \pi_1 + \cdots + \pi_\omega + \cdots + \pi_\alpha + \cdots), \\ \Sigma = M(\cdots + \sigma_\alpha + \cdots + \sigma_\omega + \cdots + \sigma_1 + \sigma_0), \end{cases}$$

where m denotes the principal element, l an arbitrary secondary element, and only a finite number of the π_α, σ_α are different from π_m, σ_m. The type M is everywhere dense; it contains $\omega\omega^*$-gaps and the gaps of the base, but otherwise no others because $\Omega(\Omega^*)$-sequences of the first kind, if they occur at all, have limits (III, p. 161) and those of the second kind belong to base-gaps [[Basislücken]] by the current choice of base.

Let us now separate the different cases for the principal element m.

(A). *Let the principal element be a middle one* (neither the first nor the last).

It then follows from (9) that
$$M = M\pi\Omega + 1 + M\sigma\Omega^*$$
$$= M\pi\omega + M\sigma\omega^* = M\pi\omega = M\sigma\omega^*,$$
or, on account of the remarks regarding π and σ,
$$M = M\pi = M\sigma.$$
The case $\pi_\alpha = 0$ occurs only with finite frequency, therefore
$$\Pi = M\pi\Omega = M\Omega, \quad \text{similarly} \quad \Sigma = M\sigma\Omega^* = M\Omega^*.$$
From this it follows easily that for a middle segment P ($=$ end segment of Π or initial segment of Σ)
$$P = M\Omega^* + M\Omega = \Sigma + \Pi.$$
Accordingly, M is isomeric, and each of its middle segments is a homogeneous type of group IV that is certainly of the cardinality of the continuum. On account of what was said about the gaps, the following species of ϱ and P belong together:

$$\varrho : I \quad \begin{cases} 1 & 2 & 3 & 4 & 5 & 6 & 7 & 8 \\ 9 & 10 & 11 & 12 & 13 & 14 & 15 & 16 \end{cases}$$
$$P : IV \quad \phantom{\{} 1 \quad 2 \quad 3 \quad 4 \quad 5 \quad 6 \quad 7 \quad 8$$

(B). *Let the principal element be the first one.*
Here
$$M = 1 + M\sigma\Omega^*$$
$$= M\pi\omega + M\sigma\omega^* = M\pi\omega = M\pi.$$
Only finitely many $\pi_\alpha \neq 0$ occur, thus
$$\Pi = M\pi\nu = M\pi = M \quad (\text{for } \nu \neq 0)$$
or $\Pi = 0$, whereas again $\Sigma = M\sigma\Omega^*$ holds. Thus apart from an initial element ($\Pi = 0$), each initial segment $\Pi = M = 1 + \Sigma = M\pi$ is cofinal with ω, and each end segment and each middle segment $= \Sigma$ is coinitial with Ω^*. If ϱ belongs to species I 1–8 or I 9–16, M is an isomeric type, and the type Σ resulting from the omission of the initial element is a homogeneous type of group III and clearly of species III 1–8.

(C). *Let the principal element be the last one.*
This is the reverse of the previous case and thus yields homogeneous types of the species II 1–8.

Types without $\omega\omega^*$-gaps

To represent these, we must take a base free from $\omega\omega^*$-gaps, and we must restrict ourselves to the species I 9 – 16 for μ; in addition, however, because of completeness μ may denote any of the isomeric types constructed in §2.

For the powers of the second class
$$M = \mu'_m(\Omega),$$

the following fundamental formulas hold:

(10)
$$\begin{cases} M = M\pi_m \Omega + 1 + M\sigma_m \Omega^*, \\ M = M\pi_l \Omega + M\sigma_l \Omega^*, \\ \Pi = M(\pi_0 + \pi_1 + \cdots + \pi_\omega + \cdots + \pi_\alpha + \cdots), \\ \Sigma = M(\cdots + \sigma_\alpha + \cdots + \sigma_\omega + \cdots + \sigma_1 + \sigma_0), \end{cases}$$

where m denotes the principal element, l a secondary element, and an at most countable set of the π_α, σ_α are different from π_m, σ_m. From the second equation, one sees that M surely has $\Omega\Omega^*$-gaps; concerning the remaining gaps, the following observation will lead us to to our goal: *each ω-sequence in M either precedes a piece of type M or else it transfers a base-gap to the type M.* Indeed, first of all let an ω-sequence of the first kind be given, formed from the coverings

$$(x_0\, x_1\, \ldots\, |\, a_\alpha \ldots\ldots\ldots\ldots\ldots),$$
$$(x_0\, x_1\, \ldots\, x_\alpha \ldots\, |\, b_\beta \ldots\ldots\ldots),$$
$$(x_0\, x_1\, \ldots\, x_\alpha \ldots x_\beta \ldots\, |\, c_\gamma \ldots),$$
$$\text{etc.,}$$

where $\alpha\,\beta\,\gamma\ldots$ is an increasing sequence of numbers from the first or second number classes, call its limit λ, that respectively denote the locations of the first differences between each pair of consecutive coverings, so that

$$a_\alpha < x_\alpha, \quad b_\beta < x_\beta, \quad c_\gamma < x_\gamma, \quad \ldots.$$

The piece all of whose elements begin with

$$(x_0\, x_1\, \ldots x_\alpha \ldots\, x_\beta \ldots x_\gamma \ldots\, |\, \lambda \ldots),$$

i.e., a piece of type $\mu'_m(\Omega) = M$, immediately follows this ω-sequence. On the other hand, if an ω-sequence of the second kind

$$(x_0\, x_1\, \ldots\, |\, a_\alpha \ldots),$$
$$(x_0\, x_1\, \ldots\, |\, b_\alpha \ldots),$$
$$(x_0\, x_1\, \ldots\, |\, c_\alpha \ldots),$$
$$\text{etc.,}$$

is given, then either the corresponding base-sequence $a_\alpha < b_\alpha < c_\alpha < \cdots$ has a limit x_α, and it follows that the totality of all permitted coverings starting with $x_0\, x_1 \ldots x_\alpha$ is again a piece of type M, or else it defines a base-gap that transfers itself to M; a third case is excluded since μ has a last element. The assertion mentioned above, which also correspondingly follows for ω^*-sequences, is thereby verified.

Let us now again separate the different cases for the principal element.

(A). *Let the principal element be a middle one.*

Here (10) implies

$$M = M\pi\Omega = M\sigma\Omega^* = M\pi\Omega + M\sigma\Omega^*,$$

and since only a countable set of $\pi_\alpha = 0$, the rest are $= \pi$, and it follows that
$$\Pi = M\pi\Omega = M, \quad \text{similarly} \quad \Sigma = M\sigma\Omega^* = M.$$
Thus all initial segments and end segments of M are $= M$ and therefore so are all middle segments; M is homogeneous of group IV.

It follows from
$$M = M\pi\omega + M = M + M\sigma\omega^* = M + M$$
that M exhibits $\omega\Omega^*$, $\Omega\omega^*$, and $\Omega\Omega^*$-gaps. On the other hand, M only has $\omega\omega^*$-gaps if they occur in the base, because a piece of type M follows an ω-sequence that does not transfer a base-gap, i.e., such a sequence produces an $\omega\Omega^*$-gap. Thus with regard to species, it is the case that

$$\begin{array}{rcc} \text{for} & \varrho: & I\,1-8, \quad I\,9-16 \\ & M \text{ is}: & IV\,8, \quad\ \ IV\,16. \end{array}$$

(B). *Let the principal element be the first one.*

Now it is the case that
$$M = 1 + M\sigma\Omega^* = M\pi\Omega = M\pi\Omega + M\sigma\Omega^*,$$
and since at most a countable set of the π_α are different from $\pi_m = 0$, $\Pi = M\pi\alpha$ (α any number from the first two number classes); thus $\Pi = M\pi$ or $\Pi = 0$. For end segments, it happens that $\Sigma = M\sigma\Omega^*$; hence $M = 1+\Sigma$. If one sets $\Pi = 1 + P$, then each initial segment of Σ, thus each middle segment of M, has type P; moreover, one finds that
$$\Pi = M\pi = M + M\rho = 1 + \Sigma + M\rho,$$
so $P = \Sigma + M\rho$. Therefore this is a homogeneous (M is only an isomeric) type of group III.

From
$$M = M\pi\Omega + M\sigma\Omega^* = M + M\sigma\omega^*,$$
it follows that M contains $\Omega\Omega^*$-gaps and $\Omega\omega^*$-gaps; however, it contains $\omega\omega^*$-gaps and $\omega\Omega^*$-gaps only in case these occur in the base since a piece of type $1 + \Sigma$ follows an ω-sequence that does not transfer a base-gap, i.e., such a sequence has a limit. Therefore the following species for P result from various choices for ϱ:

$$\begin{array}{lccccl} & \multicolumn{4}{c}{\varrho:} & P: \\ I & 1, & 2, & 4, & 5 & III\,5 \\ I & 3, & 6, & 7, & 8 & III\,8 \\ I & 9, & 10, & 12, & 13 & III\,13 \\ I & 11, & 14, & 15, & 16 & III\,16 \end{array}$$

(C). *Let the principal element be the last one.*

This yields the types inverse to those in B, thus homogeneous types of species II 6, 8, 14, 16.

So if we restrict ourselves to species I 9–16 for ϱ, the last five still missing species II 14, II 16, III 13, III 16, IV 16, i.e., the species of the last three groups without $\omega\omega^*$-gaps, are now constructed.

Let us summarize our construction methods one more time:

Species I 1–16: Powers of the 2nd class with argument ω and bases composed from the ω_{ik}.

Species II 1-8; III 1-8; IV 1-8: Powers of the 1st class with argument Ω and bases the just constructed isomeric types I and principal element the last one, the first one, or a middle one.

Species II 14, 16; III 13, 16; IV 16: Powers of the 2nd class with argument Ω and bases the just constructed isomeric types I 9–16 and principal element the last one, the first one, or a middle one.

And thus the result of this investigation is:

D. *If we disregard the five species mentioned in §1 C, then the remaining 45 species of homogeneous types with sequences of the first and second infinite cardinality are represented by types of the cardinality of the continuum.*

If we observe that at least a part of these 45 species are surely represented not only by one type but by an uncountable set of distinct types, at this point we also get a surprising insight into the boundless domain of types of higher cardinality and into the limited power of determination of the apparently significant concept of homogeneity, which we, under the spell of spatial and temporal intuitions, are inclined to take almost as a privilege of the usual continuum.

If we ask which of these homogeneous types might most likely be capable of application outside pure set theory (perhaps in analysis, function theory, or geometry), first of all, of course, the linear continuum (ID 9) and its subsets (theory of point sets) have to be mentioned; in second place, species IV 16 is a possibility since the *graduation of functions according to growth* yields "pantachie types," which, if not equal to the types of this species, are certainly similar in their basic features. The next article V shall be devoted to this subject. There we will put forward a category of types, the H-types, to which our species IV 16 (and IV 15, the one left out here[1]) belongs. For this reason, we also mention that the species IV 16 can even be represented in a simpler way than is done here—because of the uniform treatment of all species — perhaps most simply by the homogeneous type

$$M = 3'_1(\Omega), \quad 3 = \{0, 1, 2\},$$

i.e., by a power of the 2nd class with argument Ω and base 3, whose middle element figures as the principal element. Indeed, it follows from formula (10) for π_l, $\sigma_l = 0, 2$ or $2, 0$ that

$$M = M\Omega = M\Omega^*,$$

---

[1]) One can easily realize these by powers of the third class with argument Ω, thus by types of cardinality $2^{\aleph_1}$.

and since at most a countable set of the π_α are different from $\pi_m = 1$, $\Pi = M\Omega = M$, and similarly $\Sigma = M\Omega^* = M$; hence M is a homogeneous type of group IV. Since the base contains no sequences or gaps, there are only sequences of the first kind in M; a piece of type M follows an ω-sequence, and thus there is an $\omega\Omega^*$-gap; similarly, there are $\Omega\omega^*$-gaps and $\Omega\Omega^*$-gaps but no $\omega\omega^*$-gaps, i.e., M belongs to species IV 16.

Clearly, instead of 3 one could choose any finite number $\nu \geq 3$ as base and a middle element as principal element. On the other hand, if one chooses the first element as principal element for $\nu \geq 2$, then the power $M = \nu'_0(\Omega)$ is a mixed type that still contains, besides an initial element, $\omega\Omega^*$-limits and $\Omega\Omega^*$-limits; the subset of $\Omega\Omega^*$-limits is again a homogeneous type IV 16; the subset of $\omega\Omega^*$-limits is a homogeneous type $IIIA$ 13. Similar remarks hold for the last element as principal element, where types IV 16 and IIA 14 result. — Should the continuum be of the second infinite cardinality, then, as we shall see in the next note, species IV 16 is represented by only one type of this cardinality.

§4
Everywhere dense, unbounded types

As an appendix, we are going to give a general enumeration of the *everywhere dense, unbounded types* with sequences of the 1st and 2nd infinite cardinality by disregarding homogeneity and isomery. We divide these types again into species according to the kind of *elements* (limits) and *gaps* occurring; for both, it is a question of 4 kinds

$$\omega\omega^*, \quad \Omega\omega^*, \quad \omega\Omega^*, \quad \Omega\Omega^*,$$

which we again are going to abbreviate by

$$11, \quad 21, \quad 12, \quad 22,$$

and of their combinations. In order to obtain a suitable notation for species, we follow our previous numbering (III, §2) for gap combinations and utilize it also for element combinations; namely,

Number	Limits resp. Gaps	Number	Limits resp. Gaps
1	11	9	none
2	11, 21	10	21
3	11, 12	11	12
4	11, 22	12	22
5	11, 21, 22	13	21, 22
6	11, 12, 22	14	12, 22
7	11, 21, 12	15	21, 12
8	11, 21, 12, 22	16	21, 12, 22

We then give each species a limit number i and a gap number k and call it the species (i, k); so for example, species $(3, 14)$ includes the everywhere dense, unbounded types whose elements are in part 11-limits and in part 12-limits and in which 12-gaps and 22-gaps occur.

Of the 256 species that at first glance are available, some are certainly omitted; first, all 16 species $(9, k)$ are omitted since surely there are types without gaps (continuous types $(i, 9)$), but there are no types without elements. Furthermore, it has to be noted that fundamental sequences, which must produce either limits or gaps, are always present; thus ω-limits and ω-gaps cannot simultaneously be absent and similarly for ω^*-limits and ω^*-gaps. Accordingly, the following species are excluded a priori:

(α) $\left\{\begin{array}{lllll} (10,9) & (10,10) & (10,12) & (10,13) \\ (11,9) & (11,11) & (11,12) & (11,14) \\ (12,9) & (12,10) & (12,11) & (12,12) & (12,13) & (12,14) \\ (13,9) & (13,10) & (13,12) & (13,13) \\ (14,9) & (14,11) & (14,12) & (14,14). \end{array}\right.$

After the removal of these 22 species and the previous 16 species, 218 species remain. Finally, we must also exclude the species which have no 11-gaps and no 22-gaps and no 11-limits and no 22-limits since, according to Theorem B, §1, these certainly contain sequences of the third infinite cardinality; these are the 8 species

(β) $\left\{\begin{array}{llll} (10,11) & (10,15) & (11,10) & (11,15) \\ (15,9) & (15,10) & (15,11) & (15,15). \end{array}\right.$

So finally, there remain 210 species against which, as far as I can see, there are no further grounds for exclusion; hence their existence should be proved under further prescribed conditions (e.g., cardinality).

First, in respect thereof, for types of the second infinite cardinality, the 120 species with $\omega\omega^*$-gaps

(γ) $\qquad i \neq 9$, $k = 1, 2, 3, 4, 5, 6, 7, 8$

would be possible. Certainly, according to Theorem A in §1, the types without 11-limits and without 11-gaps and 22-gaps are of cardinality higher than the second infinite cardinality, i.e., in addition to the already deleted types (β), the 9 species

(δ) $\left\{\begin{array}{lllll} (12,15) & (13,11) & (13,15) & (14,10) & (14,15) \\ (16,9) & (16,10) & (16,11) & (16,15). \end{array}\right.$

So for types of the cardinality of the continuum, the existence question would next have to be restricted to 201 species.

The types with only one kind of limit, to which the homogeneous types belong, correspond to the species

(ε) $\quad\begin{array}{cccc} I & II & III & IV \\ (1,k) & (10,k) & (11,k) & (12,k). \end{array}$

Their total is 46, of which one, $(12, 15) = IV\ 15$, belongs to (δ) (in addition, four belong to (β)).

The following 9 species are *continuous* (gap-free) [[*Stetige*(lückenfreie)]] types

(ζ) $\qquad (1, 9)\ (2, 9)\ (3, 9)\ (4, 9)\ (5, 9)\ (6, 9)\ (7, 9)\ (8, 9)\ (16, 9),$

the last of which belongs to (δ). So *discontinuous* (having gaps) [[*Unstetige*(mit Lücken versehene)]] types belong to 201 species, and among these, the species (k, i) always occurs along with the species (i, k); two such species are in a kind of dual relation, which now in conclusion we are going to enlarge upon.

The complement of everywhere dense, unbounded types

Let μ be an everywhere dense, unbounded type. Then we can look upon the gaps of μ as individuals of a new order type, as well as form a third type from the elements and gaps together. To be exact, a decomposition $\mu = \alpha + \beta$ represents an element of μ if α has a last element or β has a first element (both simultaneously are excluded); it represents a gap if α has no last element and β has no first element. Thus if we decree that α shall have no last element, then all the elements and the gaps of μ are uniquely represented by *endless initial pieces* [[*endlosen Anfangsstücke*]] α. These initial pieces α can be directly transformed into an ordered set by stipulating the relation $\alpha < \alpha'$ if α is a subset of α'. On the one hand, through this the elements of μ are assigned a type that is of course μ itself, and on the other hand, the gaps are assigned a new type $\bar{\mu}$ that we call the *completion* [[*Ergänzung*]] or the *complement* [[*Komplement*]] of μ; thirdly, the elements and gaps together once again form a new type $[\mu]$ that may be called the *filling* [[*Ausfüllung*]] of μ. This last type $[\mu]$ obviously has the property of continuity since if we divide all the endless initial pieces α of μ into two classes $\{\gamma\}$ and $\{\delta\}$ so that $\gamma < \delta$ and if the class $\{\gamma\}$ has no last element, then the totality of all elements of μ that occur in some γ again forms an endless initial piece of μ, and it, as the first element of the class $\{\delta\}$, comes after each γ. The set $\{\alpha\}$ is therefore gap-free and everywhere dense, thus continuous.[1])

Of course, $\bar{\mu}$ (and similarly $[\mu]$) is uniquely determined by μ, but in general not conversely since $\bar{\mu}$ need not be an everywhere dense type; e.g., if μ has only one gap, then $\bar{\mu} = 1$. For this reason, we now define the following:

Call an everywhere dense, unbounded type *everywhere discontinuous* [[*überall unstetig*]] if none of its middle segments is continuous. Thus such a type certainly contains a gap between any two of its elements; each middle segment and, more generally, each piece of it is also everywhere discontinuous.

Here, the following theorem holds:

---

[1]) These considerations are already found in part in B. RUSSELL, *The Principles of Mathematics*, Vol. I (Cambridge, 1903), §280.

E. *The complement $\bar{\mu}$ of an everywhere dense, unbounded, everywhere discontinuous type μ is the same sort of type, and for its part, μ is the complement of $\bar{\mu}$.*

If two elements α and $\alpha' = \alpha + \beta$ of $\bar{\mu}$ represent two gaps of μ, i.e.,

$$\mu = \alpha + (\beta + \gamma) = (\alpha + \beta) + \gamma,$$

then the piece β is unbounded and so contains infinitely many elements and by hypothesis certainly a third gap, i.e., $\bar{\mu}$ is everywhere dense. In the same way, it follows that $\bar{\mu}$ is unbounded. Moreover, we show that each element of μ uniquely corresponds to a gap of $\bar{\mu}$. For if m is an element of μ that produces the decomposition $\mu = \pi + 1 + \sigma$, then π produces a decomposition of all the elements of $\bar{\mu}$ into two classes $\{\alpha\}$ and $\{\alpha'\}$, in such a way that $\alpha < \pi < \alpha'$. The class $\{\alpha\}$ cannot have a last element because then π would have a gap-free end piece and μ would have a gap-free middle piece; similarly $\{\alpha'\}$ cannot have a first element, and consequently π determines a gap of $\bar{\mu}$. Conversely, let $\{\alpha\}$ and $\{\alpha'\}$ be a decomposition of $\bar{\mu}$ that represents a gap in $\bar{\mu}$, so that $\{\alpha\}$ has no last element and $\{\alpha'\}$ has no first element. The totality π of all elements of μ that belong to some α is again an endless initial piece of μ; π however cannot determine a gap of μ since otherwise π would be an element of $\bar{\mu}$ and clearly the first element of the class $\{\alpha'\}$; consequently, π has an upper limit in μ, $\mu = \pi + 1 + \sigma$, and thus a gap of $\bar{\mu}$ corresponds to an element of μ. Therefore μ is the complement of $\bar{\mu}$. Since between any two gaps of μ there always occurs an element of μ, between any two elements of $\bar{\mu}$ there always occurs a gap of $\bar{\mu}$; i.e., $\bar{\mu}$ is everywhere discontinuous.

Hence the relationship between μ and $\bar{\mu}$ is mutual; each of these types is the completion (the type of the gaps) of the other, and both have the same filling $[\mu]$.

In particular, if one considers the sequences occurring in our types and one again uses the result just found that between two elements of μ there always lies an element of $\bar{\mu}$ and conversely, then it follows: each $\omega_\nu \omega_\pi^*$-gap in μ corresponds to an $\omega_\nu \omega_\pi^*$-limit in $\bar{\mu}$ and an $\omega_\nu \omega_\pi^*$-limit in $[\mu]$; each $\omega_\nu \omega_\pi^*$-limit in μ corresponds to an $\omega_\nu \omega_\pi^*$-gap in $\bar{\mu}$ and an $\omega_\nu \omega_\pi^*$-limit in $[\mu]$.

So if we restrict ourselves to everywhere discontinuous types, of our species, those denoted by (i, k) and (k, i) belong together as complements, and a continuous type that contains the combined limits corresponding to the numbers i and k belongs to each pair of complementary types as filling. For example, to the species $(1, 1)$, with $\omega\omega^*$-limits and $\omega\omega^*$-gaps, belongs the same species as complement and the species $(1, 9)$ as filling; the simplest case of this kind is present in the type η of the rational numbers, whose complement is the type of the set of irrational numbers and whose filling is the type of the set of real numbers (the unbounded linear continuum).

V. On Pantachie Types

§1
Infinitary Rank Ordering

The idea of *graduating* [[zu *graduieren*]] the convergence of functions to a limit, e.g., their becoming zero or infinite, has been carried out in a systematic way by P. Du Bois-Reymond[1]) in particular and has been developed as a so-called "infinitary calculus" [["Infinitärkalkül"]]. In order to select a particular one from among the different versions of the problem that are possible in principle, we consider monotonically increasing functions of a positive real variable for which

$$\lim_{x=+\infty} f(x) = +\infty.$$

If the quotient $f(x):g(x)$ of two such functions converges for $\lim x = +\infty$ to a finite, non-zero limit, then it is said that the functions have equal infinity; if the limit in question is 0 or $+\infty$, then $f(x)$ has a smaller infinity, respectively, greater infinity.[2]) This is also expressed as follows: $f(x)$ is *infinitarily equal to, smaller than, greater than* $g(x)$, and in these three cases both functions may be called *infinitarily comparable*. "In general," none of these three cases occurs, but rather a fourth does: namely, the quotient has no limit at all; we then call both functions *infinitarily incomparable*. It is quite desirable to extend the previously existing symbolism with a notation for this fourth case (which is actually the rule and only for the usually considered simple

---

[1]) P. Du Bois-Reymond, Sur la grandeur relative des infinis des functions, Ann. di Mat. (2) **4** (1870), 338–353.

Über die Paradoxen des Infinitärkalküls, Math. Ann. **11** (1877), 149–167.

Die allgemeine Funktionentheorie I (Tübingen 1882), Chapter 5.

O. Stolz, Vorlesungen über allgemeine Arithmetik I (Leipzig 1885), 205–215.

Über zwei Arten von unendlich kleinen und von unendlich großen Größen, Math. Ann. **31** (1888), 601–604.

Here and in the upcoming work, I confine myself to mentioning that which deals with the fundamental side of matters; the technique of the infinitary calculus, i.e., the computation of limits or in special cases the proof of their existence, is a chapter in itself. The latest reports on the state of the pantachie question are:

A. Schoenflies, Die Entwickelung der Lehre von den Punktmannigfältigkeiten, Jahresber. d. Deutschen Math. Ver. **8** (1900), 53–56.

E. Borel, Leçons sur la Théorie des Fonctions (Paris 1898), Note II (111–122).

[2]) [[H 1907a, 106n1]] Of course, these stipulations are somewhat arbitrary. For example, they could be extended so that it is not the existence of $\lim f(x):g(x)$ that is called for, but lim inf and lim sup are to be considered; $f(x)$ could then be called infinitarily equal to, smaller than, or greater than $g(x)$ according to whether for the quotient $f(x):g(x)$

$$0 < \liminf < \limsup < +\infty,$$
$$0 = \liminf \leqq \limsup < +\infty,$$
$$0 < \liminf \leqq \limsup = +\infty.$$

Then incomparability would remain confined to the case that $\liminf = 0$ and $\limsup = +\infty$. Cf. p. 114 about further arbitrariness in the preference of quotients.

functions an exception); we shall write $f \parallel g$. Accordingly, here are the symbols and what they stand for:

(α) $f(x) \sim g(x)$: $f(x)$ is infinitarily equal to $g(x)$,
(β) $f(x) < g(x)$: $f(x)$ is infinitarily less than $g(x)$,
(γ) $f(x) > g(x)$: $f(x)$ is infinitarily greater than $g(x)$,

$\left.\right\}$ $f(x)$ is infinitarily comparable with $g(x)$,

(δ) $f(x) \parallel g(x)$: $f(x)$ is infinitarily incomparable with $g(x)$,

and these symbols and expressions are defined by:

(α) $\lim\{f(x):g(x)\} = k$, finite and $\neq 0$,
(β) $\lim\{f(x):g(x)\} = 0$,
(γ) $\lim\{f(x):g(x)\} = +\infty$,
(δ) $f(x):g(x)$ has no limit.

So for any two functions $f(x)$ and $g(x)$ there exists a relation $f(x)\varrho g(x)$, where the relation symbol ϱ is one of the above four symbols; for any totality of functions, the aggregate of these four relations is called its *infinitary rank ordering*.

One sees immediately that the first three relations clearly have the properties of the usual relations "equal to, smaller than, greater than"; for example,

$g \sim f$ follows from $f \sim g$,
$g > f$ follows from $f < g$,
$f \sim h$ follows from $f \sim g$, $g \sim h$,
$f < h$ follows from $f < g$, $g < h$,

etc. Therefore a set of pairwise comparable functions can be ordered rather like a set of ordinary numbers "according to size," and when one identifies infinitarily equal functions with a single element, it has a definite order type in the sense of G. CANTOR. Certainly through the fourth relation, which allows no other inferences besides the two propositions

$g \parallel f$ follows from $f \parallel g$,
$f \parallel h$ follows from $f \parallel g$, $g \sim h$

and which has no analogue in the domain of ordinary numbers, the relationship of the infinitary rank ordering to a simple ordering by magnitude is completely destroyed. For this reason, all attempts to produce a simple (linearly) ordered set of elements in which each infinity occupies its specific place had to fail: *the infinitary pantachie* in the sense of Du BOIS-REYMOND *does not exist*.[1]) Of course, this fact has not escaped the notice of those who

---

[1]) There is no reason to reject the entire theory because of the possibility of incomparable functions as G. CANTOR has done (Sui numeri transfiniti, Riv. di Mat. **5** (1895),

deal with this subject, and besides isolated, unclear, and fanciful opinions about that illusory set of all infinities, in general only constructions of *restricted* sets of pairwise comparable functions are found, mostly sets *that are countable or that at least contain only countable sequences.* What is more, the limited means of analysis make sure that one does not essentially step beyond the boundary of the elementary functions x^α, e^x, $\mathrm{l}x$ and their finite combinations. Within this territory lie both the logarithmic scales, which are applied in the investigation of the convergence of series or of definite integrals (and which can just fail for the very reason that they do not overcome the countable), and the newer investigations into the growth of entire functions, and much else. In addition, starting out from the mode of expression that x^α is infinite of order α, people have repeatedly endeavored to associate to the elements of such restricted function classes magnitude-like symbols with corresponding laws of combination; among these are STOLZ's moments, THOMAE's complex numbers, and the symbols of PINCHERLE, BOREL, BORTOLOTTI et al.[1]) These magnitudes do not satisfy the Archimedean axiom and are thus "actually" infinitely large or infinitely small relative to each other, which says nothing against their logical admissibility; their usefulness can be debated since, on the one hand, they break down with respect to the intermediate levels of the scale, and, on the other hand, they are more complicated than the functions whose infinity they are supposed to express.

Thus all these investigations are still far from leading out from under the spell of the exponential-logarithmic and out of the domain of the countable. Now, neither the interest nor the difficulty of these problems with which analysis and function theory have to contend should be underestimated. But from the point of view of set theory, which we adopt here, all these special function classes have only secondary significance, precisely because in this case the countable fails and because one cannot learn anything about the totality of the infinitary rank ordering through subsets to which one can, within certain limits, give an arbitrary structure. In order to illustrate this with a simple example, we consider the well-known scale of BONNET (Journ. de Mathém. **8** (1843), p. 78), which is gone into in all the textbooks. We set

$$x_1 = \mathrm{l}x, \quad x_2 = \mathrm{ll}x, \quad x_3 = \mathrm{lll}x, \quad \ldots,$$

104–108); the concept of pantachie, defined in what follows and certainly represented by innumerable individual examples, removes the difficulty.

[1]) O. STOLZ, Allg. Arithmetik, loc. cit.
   J. THOMAE, Elementare Theorie der analytischen Funktionen (Halle 1880),
      §§141–145.
   S. PINCHERLE, Alcune osservazioni sugli ordini d'infinito delle funzioni,
      Mem. Acc. Bologna (4) **5** (1884), 739–750.
   E. BOREL, Leçons sur les fonctions entières (Paris 1900),
      Leçons sur les séries à termes positifs (Paris 1902).
   E. BORTOLOTTI, Sulla determinazione dell' ordine di infinito, Atti Soc.
      Natur. Modena (4) **3** (1901), 13–77, and Mem. Acc. Modena (3)
      5 (1905), LXII–LXXI.

$$f(x) = x^\alpha \, x_1^{\alpha_1} \, x_2^{\alpha_2} \cdots x_n^{\alpha_n},$$

so that this function is characterized by the exponent system

$$\alpha_f = (\alpha \, \alpha_1 \, \alpha_2 \cdots \alpha_n) = (\alpha \, \alpha_1 \, \alpha_2 \cdots \alpha_n \, 0\,0\,0 \cdots),$$

where the α are real numbers. Each $f(x)$ is real and positive from a certain x on; at infinity, it becomes 0 or ∞ according to whether the first nonvanishing α_m is negative or positive. The quotient of two such functions is a function of the same kind; all of these are pairwise comparable, and $f \lessgtr f'$ according to whether the first nonvanishing difference $\alpha_m - \alpha'_m$ is $\lessgtr 0$. The rank ordering of the f is thus identical with the rank ordering of the exponent system α_f by first differences, and in our terminology the set $\{f(x)\}$ has the type of the power of first class $\lambda_0(\omega)$ with base the unbounded linear continuum (the set of all real numbers) and with an arbitrary principal element (0) (cf. I. i. O. T. I, p. 108). The improper definite integral $\int^\infty f(x)\,dx$ or the series $f(n) + f(n+1) + \cdots$ is convergent or divergent according to whether the first exponent α_m differing from -1 is < -1 or > -1. Consequently, we have, for example, an ω-sequence of convergent elements

$$(-2) \quad (-1,-2) \quad (-1,-1,-2) \quad \cdots$$

and an ω^*-sequence of divergent elements

$$(-1) \quad (-1,-1) \quad (-1,-1,-1) \quad \cdots ;$$

together they determine an $\omega\omega^*$-gap, i.e., there is no function $f(x)$ in our set which lies between these two sequences; however, infinitely many other convergent as well as divergent elements[1] lie there, and therefore this scale must break down for them.

Thus if we designate it as our task to connect the infinitary rank ordering as a whole with CANTOR's theory of order types, then nothing remains but to investigate *the sets of pairwise comparable functions that are as comprehensive as possible* [[möglichst umfassend]]: as comprehensive as possible in the sense that such a set should not be extendible by functions that are comparable to all the functions in the set. Retaining the term of Du BOIS, but abandoning the unsuccessful concept, we are going to call such a class of functions—for which first of all, of course, an existence proof must be furnished—*an infinitary pantachie* (so not "the" pantachie, but instead "a" pantachie), noting immediately that two incomparable functions belong to different pantachies in any case. Since the attempt to actually legitimately construct a pantachie seems completely hopeless, it would now be a matter of gathering information *without further assumptions* about the order type of

---

[1] This was proved by A. PRINGSHEIM (Allgemeine Theorie der Divergenz und Konvergenz von Reihen mit positiven Gliedern, Math. Ann. **35** (1890), 297–394, in particular p. 353). A convergent sequence of this kind had already been given by DU BOIS-REYMOND (Eine neue Theorie der Konvergenz und Divergenz von Reihen mit positiven Gliedern, Journ. f. Math. **76** (1873), 61–99, Appendix). Also cf. §2, Theorem P.

any pantachie — understanding that infinitarily equal functions are considered as only one function; and with some yet to be discussed modifications, this procedure shall be pursued in what follows.

Next, so as not to have to continually reiterate the additional remark regarding infinitarily equal functions, we are going to strengthen our definition to this effect. So that we say:

A set of functions that are pairwise in the relations $\lessgtr$ is called an ordered domain [[Bereich]]. An ordered domain which is not contained in any more comprehensive ordered domain is called a pantachie.

Thus an ordered domain or a pantachie contains neither infinitarily equal nor infinitarily incomparable functions.

The first question that should be asked concerns the *cardinality* of the pantachie types; it is easy to answer under the hypotheses imposed so far since the set of all monotonic functions has, like the set of continuous functions, only the cardinality of the continuum ($\aleph$).[1]) Each pantachie of monotonic functions is thus of cardinality at most that of the continuum; since, as will be shown later (§3), it cannot also be of smaller cardinality, it has exactly the cardinality of the continuum.

Upon further scrutiny, a certain simplification is permitted (as is also usually done in the application of the infinitary calculus to the convergence of series and to the behavior of the coefficients of entire functions, and so on), instead of functions of a continuous variable x, to consider functions of a positive integer index n, thus *numerical sequences* [[Zahlenreihen]] of the form

$$A = (a_1 \, a_2 \, a_3 \, \cdots \, a_n \, \cdots).$$

For these, infinitary relations can be defined by the quotient $a_n : b_n$ exactly as for functions of x, giving the criteria for $A \varrho B$ according to its behavior for $\lim n = \infty$. The notions of "ordered domain of numerical sequences" and "pantachie of numerical sequences" are thereby also explained. Now in each function $f(x)$ there is contained a specific numerical sequence

$$F = (f(1) \, f(2) \, f(3) \, \cdots \, f(n) \, \cdots),$$

which grows monotonically to infinity simultaneously with $f(x)$, while conversely, infinitely many ($\aleph$) functions $f(x)$ belong to a numerical sequence F.

Now the following holds:

$$\begin{aligned}
F \sim G & \quad \text{follows from} \quad f(x) \sim g(x), \\
F < G & \quad \text{follows from} \quad f(x) < g(x), \\
F > G & \quad \text{follows from} \quad f(x) > g(x),
\end{aligned}$$

---

[1]) For a monotonic, nondecreasing function $f(x)$ both limits $f(x+0)$ exist for each point, and indeed $f(x-0) \leq f(x+0)$; if both are equal, then x is a point of continuity, and if they are different, then x is a point of discontinuity and indeed a jump discontinuity. In each interval, the set of all jumps $\geq \sigma$ is finite, and so the set of points of discontinuity is at most countable; and $f(x)$ is determined by a countable set of function values (for instance, at all rational points and at the irrational jumps), from which the statement of the text follows.

112 while $f(x) \| g(x)$ permits none of the four possible relations between F and G to be inferred or excluded. With that, we immediately find:

(A) *Each ordered domain and each pantachie of functions is similar to an ordered domain of numerical sequences. An ordered domain $\{f(x)\}$ of functions is certainly a pantachie when the corresponding ordered domain $\{F\}$ of numerical sequences is a pantachie.*

In addition, a *continuous* function $\alpha(x)$ can be associated with each numerical sequence A in infinitely many ways so that the infinitary relations are preserved. The most obvious way is to define a piecewise linear function by

$$\alpha(x) = a_n(n+1-x) + a_{n+1}(x-n) \quad \text{for } n \leqq x \leqq n+1,$$

i.e., by connecting the points with rectangular coordinates (n, a_n) by a linear graph [[Streckenzug]]. The simple proof that with $A \varrho B$ also $\alpha(x) \varrho \beta(x)$ (where ϱ is one of the four symbols $\sim < > \|$) can be omitted. If one assigns to an arbitrary function $f(x)$ a numerical sequence F and to this sequence one assigns a piecewise linear function $\varphi(x)$, then the pantachie $\{f(x)\}$ is similar to the ordered domain $\{\varphi(x)\}$ that consists of only continuous functions; if $\{F\}$ is a pantachie, then $\{\varphi(x)\}$ is a pantachie formed from only continuous functions. Thus:

(B) *There are pantachies consisting of only continuous functions; each pantachie of arbitrary functions is similar to an ordered domain or a pantachie of only continuous functions.*

These theorems must of course not be understood *as if for any discontinuous function there would be a continuous one that is infinitarily equal to it.* Functions that jump at all integer arguments at a constant or an increasing rate show that this is not the case, as for example

$$f(n+0) = 2 \cdot f(n-0),$$
$$f(n+0) = n \cdot f(n-0).$$

Obviously, such a function cannot be infinitarily equal to a continuous function, in the second case, not even according to the more general definition of infinitary equality (cf. the remarks on p. 106).

From now on, we will even drop the previous requirement of monotonicity and infinite growth for our functions and numerical sequences since these 113 hypotheses, useful for analytic purposes, unnecessarily hinder the order theoretic investigation. Thus $f(x)$ and a_n shall have arbitrary *positive* (> 0) values, which implies that the sum, product and quotient of two such functions (numerical sequences) is again a function (numerical sequence) with the same property. Because of this, something has changed. The totality of all functions $f(x)$ is now of cardinality $2^{\aleph}$, whereas the totality of all numerical sequences continues to be of cardinality $\aleph$. The relation between functions $f(x)$ and numerical sequences F remains; only now, $2^{\aleph}$ functions belong to each numerical sequence; also theorems (A) and (B) remain valid, so as before, *each pantachie is of the cardinality of the continuum*. With respect to discontinuous functions, the scope is now so enlarged that we can even give such functions that are *comparable to no continuous function*: this produces functions that take on arbitrarily large and arbitrarily small values in each interval, e.g., the function that is equal to q or $1/q$ for the rational value $x = p/q$ according to whether the numerator of the reduced fraction p/q is even or odd. Consequently, a pantachie to which one such function $f(x)$ belongs contains no continuous function whatever: *there are pantachies consisting of only discontinuous functions*.

Moreover, to each function there certainly now corresponds $2^\aleph$ infinitarily equal functions, but when we collect these functions into one class, there are still $2^\aleph$ distinct classes; or said more concisely, there are $2^\aleph$ infinitarily distinct functions. The first assertion is proved by considering a function $g(x)$ that has one of the two values

$$f(x) \quad \text{or} \quad f(x) \cdot \left(1 + \frac{1}{x}\right),$$

so that $g(x) \sim f(x)$. There are as many such functions $g(x)$ as there are coverings of the continuum with both values 0 and 1, thus $2^\aleph$. To prove the second assertion, we restrict ourselves to those functions which can take on only the values n, $1/n$ in each interval $n - 1 < x \leq n$; two such functions are obviously infinitarily equal only if $f(x) = g(x)$ from some x on. Among those, we take only the quasiperiodic ones whose values are determined by their values in the interval $(1, 2)$, i.e., so that for $1 < x \leq 2$, $n = 1, 2, 3, \ldots$

$$f(x + n) \lessgtr 1, \quad \text{according to } f(x) \lessgtr 1.$$

Two such functions are then infinitarily equal only if they are altogether identical. Again, their totality is in one-to-one correspondence with the set of coverings of the continuum with two values, so of cardinality $2^\aleph$. Whereby, it is proved that there are $2^\aleph$ infinitarily distinct functions.

It follows from this (since $2^\aleph > \aleph$) that for each set of functions of cardinality $\leq \aleph$ there exists a new function that is infinitarily different from each of its elements, which assertion one can still strengthen to the effect that a function exists that is *incomparable with each of its elements* (e.g., a comparable function need not exist when the aforementioned set is a pantachie). For if one assigns the functions of the set, which may be of cardinality $\aleph$, to the numbers ξ of the interval $0 < \xi \leq 1$ so that a function $f_\xi(x)$ corresponds to each number ξ, then one would choose a function $g(x)$ in such a way that the quotient $g(\xi + n) : f_\xi(\xi + n)$ has no limit for $\lim n = \infty$; then $g(x) \parallel f_\xi(x)$, and this can happen for each ξ since none of the point-sequences ξ, $\xi + 1$, $\xi + 2$, $\ldots$ has points in common with the others.

Finally, we allow another fundamental modification to occur, which once again removes a useless restraint from set theoretic considerations. The infinitary rank ordering defined by Du Bois-Reymond singles out the *quotient* of two functions as the standard of comparison. As has been pointed out by S. Pincherle in particular (loc. cit.), this is an arbitrary choice since

instead of $f(x) : g(x)$ or, what amounts to the same thing, the difference $\lg f(x) - \lg g(x)$, one can even take as a starting point, with appropriate stipulations, the difference of the functions themselves or more generally the difference
$$\Phi(f(x)) - \Phi(g(x)),$$
where $\Phi(z)$ denotes a suitably chosen function, for instance, one that is continuous and monotonically increasing. The former arbitrariness manifests itself through this because the infinitary rank ordering is not invariant with respect to such transformations, e.g.,
$$2x \sim x, \quad \text{but} \quad e^{2x} > e^x,$$
$$xE(x) \sim x^2, \quad \text{but} \quad e^{xE(x)} \parallel e^{x^2},$$
where $E(x)$ denotes the greatest integer contained in x. Only a few very simple transformations such as
$$z' = mz, \quad z' = z^\mu$$
have the property that they leave the infinitary rank ordering relations unchanged; in addition to which, the transformation $z' = 1 : z$ reverses all rank ordering relations, i.e., it exchanges the signs $<$ and $>$ and leaves the signs $\sim$, $\parallel$ invariant. With regards to a fundamental means of investigation, namely, the transformation of one pantachie into another pantachie or into a piece of another pantachie, we are surely subject to strong restrictions by the Du Bois formulation; e.g., I have not succeeded in mapping an infinitary pantachie similarly onto a middle segment of a pantachie. In this connection, it should also be stressed (since the contrary assertion is repeatedly made or tacitly assumed) that passing to *inverse functions* does not produce a single-valued inversion [[eindeutige Umkehrung]] of the infinitary rank ordering, and it cannot be concluded from $f(x) < g(x)$ that the inverse functions are in the relation $\varphi(x) > \psi(x)$; the example
$$f(x) = e^{x^2}, \quad g(x) = e^{(x+1)^2}$$
shows this, as the inverse functions differ from each other by only a unit and so $\varphi(x) \sim \psi(x)$.[1]) — For all these reasons, we prefer simply to choose, instead of the quotient $f(x) : g(x)$, the eventual *sign of the difference* $f(x) - g(x)$ as the basis for a rank ordering[2]), which we wish to call the *final rank ordering* [[finale Rangordnung]]. Thus we say that $f(x)$ is *finally equal to, smaller than, or greater than* $g(x)$ according to whether from some x on $f(x)$ is always *equal to, smaller than, or greater than* $g(x)$; in any other case, we say that $f(x)$ is *finally incomparable* to $g(x)$.[3]) We make the

---

[1]) The Borel symbols allow these to be expressed, but then they are no longer simply equivalent to the infinitary rank ordering.

[2]) This has also already occurred on occasion, as in Du Bois-Reymond (loc. cit. Ann. di Mat. (2) **4**).

[3]) [[H 1907a, 116n1]] The infinitary relations $\sim < >$ each represent a countable set of final relations, namely: for each $n = 1, 2, 3, \ldots,$

analogous stipulations for numerical sequences. The relationships between functions and numerical sequences and theorems (A) (B) also remain valid. With the name "final rank ordering," we express what, in our view of the theory of order types, is the essence of DU BOIS's idea, namely, the graduation by *final behavior* [[Graduierung nach dem *Endverlauf*]] of functions and numerical sequences, by which we have surely stripped the idea of its *metric* specialization as far as possible. It will now also be possible to generalize this principle of final ordering to arbitrary order types (§3).

To summarize what has been said, we have modified DU BOIS's problem in the following aspects:

the definition of the pantachie concept and the investigation of the common properties of all pantachies;

the investigation of numerical sequences instead of functions;

the removal of the monotonicity hypothesis;

the replacement of the infinitary rank ordering by the final one.

§2

The Final Rank Ordering

We understand by $x_1\, x_2\, x_3 \ldots x_n \ldots$ positive real numbers, and we call a sequence (of type ω) of such numbers

$$X = (x_1\, x_2\, x_3 \ldots x_n \ldots) \qquad (x_n > 0)$$

a *numerical sequence* [[*Zahlenreihen oder Zahlenfolge*]]. For two numerical sequences X, Y, if *from some index on* $(n \geqq n_0)$

(α) $x_n = y_n$ throughout, then X is said to be finally equal to Y,

(β) $x_n < y_n$ throughout, then X is said to be finally smaller than Y,

(γ) $x_n > y_n$ throughout, then X is said to be finally greater than Y,

(δ) in every other case X is said to be finally incomparable with Y.

In symbols:

(α) $\qquad\qquad X \sim Y,$
(β) $\qquad\qquad X < Y,$
(γ) $\qquad\qquad X > Y,$
(δ) $\qquad\qquad X \parallel Y.$

The aggregate of the relations $X \varrho Y$ existing between numerical sequences is called their *final rank ordering*.

The four relations have the following properties that one can express in the language of the "relational calculus," so that $\sim$ is a reflexive relation,

---

$$\begin{array}{cc} \textit{infinitary} & \textit{final} \\ f \sim g & (k - \frac{1}{n})g < f < (k + \frac{1}{n})g \\ f < g & nf < g \\ f > g & f > ng \end{array}$$

$\sim$ and $\parallel$ are symmetric relations, $<$ and $>$ are asymmetric relations, $\sim$ is a conservative relation, and $\sim\ <\ >$ are transitive relations:

I. $\qquad X \sim X,$

II. $\begin{cases} Y \sim X \text{ follows from } X \sim Y, \\ Y > X \text{ follows from } X < Y, \\ Y < X \text{ follows from } X > Y, \\ Y \parallel X \text{ follows from } X \parallel Y, \end{cases}$

III. $\begin{cases} X \varrho Z \text{ follows from } X \sim Y, Y \varrho Z, \\ X \varrho Z \text{ follows from } X \varrho Y, Y \sim Z, \\ X < Z \text{ follows from } X < Y, Y < Z, \\ X > Z \text{ follows from } X > Y, Y > Z. \end{cases}$

A set of numerical sequences that are pairwise in the relations $\lessgtr$ is called an ordered domain; an ordered domain that is not contained in a more comprehensive ordered domain is called a pantachie.

Here first of all, the *existence question* (postponed in §1) for pantachies raises itself. The existence proof is quite simple if one bases it on the *well-ordering of the continuum*.[1]) If one brings all $\aleph$ numerical sequences into the form of a well-ordered set

$$A^0\, A^1\, A^2\, \ldots\, A^\omega\, \ldots\, A^\alpha\, \ldots,$$

one would first delete all elements that are finally equal to A^0 or finally incomparable with it. Let A^α be the first among those remaining. Afterwards, delete all elements $\overset{\parallel}{\sim} A^\alpha$ and keep A^β as the first left over; further, delete all elements $\overset{\parallel}{\sim} A^\beta$ and find A^γ, the first among those remaining, and so forth. By this means, one next arrives at a set of as yet undeleted elements that begins with the fundamental sequence

$$A^0\, A^\alpha\, A^\beta\, A^\gamma\, \ldots \quad (0 < \alpha < \beta < \gamma < \cdots)$$

and the rest of which contains only elements that are in the relation $\lessgtr$ with the very elements mentioned above; let A^λ be the first among these, whereupon again all elements $\overset{\parallel}{\sim} A^\lambda$ are to be deleted, and so forth. The process continues in this way as long as the domain obtained $\{A^0\, A^\alpha\, A^\beta\, A^\gamma\, \ldots\, A^\lambda\, A^\mu \ldots\}$ is still extendible, and it either ends together with the whole set $\{A^\alpha\}$ or in one of its initial segments [[Abschnitte]]. — One can immediately generalize this argument in such a way that it always gives pantachies that contain any arbitrary prescribed ordered domain $\mathfrak{B}$. Of course, it can also be given in a different form; e.g., argue as follows: if there were no pantachie, thus each ordered domain would be part of a more comprehensive

---

[1]) E. ZERMELO, Beweis, daß jede Menge wohlgeordnet werden kann. Math. Ann. **59** (1904), 514–516.

ordered domain, then for each ordered domain $\mathfrak{B}_\alpha$ one could give a more comprehensive ordered domain $\mathfrak{B}_{\alpha+1}$, and for each well-ordered sequence (countable or uncountable) of ordered domains, each of which contains all its predecessors, one could give an ordered domain that contains all of them. So one could form a well-ordered sequence of ordered domains

$$\mathfrak{B}_0\, \mathfrak{B}_1\, \mathfrak{B}_2\, \ldots\, \mathfrak{B}_\omega\, \ldots\, \mathfrak{B}_\alpha\, \ldots$$

up to any arbitrary index α, and the cardinality of the continuum would have to be greater than each aleph.

We now seek to determine the common properties of all pantachies. The following simple theorems serve as foundation for this investigation.

(C). *If $\{U\}$ is a finite or countable set of numerical sequences, then there always exists a numerical sequence that is finally greater than each U and a numerical sequence that is finally smaller than each U.*

Let $\{U\} = A, B, C, D, \ldots$ be the set mentioned above. We presume nothing about its final rank ordering. If one then chooses

$$x_1 > a_1,$$
$$x_2 > a_2, b_2,$$
$$x_3 > a_3, b_3, c_3, \quad \text{etc.},$$

then $X > A$, $X > B$, $X > C$, $\ldots$. The second part of the statement is proved analogously. In particular, it follows that:

(C′). *For each ω-sequence of numerical sequences, there is a finally greater numerical sequence; for each ω^*-sequence, there is a finally smaller numerical sequence.*

(D). *If $\{U\}$ and $\{V\}$ are two at most countable sets of numerical sequences and throughout $U < V$, then there always exists a numerical sequence X that finally lies between both sets, so that $U < X < V$.*

Let $\{U\} = A, B, C, \ldots$ and $\{V\} = P, Q, R, \ldots$ be the two sets about whose internal rank ordering nothing is presumed; only each U is supposed to be finally smaller than each V. Accordingly, there exists for every pair U, V an index κ such that $u_n < v_n$ for $n \geq \kappa$. Consequently, one can define an increasing sequence of numbers $1 < \lambda < \mu < \nu < \cdots$ such that

$$\begin{aligned} a_n &< p_n & &\text{for} \quad n \geqq \lambda, \\ a_n, b_n &< p_n, q_n & &\text{for} \quad n \geqq \mu, \\ a_n, b_n, c_n &< p_n, q_n, r_n & &\text{for} \quad n \geqq \nu, \end{aligned}$$

etc. Accordingly, if one chooses X so that

$$\begin{aligned} a_n &< x_n < p_n & &\text{for} \quad \lambda \leqq n < \mu, \\ a_n, b_n &< x_n < p_n, q_n & &\text{for} \quad \mu \leqq n < \nu, \\ a_n, b_n, c_n &< x_n < p_n, q_n, r_n & &\text{for} \quad \nu \leqq n < \pi, \end{aligned}$$

etc., then

$$\begin{aligned} a_n &< x_n < p_n & &\text{for} \quad n \geqq \lambda, \\ b_n &< x_n < q_n & &\text{for} \quad n \geqq \mu, \\ c_n &< x_n < r_n & &\text{for} \quad n \geqq \nu, \end{aligned}$$

etc.; thus $A < X < P, B < X < Q, C < X < R, \ldots$.

The following special cases are contained in this theorem:

(D'). *Between an ω-sequence of numerical sequences and a later numerical sequence, as well as between an ω^*-sequence of numerical sequences and an earlier numerical sequence, there always lies yet a further numerical sequence.*

(D''). *Between an ω-sequence and a later ω^*-sequence of numerical sequences there always lies a further numerical sequence.*

After this, the following properties of each pantachie are easily established:

(E). *Each pantachie is unbounded, i.e., it has no first or last element.*

For if an ordered domain $\mathfrak{B}$ contains a first element A, so that all its remaining elements satisfy $U > A$, then by (C) there exists an element $X < A$, consequently, also $X < U$. Thus $\{X, \mathfrak{B}\}$ is an ordered domain as well, and as a result $\mathfrak{B}$ is not a pantachie. The second part of the statement follows analogously.

(F). *A pantachie is neither cofinal with ω nor coinitial with ω^*.*

Let $\mathfrak{B}$ be an ordered domain cofinal with ω, and let $A < B < C < \cdots$ be a fundamental sequence contained in it that is not followed by any element of $\mathfrak{B}$. Thus each element U of $\mathfrak{B}$ is surpassed by some element of this fundamental sequence (and all that follow it). Now according to (C) and (C'), for this ω-sequence there exists an element X that surpasses all its elements and so also each U. Thus $\{\mathfrak{B}, X\}$ is again an ordered domain, and $\mathfrak{B}$ is not a pantachie. The second part of the statement follows analogously.

(G). *Each pantachie is everywhere dense, i.e, between any two of its elements it always contains a further element.*

Let $\mathfrak{B}$ be an ordered domain with two elements $A < B$ that follow immediately upon each other (consecutive), so that for each other element, either $U < A$ or $V > B$. According to (D), there is an element X so that $A < X < B$, thus also $U < X < V$. Consequently, $\{\mathfrak{B}, X\}$ is a larger ordered domain, and $\mathfrak{B}$ is not a pantachie.

(H). *In a pantachie no fundamental sequence has a limit.*

Let an ω-sequence $A < B < C < \cdots$ and its limit P be present in an ordered domain $\mathfrak{B}$, so that each element U of $\mathfrak{B}$ that is $< P$, is surpassed by an element of the sequence and each other element V is $> P$. According to (D) and (D'), there exists a further element X between the ω-sequence and P, so $U < X < P < V$. Consequently, $\{\mathfrak{B}, X\}$ is an ordered domain, and $\mathfrak{B}$ is not a pantachie. The same holds for ω^*-sequences.

(J). *In a pantachie there are no $\omega\omega^*$-gaps.*

Let $\mathfrak{B}$ be an ordered domain with an $\omega\omega^*$-gap, i.e., with an ω-sequence $A < B < C < \cdots$ and with a later ω^*-sequence $P > Q > R > \cdots$ between which lies no element of $\mathfrak{B}$. Hence the elements of $\mathfrak{B}$ fall in two categories U and V; each U is finally smaller than an element of the ω-sequence, and each V is finally greater than an element of the ω^*-sequence. Then by (D'') one finds an element X that lies between both sequences, so $U < X < V$; thus $\{\mathfrak{B}, X\}$ is an ordered domain, and $\mathfrak{B}$ is not a pantachie.

We are going to name the types of the category found here H-types. Thus we define:

An order type that is unbounded, everywhere dense and neither cofinal with ω nor coinitial with ω^* and that contains no fundamental limits and no $\omega\omega^*$-gaps is called an H-type.

Consequently, we could express our results as follows:

(K). *Each pantachie and each segment of a pantachie is an H-type.*

In addition, I note that the results obtained here prove also to be valid for the infinitary rank ordering and for transference to functions $f(x)$. However, then the simple proofs given here need various modifications. First touching upon the case of the infinitary rank ordering for numerical sequences, for instance, one would now have to adjust the choice of X in the proof of theorem (C) through the following strengthened inequalities

$$x_1 > a_1,$$
$$\tfrac{1}{2}x_2 > a_2, b_2,$$
$$\tfrac{1}{3}x_3 > a_3, b_3, c_3,$$

etc. and correspondingly for the second part of Theorem (C) through the inequalities

$$x_1 < a_1,$$
$$2x_2 < a_2, b_2,$$
$$3x_3 < a_3, b_3, c_3,$$

etc. For theorem (D), the numbers λ, μ, ν, ... should be chosen so that, say,

$$a_n < p_n \qquad \text{for} \quad n \geqq \lambda,$$
$$2a_n, 2b_n < \tfrac{1}{2}p_n, \tfrac{1}{2}q_n \qquad \text{for} \quad n \geqq \mu,$$
$$3a_n, 3b_n, 3c_n < \tfrac{1}{3}p_n, \tfrac{1}{3}q_n, \tfrac{1}{3}r_n \qquad \text{for} \quad n \geqq \nu,$$

etc., and the x_n should be put in the intervals thereby left open; meanwhile, in the special case (D''), one manages with the valid argument for the final ordering.

The case of functions $f(x)$, for instance, is dealt with by a decomposition into intervals (0 1), (1 2), (2 3), ...

The effect of the monotonicity requirement is not quite so simple; however, we abstain from entering into it.

On the other hand, it is of interest to carry out the conversion of theorems (C) and (D) for the case in which one is supposed to distinguish

122 between *convergent and divergent series*.[1]) Call the numerical sequence $X = (x_1\, x_2\, x_3\, \ldots)$ convergent or divergent according to whether the sequence taken additively $x_1 + x_2 + x_3 + \cdots$ is convergent or divergent. We put
$$X^n = x_1 + x_2 + \cdots + x_n,$$
and in the case of a convergent sequence, we put
$$X_n = x_{n+1} + x_{n+2} + \cdots;$$
so in the cases of

convergence: $\lim X^n = X_0$, $\lim X_n = 0$,
divergence: $\lim X^n = +\infty$.

If C is a convergent sequence and $X \lesssim C$, then X is also convergent; if D is divergent and $Y \gtrsim D$, then Y is also divergent[2]); a convergent and a divergent sequence can only have the relations $C < D$ or $C \parallel D$. If we combine this with theorems (C) and (D), some theorems follow *immediately*, such as this: for an at most countable set of elements there exists a finally greater divergent element, etc. Of greater importance are the following statements, which are the only ones that still need proof.

(L). *If $\{U\}$ is an at most countable set of convergent sequences, then there exists a convergent sequence that finally surpasses each U.*

123 Let $\{U\} = A, B, C, \ldots$ be this set whose terms are permitted to stand in an arbitrary final rank order with respect to each other (e.g., they could be incomparable). Understand by $\varepsilon_1 + \varepsilon_2 + \varepsilon_3 + \cdots$ a convergent series of positive terms and determine a sequence of increasing integers $\mu, \nu, \ldots$ so that
$$B_\mu < \varepsilon_1, \quad C_\nu < \varepsilon_2, \quad \ldots.$$
Thereupon set
$$X = (a_1, \ldots, a_\mu, a_{\mu+1} + b_{\mu+1}, \ldots, a_\nu + b_\nu, a_{\nu+1} + b_{\nu+1} + c_{\nu+1}, \ldots).$$
Then X is convergent and
$$X_0 = A_0 + B_\mu + C_\nu + \cdots < A_0 + \varepsilon_1 + \varepsilon_2 + \cdots;$$

---

[1]) Besides the already cited paper of PRINGSHEIM (Math. Ann. **35**), cf. J. HADAMARD, Sur les charactères de convergence des séries ..., Acta. Math. **18** (1894), 319–336. Theorems (C)(D)(L)–(P) of the present § are for the most part, of course, well known, at least implicitly; but one will find that through the final ordering they have gained in simplicity and generality.

[2]) The graduation of convergence and divergence (weaker and stronger convergence or divergence) is different among different authors; cf. O. STOLZ and A. GMEINER, Einleitung in die Funktionentheorie II (Leipzig 1905), p. 246. One usually takes as fundamental the infinitary rank ordering of the sequences $(X_1\, X_2\, \ldots)$ for convergent X and that of the sequences $(X^1\, X^2\, \ldots)$ for divergent X (like STOLZ, CESÀRO, BOREL); PRINGSHEIM takes into consideration the final rank ordering of these sequences and sometimes (Ezyklop. d. math. WISS. *IA* 3, No. 26) even considers the infinitary rank ordering of the sequences $X = (x_1\, x_2\, \ldots)$ themselves. We do not want to add yet one more to these different definitions; however, we see no reason, just this last one, to completely omit the final rank ordering of the sequences X themselves from consideration.

in addition, $X > A, B, C, \ldots$.

(M). *If $\{V\}$ is an at most countable ordered domain*[1]) *of divergent sequences, then there exists another divergent sequence that is finally smaller than each V.*

If the ordered domain $\{V\}$ has a smallest element P, then the sequence that we seek is $\frac{1}{2}P = (\frac{1}{2}p_1, \frac{1}{2}p_2, \ldots)$. If it has no smallest element, then it is coinitial with an ω^*-sequence, and it is then a matter of finding a divergent element that finally precedes this ω^*-sequence. Let $P > Q > R > \cdots$ be the elements of this sequence. We assume for the sake of simplicity that for each n, $p_n > q_n > r_n > \cdots$; if it is not the case, then this can be attained by the modification of a finite number of values $q_n, r_n, \ldots$, by which $Q, R, \ldots$ are replaced by finally equal elements.[2]) Now determine a sequence of increasing integers $\lambda, \mu, \nu, \ldots$ so that

$$P^\lambda > \varrho, \quad Q^\mu > \varrho_1, \quad R^\nu > \varrho_2, \quad \ldots,$$

where $\varrho \; \varrho_1 \; \varrho_2 \ldots$ can be arbitrary positive numbers with $\lim \varrho_n = +\infty$. Then one sets

$$X = (p_1 \ldots p_\lambda \, q_{\lambda+1} \ldots q_\mu \, r_{\mu+1} \ldots r_\nu \ldots)$$

and finds

$$X^\lambda = P^\lambda > \varrho,$$
$$X^\mu > Q^\mu > \varrho_1,$$
$$X^\nu > R^\nu > \varrho_2, \quad \text{etc.}$$

So X is divergent and at the same time $X < P, Q, R, \ldots$.

In addition, we prove two special theorems (N) and (O) before putting forward the general analogue for (D).

(N). *If $\{U\}$ is an at most countable set of convergent sequences and P is an arbitrary sequence $> U$, then there exists a convergent sequence X between $\{U\}$ and P ($U < X < P$).*

Let $\{U\} = A, B, C, \ldots$ and construct a convergent sequence $Y > U$ by (L). If $Y < P$, then the task is solved. If not, then there are infinitely many places $n = \alpha, \beta, \gamma, \ldots$ where $y_n \geqq p_n$. If one denotes the sequence $(y_\alpha \, y_\beta \, y_\gamma \ldots)$ by Y' and similarly for the remaining letters, then Y', and consequently P', is a convergent sequence. According to (D), one can insert

---

[1]) This restrictive addendum or another equally strong one is essential. If P and Q are two divergent sequences, while $p_1 + p_3 + p_5 + \cdots$ and $q_2 + q_4 + q_6 + \cdots$ converge, then each sequence that is finally smaller than P and Q is convergent. It does not interest us which minimum of conditions suffice to eliminate this case since, in any case, the above sufficient formulation of Theorem (M) is commended by its simplicity. For Theorem (L) the analogous restriction is superfluous.

[2]) In each countable ordered domain, one can replace elements with finally equal elements $P, Q, R \ldots$ in such a way that *for each n* the order type of the set of numbers $p_n, q_n, r_n, \ldots$ is equal to the order type of the domain; this is no longer valid for an ordered domain of higher cardinality.

a sequence $X' = (x_\alpha\, x_\beta\, x_\gamma\, \ldots)$ between P' and the convergent sequence U' $(U' < X' < P')$, which is necessarily also convergent. Then set

$$X = (y_1\, \ldots\, y_{\alpha-1}\, x_\alpha\, y_{\alpha+1}\, \ldots\, y_{\beta-1}\, x_\beta\, y_{\beta+1}\, \ldots),$$

i.e., let X coincide with Y in the places different from $\alpha\ \beta\ \gamma\ldots$. So $U < X < P$, and X is convergent.

(O). *If $\{V\}$ is an at most countable ordered domain of divergent sequences and A is an arbitrary sequence $< V$, then there exists a divergent sequence X between A and $\{V\}$ $(A < X < V)$.*

If $\{V\}$ has a smallest element P, then put

$$X = \frac{A+P}{2} = \left(\frac{a_1 + p_1}{2}, \frac{a_2 + p_2}{2}, \ldots\right).$$

If $\{V\}$ does not have such an element, then, as in (M), it is a matter of inserting a divergent element between A and an ω^*-sequence $(P > Q > R > \cdots > A)$. By the possible modification of a finite set of values $p_n, q_n, \ldots$, one can again ensure that for each n, $p_n > q_n > r_n > \cdots > a_n$, and one can then apply the same method as in (M).

Now the general theorem parallel to (D) says:

(P). *If $\{U\}$ is an at most countable set of convergent sequences and $\{V\}$ an at most countable ordered domain of divergent sequences and $U < V$ throughout, then there are both a convergent sequence and a divergent sequence X between both sets $(U < X < V)$.*

Namely: by (D) we form some element Y between these two sets. Then according to (N) there is a convergent element X so that $U < X < Y$; and according to (O) there is a divergent element Z so that $Y < Z < V$.

In particular, between an ω-sequence of convergent elements and a following ω^*-sequence of divergent elements there always exists further convergent as well as divergent sequences; so that it seems totally wrong to want to fill in any such $\omega\omega^*$-gap, in which by right there belong infinitely many real elements, by *a single* "ideal" element.[1]

---

[1]) The "ideal function" of Du Bois-Reymond, which was also adopted by Borel, has undergone a vigorous and pointed critique by A. Pringsheim (Über die sogenannte Grenze und die Grenzgebiete zwischen Konvergenz und Divergenz, Münch. Ber. **26** (1896), 605–624; Über die Du Bois-Reymondsche Konvergenzgrenze ... ibid. **27** (1897), 303–334). The analogy with the irrational numbers is a logical misconception; one can insert a new thing corresponding to the relation $x^2 = 2$ between the elements of the two classes defined by $x^2 < 2$ and $x^2 > 2$, but between convergence and divergence of positive series there is no third alternative. If one then wants to assign new "ideal" elements to the actual gaps of a pantachic (which certainly according to (J) are not $\omega\omega^*$-gaps), which is of course permitted, then these are no longer numerical sequences or functions, and to let such an element appear, for instance, as a function under the integral sign is to dispense with any meaning. Cf. §5 also regarding the decomposition $\mathfrak{P} = \mathfrak{P}_c + \mathfrak{P}_d$.

§3
About H-types

The category of H-types can be viewed as the second term in a sequence of type categories for which the first term is the category of *everywhere dense, unbounded* types and which can be continued in a systematic way.

An everywhere dense, unbounded type is so defined because for each element there can be found a preceding one as well as following one and for each pair of elements there can be found an element lying between them. From this, it can be immediately concluded: if A and B are finite subsets $(A < B)$ of such a type, then there are always elements x, y, z that satisfy the conditions

(1) $\qquad x < A, \quad A < y < B, \quad B < z.$

So to the left and right of each finite subset and between any two finite subsets, there are always further elements.

Let us take the liberty of designating the everywhere dense, unbounded types as η-types after the simplest type of this category, the type η of the set of rational numbers. *Any η-type (e.g., η itself) contains each arbitrary countable type as a subset; there is only one countable η-type, namely, η itself.* The proof of these well-known theorems depends exclusively on the satisfaction of the conditions (1); I need not go into this since the basic idea will again appear in the following proofs.

Now, for instance, the conditions (1) can be strengthened so that the existence of such elements x, y, z is required even when A and B *are at most countable subsets* of the type in question. For the present, we call the types that satisfy these strengthened conditions, and thus in which there exist yet further elements to the left and right of any at most countable subset or between any two separated, at most countable subsets, η_1-types. If we then apply the conditions (1) to the special case when A, B are either single elements or fundamental sequences, we find:

An η_1-type is unbounded and everywhere dense; it is neither cofinal with ω nor coinitial with ω^*; no fundamental sequence in it has a limit; it contains no $\omega\omega^*$-gaps.

Conversely, these single conditions taken together again suffice to characterize a type as an η_1-type. For if μ is a type that satisfies these conditions and if A is an at most countable subset, then A has either a last element or is cofinal with ω, and in both cases, there are elements $z > A$ in μ (since μ has no last element and is not cofinal with ω). Similarly, there are always elements $x < A$. Finally, if A and B $(A < B)$ are two at most countable subsets, then there are 4 possible cases in that A can have a last element or be cofinal with ω and B can have a first element or be coinitial with ω^*. In all cases, there are always elements $A < y < B$ since by hypothesis μ is everywhere dense, contains no limits of ω-sequences or of ω^*-sequences, and contains no $\omega\omega^*$-gaps. — Consequently, η_1 types and H-types are identical (the notation was chosen by analogy with ω, Ω).

At this point, we are going to study some interesting properties of H-types.

I. Each H-type contains any arbitrary type of the second infinite cardinality as a subset.

Let $M = \{m\}$ be an H-type, and let A be a type of the second infinite cardinality, thus whose elements we can write in the form of an Ω-sequence

$$a_0 \, a_1 \, a_2 \, \ldots \, a_\omega \, \ldots \, a_\alpha \, \ldots$$

To these $\aleph_1$ elements, it is now a matter of assigning $\aleph_1$ elements from M

$$m_0 \, m_1 \, m_2 \, \ldots \, m_\omega \, \ldots \, m_\alpha \, \ldots$$

in such a way that m_α has the same rank ordering with respect to m_β in M as a_α has with respect to a_β in A. We show by recursion [[Rekursion]] that this surely is possible: if elements m_β have already been assigned to all elements a_β ($\beta < \alpha$), then it surely is possible to assign an m_α to a_α. The sets $\{a_\beta\}$ and $\{m_\beta\}$ are at most countable; if the set $\{a_\beta\}$ is now split by the element a_α into the subsets A' and A'' of elements lying to the left and to the right of a_α ($A' < a_\alpha < A''$), one of which by the way can be empty, and if M' and M'' are the corresponding subsets from $\{m_\beta\}$, then it is a matter of picking an element m_α from M in accordance with the condition $M' < m_\alpha < M''$. Thus m_α has to be chosen between two separated, at most countable subsets of M or possibly to the left or right of such a set, and this is always possible for an H-type. Consequently, understanding that m_0 is an arbitrary element, one can according to the stated requirement determine in turn the sequence $m_1 \, m_2 \, \ldots \, m_\omega \, \ldots \, m_\alpha \, \ldots$ up to each index of the second number class, and certainly one can determine it in infinitely many ways.

II. If an H-type of the second infinite cardinality exists at all, then there is only one.

In particular, let A and B be two H-types of the second infinite cardinality, whose elements we can thus write in the form of well-ordered sets

$$a_0 \, a_1 \, a_2 \, \ldots \, a_\omega \, \ldots \, a_\alpha \, \ldots,$$

$$b_0 \, b_1 \, b_2 \, \ldots \, b_\omega \, \ldots \, b_\alpha \, \ldots.$$

We already know that A can be similarly mapped onto a subset of B and B can be similarly mapped onto a subset of A. Now we only need to show that the previously given method can be adapted in such a way that *each* element a_α and *each* element b_β get mapped; thus A is mapped similarly onto B itself. The simplest way is to carry out the mapping *alternatingly*: first let b_0 correspond to a_0, then let the *lowest, suitable* a_α (i.e., the a_α with smallest index that has the same rank ordering with respect to a_0 that b_1 has with respect to b_0) correspond to b_1, then again let the *lowest, suitable* b_β correspond to the *lowest, still free* a_α (in this case a_1 or a_2), etc. In this way, element pairs from A and B

$$(a^0, b^0) \, (a^1, b^1) \, (a^2, b^2) \, \ldots \, (a^\omega, b^\omega) \, \ldots \, (a^\gamma, b^\gamma) \, \ldots$$

are formed in such a way that for $\gamma = 0, 2, 4, \ldots, \omega, \omega+2, \ldots, 2\xi, \ldots$ a^γ is always the lowest, still free a_α and b^γ is the lowest, suitable b_β (which has the same rank ordering with respect to all preceding $b^0\, b^1\, b^2\, \ldots\, b^\omega \ldots$ that a^γ has with respect to $a^0\, a^1\, a^2 \ldots a^\omega \ldots$), and conversely in the case $\gamma = 1, 3, 5, \ldots, \omega+1, \omega+3, \ldots, 2\xi+1, \ldots$ b^γ is the lowest, still free b_β and a^γ is the lowest, suitable a_α; in this way, it is obviously impossible that any a or b could be omitted. Hence there cannot be two distinct H-types of the second infinite cardinality.

III. *Each H-type is at least of the cardinality of the continuum $(2^{\aleph_0})$, and in case it contains no $\Omega\Omega^*$-gaps, it is at least of cardinality $2^{\aleph_1}$.*

For a proof, we shall show that such a type μ contains certain *powers* as subsets (cf. Note I: *The Powers of Order Types*). In a certain sense, the choice of base is free; for simplicity's sake, we prefer the type $\varphi = \omega^* + \omega$ of the two-sided unbounded numerical sequence

$$\ldots, -3, -2, -1, 0, 1, 2, 3, \ldots,$$

and in case any such is considered, we prefer 0 as the principal element (which, by the way, in the sense of the remark there, E, p. 120, has no effect because of the isotomy of φ). The numbers α of the first two number classes and ultimately Ω serve as arguments, and powers of the first, second, and third classes are dealt with.

We write a covering of α by φ in the form

$$X_\alpha = (x_0\, x_1\, \ldots\, x_\beta\, \ldots)_\alpha \quad (\beta < \alpha)$$

or, in case α is not a limit number, in the form

$$X_\alpha = (x_0\, x_1\, \ldots\, x_{\alpha-1}),$$

where the x_β are integers $\gtreqless 0$. This covering of α by φ will be simultaneously interpreted as a covering of $\alpha+1, \alpha+2, \ldots, \Omega$ by φ, in the sense that the principal element is to be put for $x_\alpha, x_{\alpha+1}, \ldots$; for this, we write the easily understandable form

$$X_\alpha = (x_0\, x_1\, \ldots\, |_\alpha 0) = (x_0\, x_1\, \ldots\, |_\alpha 0\, 0\, 0\, \ldots)$$

or, more concisely,

$$X_\alpha = (X_\alpha, 0) = (X_\alpha, 0\, 0\, 0\, \ldots)$$

by generally setting

$$X_{\alpha+1} = (x_0\, x_1\, \ldots\, x_\alpha) = (X_\alpha, x_\alpha).$$

Correspondingly, we denote an arbitrary covering of Ω by φ, thus where secondary elements ($\gtreqless 0$) can even occur in an uncountable set, by

$$X_\Omega = (x_0\, x_1\, \ldots\, x_\omega\, \ldots\, x_\alpha\, \ldots).$$

These coverings, ordered by first differences, form the following power types (where among the coverings X_α, those that contain an end piece of nothing but zeroes and that thus reduce to coverings with a smaller index are always to be included):

(1) For a fixed finite ν, all the X_ν form the power $\varphi(\nu) = \varphi^\nu$ of the first class.

(2) All the X_ν with finite index together form the power $\varphi_0(\omega)$ of the first class.

(3) All the X_ω form the power $\varphi'_0(\omega) = \varphi'(\omega)$ of the second class, in which the principal element has lost its special place and which contains all the coverings of ω by φ and so has cardinality $\aleph_0^{\aleph_0} = 2^{\aleph_0}$.

(4) For a fixed number α of the second number class, all the X_α form the power $\varphi'(\alpha)$ of the second class.

(5) All the X_ν and X_α together form the power $\varphi'_0(\Omega)$ of the second class.

(6) All the X_Ω form the power of the third class $\varphi''_0(\Omega) = \varphi''(\Omega)$, which again, through the irrelevancy of principal elements, contains all coverings of Ω by φ and has cardinality $2^{\aleph_1}$.

This scale of powers of increasing class with increasing arguments can of course be extended; we temporarily pause here, and we want to satisfy ourselves that an H-type μ contains all these powers as subsets, excluding or including the last one according to whether or not it contains $\Omega\Omega^*$-gaps. First, we can construct in μ a subset of type φ by designating any element by (0), a following one by (1), a still later one by (2), similarly a preceding one by (-1), a still earlier one by (-2), etc. This is always possible on account of the unboundedness of μ; so we have a first subset of elements $X_1 = (x_0)$. Now based on the everywhere denseness of μ, we place between the elements

$$(x_0) = (x_0, 0) \quad \text{and} \quad (x_0 + 1) = (x_0 + 1, 0)$$

two others, which we denote by $(x_0, 1)$ and $(x_0 + 1, -1)$, again between these, two others $(x_0, 2)$ and $(x_0 + 1, -2)$, etc.; so finally, each element $(x_0) = (x_0, 0)$ appears as the center of a φ-sequence whose elements bear the double numbering $X_2 = (x_0, x_1)$ and which neither reach nor exceed both neighbors; thus μ contains a subset of type φ^2. Again, by placing another φ-sequence about each of these elements $X_2 = (x_0, x_1, 0)$, we arrive at the three-placed elements $X_3 = (x_0, x_1, x_2)$, and continuing analogously, we arrive at the totality of all elements X_ν with finite ν; this totality of type $\varphi_0(\omega)$ is, by the way, already contained in each η-type as a subset since we have only made use of unboundedness and everywhere denseness up to this point (incidentally, it is the case that $\varphi_0(\omega) = \eta$).

Now, however, μ is supposed to contain no $\omega\omega^*$-gaps, and thus we can assign further elements of μ to all the $\omega\omega^*$-gaps of $\varphi_0(\omega)$ or to a part of them; we restrict ourselves to those gaps that are enclosed by fundamental sequences of the form

$$(x_0 \pm 1) \quad (x_0, x_1 \pm 1) \quad (x_0, x_1, x_2 \pm 1) \quad (x_0, x_1, x_2, x_3 \pm 1) \quad \ldots$$

Here in particular, we have a set of the type $\omega + \omega^*$; if the covering

$$X_\omega = (x_0\, x_1\, x_2\, \ldots\, x_\nu\, \ldots)$$

reduces to an X_ν, then both fundamental sequences enclose an element of $\varphi_0(\omega)$; if not, then they determine an $\omega\omega^*$-gap of $\varphi_0(\omega)$. We assign to the covering X_ω an element of μ that lies in this gap, and μ thus contains the totality of all coverings X_ω, which is of type $\varphi'(\omega)$ and has cardinality $2^{\aleph_0}$. Since we have not yet made use here of all properties of H-types, the theorem, already used earlier (I. i. O. T. III, IV), that an everywhere dense type without $\omega\omega^*$-gaps has cardinality at least $2^{\aleph_0}$ follows in passing.

However, in addition we conclude the following: each element X_ω is a two-sided limit of fundamental sequences in $\varphi'(\omega)$, whereas in μ no fundamental sequences have limits; thus there exist further elements of μ between X_ω and its neighbors in $\varphi'(\omega)$. From here, we again place about each element $X_\omega = (X_\omega, 0)$ a φ-sequence whose elements are to be denoted by $X_{\omega+1} = (X_\omega, x_\omega) = (x_0\, x_1\, \ldots\, x_\omega)$; by repeated application of the same procedure, we obtain the elements $X_{\omega+2}$, $X_{\omega+3}, \ldots$ and then again by filling certain $\omega\omega^*$-gaps the elements $X_{\omega 2}$; so continuing, we obtain the element X_α for each number of the second number class. Corresponding to both of Cantor's generation principles, two kinds of advancement towards new elements can be distinguished: the elements $X_{\alpha+1} = (X_\alpha, x_\alpha)$ arise by surrounding each element $X_\alpha = (X_\alpha, 0)$ with a φ-sequence; the elements X_λ with limit number index (λ a limit of the sequence $\alpha < \beta < \gamma < \cdots$) arise through the filling of those gaps that are enclosed by certain pairs of fundamental sequences,

$$(X_\alpha,\ x_\alpha \pm 1) \quad (X_\beta,\ x_\beta \pm 1) \quad (X_\gamma,\ x_\gamma \pm 1) \quad \ldots$$

With this, it has been proved that an H-type contains all the powers $\varphi'(\alpha)$ and their totality $\varphi'(\Omega)$ as a subset. In general, one cannot go further since an *arbitrary* covering X_Ω that does not reduce to an X_α (thus in which secondary elements occur in an uncountable set) represents an $\Omega\Omega^*$-gap of $\varphi'_0(\Omega)$. Only if we know that μ, too, does not contain $\Omega\Omega^*$-gaps, can elements of μ also be assigned to all coverings X_Ω, and in this case our H-type contains the power of third class $\varphi''(\Omega)$. Theorem *III* has thereby been proved.

The type $\varphi'_0(\Omega)$, which is contained in each H-type, is itself clearly not an H-type (since it still contains $\omega\omega^*$-gaps and is coextensive with $\omega^* + \omega$), but for its own part, it again contains H-types. For example, if we restrict ourselves to coverings with the three numbers -1, 0, 1, then to each power of φ there corresponds a power with base 3 of the same class and with the same argument and with the middle element as principal element; thus each H-type contains the power of second class

$$H = 3'_0(\Omega), \quad 3 = \{-1,\ 0,\ 1\}$$

as a subset, which is itself (I. i. O. T. IV, §3, p. 99) an H-type that belongs to the species *IV* 16 of homogeneous types.

Now, since our pantachie types, as well as their segments, are H-types and *at most* of the cardinality of the continuum, we can record the following result:

IV. All pantachie types are of the cardinality of the continuum; the homogeneous H-type $H = 3'_0(\Omega)$, as well as each arbitrary type of the second infinite cardinality, is contained as a subset in each pantachie; if the continuum is of the second infinite cardinality, then all pantachie types are similar, homogeneous, and of type H.

Our considerations regarding H-types have been recorded in a form that immediately yields their generalization. As before, we understand by $\omega \ \omega_1 \ \omega_2 \ \ldots \ \omega_\nu$ the smallest ordinal numbers of the 2nd, 3rd, 4th, ..., $(\nu + 2)$th number classes ($\omega_1 = \Omega$). Moreover, we are going to strengthen the conditions (1) that are supposed to hold for finite subsets A and B in η-types and for at most countable subsets A and B in H-types by postulating them for subsets A and B of at most the second infinite cardinality, third infinite cardinality, etc.; so we arrive at types that we designate as η_2-types, η_3-types, etc. Our definition thus reads (with restriction to finite indices ν):

A type is called an η_ν-type if each subset in it of cardinality $< \aleph_\nu$ has both a following and a preceding element, and for each two separated subsets of cardinality $< \aleph_\nu$, there always exists an element lying between them.

By this we obtain a sequence of type families [[Typengattungen]], the η-types, η_1-types, η_2-types, ..., where each family includes the next, i.e., each $\eta_{\nu+1}$-type is at the same time an η_ν-type; the η-types are the everywhere dense, unbounded types, and the η_1-types are identical to the H-types.

The theorems about H-types can be carried over without any difficulty to η_ν-types, and they give:

V. Each η_ν-type contains each arbitrary type of cardinality $\aleph_\nu$ as a subset.

VI. If there exists an η_ν-type of cardinality $\aleph_\nu$ at all, then there is only one.

VII. Each η_ν-type is of cardinality at least $2^{\aleph_{\nu-1}}$ and contains as a subset the power of $(\nu+1)$st class

$$\eta_\nu = 3_0^{(\nu)}(\omega_\nu), \quad 3 = \{-1, 0, 1\}$$

that is itself an η_ν-type; each η_ν-type without $\omega_\nu \omega_\nu^*$-gaps is at least of cardinality $2^{\aleph_\nu}$.

Again as for *III*, one will prove this by showing the existence of a subset $\varphi_0^{(\nu)}(\omega_\nu)$ in each η_ν-type; so such a type also contains the subset $\varphi_0^{(\nu)}(\omega_{\nu-1}) = \varphi^{(\nu)}(\omega_{\nu-1})$, which because of the irrelevant principal element contains all coverings of $\omega_{\nu-1}$ by φ and has cardinality $\aleph_0^{\aleph_{\nu-1}} = 2^{\aleph_{\nu-1}}$; on the other hand, it also contains the subset η_ν from which, analogously as in case $\nu = 1$, it can be proved that it is itself an η_ν-type. An η_ν-type without $\omega_\nu \omega_\nu^*$-gaps contains the subset $\varphi_0^{(\nu+1)}(\omega_\nu) = \varphi^{(\nu+1)}(\omega_\nu)$, i.e., the set of all coverings of ω_ν by φ, which is of cardinality $2^{\aleph_\nu}$.

Should CANTOR's hypothesis $2^{\aleph_0} = \aleph_1$ turn out to be true in an extended range, that is, $2^{\aleph_\nu} = \aleph_{\nu+1}$ also holds for each finite ν, then there exists one and only one η_ν-type of cardinality $\aleph_\nu$, namely, η_ν itself.

It may still be pointed out that the scale of η_ν-types is not as arbitrary a construction as it might appear; for order types, its generation principle is basically the same as the one that for cardinal numbers raises us over each already attained level to a higher one. Apart from rank ordering, if I ask that for each subset of cardinality $< \aleph_\nu$ another element should exist, thus that the cardinalities are $\geq \aleph_\nu$ and $\aleph_\nu$ is defined as their minimum, and if I ask that even ordinally no such subset should exhaust our set, rather that each cut $A + B$ of such a subset should still allow place for at least one element of the whole set, then the set as defined is an η_ν-type.

This might be the most appropriate place to also briefly indicate the *generalization of the final rank ordering to arbitrary covering sets*. Let $A = \{a\}$ and $M = \{m\}$ be arbitrary sets of types α and μ; a covering of A by M

$$x = \{x_a\} = \sum_a x_a$$

arises[1]) by assigning to each "place" a an element x_a from M. We now define the final rank relationship [[Rangbeziehung]] of two coverings x, y as follows:

If there is an end piece $C = \{c\}$ of A for all of whose places

$$x_c = y_c, \quad \text{or} \quad x_c < y_c, \quad \text{or} \quad x_c > y_c \quad (\text{in } M),$$

then the rank ordering shall be

$$x \sim y, \quad \text{or} \quad x < y, \quad \text{or} \quad x > y$$

(x is finally equal to, smaller than, greater than y); when none of these cases occurs, then $x \| y$ (x is finally incomparable with y).

Hereby, the final rank ordering is defined within covering sets, and as before, we can speak of ordered domains and pantachies, the last of which we wish to denote by $\mathfrak{P}(\mu, \alpha)$.[2]) For the usual pantachies considered up to now, M was the set of all positive real numbers, thus μ was the unbounded linear continuum λ; for functions $f(x)$, A was likewise the set of positive real numbers, and for numerical sequences, it was the set of positive integers (type ω), so that our pantachies of functions would be denoted by $\mathfrak{P}(\lambda, \lambda)$ and those of numerical sequences by $\mathfrak{P}(\lambda, \omega)$. However, were one, for example, to consider numerical sequences consisting of only rational numbers or of only positive integers, then one would obtain the pantachies $\mathfrak{P}(\eta, \omega)$ and $\mathfrak{P}(\omega, \omega)$. Now as far as the characteristic properties of our pantachies are concerned, namely, their membership in the category of H-types or η_1-types, an examination of Theorems (C) and (D) of the previous § shows that only the following properties were really made use of there: the base μ is everywhere dense and unbounded, thus an η-type, and the exponent α is cofinal with ω. By the corresponding generalization, one finds that $\mathfrak{P}(\mu, \alpha)$ is an

---

[1]) Cf. Note I, Die Potenzen von Ordnungstypen, p. 109. Also for each a, a particular set M_a can be chosen to which x_a is supposed to belong; we then obtain the coverings of A by the complex of sets M_a.

[2]) We are going to call μ the base and α the exponent.

$\eta_{\nu+1}$-type if μ is an η_ν-type and α is cofinal with ω_ν; so in particular, each pantachie $\mathfrak{P}(\eta_\nu,\omega_\nu)$ is an $\eta_{\nu+1}$-type; thus one can construct these types not only by raising to powers but also by the final rank ordering of higher numerical sequences. But in contrast to the clearly defined formation of powers, the pantachie concept is extraordinarily more broad and more indefinite, so already in the simplest cases, the complete exploration of pantachie types such as $\mathfrak{P}(\eta,\omega)$ or $\mathfrak{P}(\lambda,\omega)$ still remains an unsolved problem.[1])

Before we apply ourselves again to pantachies, we observe that one can prove Theorems *III* and *VII* of this § under somewhat extended hypotheses, and thereby one can achieve a strengthening of Theorems A and B in §1 of the previous Article IV. We saw in the proof of Theorem *III* that an everywhere dense type without $\omega\omega^*$-gaps already has cardinality at least $2^{\aleph_0}$; however, it also holds more widely:

VIII. An everywhere dense type without $\omega\omega^*$-limits as well as without $\omega\omega^*$-gaps and $\Omega\Omega^*$-gaps is of cardinality at least $2^{\aleph_1}$,
and in general:

IX. An everywhere dense type
without $\omega\omega^*, \omega_1\omega_1^*, \ldots, \omega_{\nu-1}\omega_{\nu-1}^*$-limits
and without $\omega\omega^*, \omega_1\omega_1^*, \ldots, \omega_{\nu-1}\omega_{\nu-1}^*, \omega_\nu\omega_\nu^*$-gaps
is at least of cardinality $2^{\aleph_\nu}$.

In order to recognize which modifications are necessary to adapt the proof of Theorem *III* to the reduced hypotheses of Theorem *VIII*, we first note that if a φ-sequence is to be constructed in μ, the unboundedness of μ plays no role and only its everywhere denseness does; in the choice of the element (x_0), we only have to avoid the possible boundary elements of μ — to put it another way, instead of μ we use the unbounded type that in any case arises from μ after the removal of possible boundary elements. Now exactly as there, we can assign elements of μ to all coverings X_ν and we can place new elements X_ω in certain $\omega\omega^*$-gaps, so that μ again contains the power $\varphi'(\omega)$ of the second class. However, a difference now arises. In $\varphi'(\omega)$, the elements X_ω are two-sided fundamental limits ($\omega\omega^*$-limits); they are certainly not such in μ, but they could be one-sided (ω-limits or ω^*-limits), and then we cannot legitimately make them centers of new φ-sequences. However, now it is of importance that the *actual* coverings X_ω (those that do not reduce to X_ν) can surely be assigned elements of μ that one can surround by new φ-sequences; for we had filled an $\omega\omega^*$-gap of $\varphi_0(\omega)$ with X_ω, and by assumption, into this gap falls a piece μ_1 of μ that contains at

---

[1]) For rank ordering by first differences, which for the moment offers the possibility of incomparable elements as well, we could, in a simple way, so restrict the coverings that certain ordered domains (just our power types) arise; here one could also study pantachies, i.e., non-extendible ordered domains, as such with no further hypotheses. I also note that the final rank ordering itself, with a certainly not insubstantial modification, can be so formulated that it includes rank ordering by *last* differences (which differs from ordering by first differences only by inversion of the exponent) as a special case; perhaps an opportunity to go into this will present itself later.

least two and therefore infinitely many members and that has exactly the same properties as μ. So by avoiding the possible boundary elements of μ_1, we can certainly insert in μ_1, not only the single element X_ω, but an entire φ-sequence $(X_\omega, x_\omega) = X_{\omega+1}$. The step from the covering $X_\alpha = (X_\alpha, 0)$ to the covering $X_{\alpha+1} = (X_\alpha, x_\alpha)$, which is analogous to the first generation principle, is now only possibly hindered if α is a limit ordinal and X_α reduces to a covering with a smaller index; if α is not a limit ordinal or if X_α is an actual cover with limit index, i.e., an element newly placed in a hitherto existing $\omega\omega^*$-gap, then X_α is usable as center of a φ-sequence of elements $X_{\alpha+1}$.

Under these circumstances, μ certainly still contains a set of coverings of sufficiently high cardinality. For if we now restrict ourselves, say, to such coverings
$$Y_\alpha = (y_0\, y_1\, \ldots)_\alpha = (y_0\, y_1\, \ldots |_\alpha 0\, 0\, 0\, \ldots)$$
in which all elements y_β ($\beta < \alpha$) are secondary, i.e., different from the principal element 0, then μ contains the set of all these coverings for all indices of the first and second number classes and, in addition, all coverings Y_Ω because of the absence of $\Omega\Omega^*$-gaps; the totality of these coverings (of Ω by the sequence of numbers $\ldots, -2, -1, 1, 2, \ldots$) still has cardinality $\aleph_0^{\aleph_1} = 2^{\aleph_1}$; whereby, Theorem *VIII* is proved. Correspondingly, in the case of Theorem *IX*, one shows that μ contains all coverings Y_α for $\alpha \leqq \omega_\nu$.

§4
Transformations

One obtains some further information regarding the structure of pantachie types by considering transformations that leave the final rank ordering invariant. For example, such a transformation is given by
$$X = (x_1\, x_2\, \ldots\, x_n\, \ldots),$$
$$X' = (f(x_1)\, f(x_2)\, \ldots\, f(x_n)\, \ldots),$$
where $f(x)$ denotes a single-valued, positive, monotonically increasing function of the positive variable x, so that for $x < y$, always $f(x) < f(y)$; in fact, from $X\varrho Y$ it immediately follows that $X'\varrho Y'$ if ϱ is any of the four symbols $\sim\, <\, >\, \|$. In this way, the totality of all numerical sequences goes into itself or into a subset, and to each pantachie there corresponds a *similar* ordered domain, which could possibly again be a pantachie. The same still holds even if the transforming function varies with the index n, thus for
$$X' = (f_1(x_1)\, f_2(x_2)\, \ldots\, f_n(x_n)\, \ldots)$$
if each f_n is a function with the properties mentioned above. Correspondingly, one can also produce a transformation with monotonic *decreasing* functions that *reverses* the final ordering, i.e., that switches the symbols $<$ and $>$.

For our purposes, certain rational transformations suffice. For the representation of these transformations, it is advisable to lay down appropriate rules for arithmetic operations on numerical sequences. We are going to define the sum, product and quotient of two numerical sequences as a new numerical sequence by

$$X + Y = (x_1 + y_1, x_2 + y_2, \ldots, x_n + y_n, \ldots),$$
$$X\,Y = (x_1 y_1, x_2 y_2, \ldots, x_n y_n, \ldots),$$
$$\frac{X}{Y} = (\frac{x_1}{y_1}, \frac{x_2}{y_2}, \ldots, \frac{x_n}{y_n}, \ldots),$$

so that numerical sequences behave like complex numbers $x_1 \varepsilon_1 + x_2 \varepsilon_2 + \cdots$ with infinitely many units ε_i and with the multiplication rules

$$\varepsilon_i^2 = \varepsilon_i, \quad \varepsilon_i \varepsilon_k = 0 \quad (i \neq k).$$

Accordingly, in case λ is a positive number, λ is also to be understood as the numerical sequence

$$\lambda = (\lambda, \lambda, \ldots, \lambda, \ldots)$$

and

$$\lambda X = (\lambda x_1, \lambda x_2, \ldots, \lambda x_n, \ldots).$$

With this, rational functions of one or more numerical sequences with coefficients that are positive numbers or numerical sequences are also defined, e.g.,

$$AX^2 + 2BXY + CY^2 = (\ldots, a_n x_n^2 + 2 b_n x_n y_n + c y_n^2, \ldots).$$

Finally, for $X > Y$ one can also define the difference to be

$$X - Y = (x_1 - y_1, x_2 - y_2, \ldots, x_n - y_n, \ldots)$$

by agreeing to insert any positive numbers for the negative or vanishing values $x_n - y_n$ (occurring only with finite frequency, according to hypothesis); essentially, that is, if one considers finally equal numerical sequences as identical, subtraction is unambiguously defined by this.

Let us now consider some "projective" transformations that completely suffice for our purpose. The *dilatation*

(1) $$X' = MX, \quad X = \frac{1}{M} \cdot X'$$

transforms the totality of numerical sequences into itself and each pantachie $\mathfrak{P}$ into a similar pantachie $\mathfrak{P}'$. If the multiplier $M \sim 1$, then $X' \sim X$; so $\mathfrak{P}'$ is identical with $\mathfrak{P}$ if finally equal elements are considered as identical. If $M < 1$ or $M > 1$, then, in general, $\mathfrak{P}$ does not go into itself but can have elements in common with $\mathfrak{P}'$. If $M \parallel 1$, then it is the case that $X' \parallel X$, and the two pantachies $\mathfrak{P}, \mathfrak{P}'$ have no common elements.

The *involution*

(2) $$XX' = 1$$

transforms each pantachie $\mathfrak{P}$ to a pantachie $\mathfrak{P}'$ of inverse type.

The *translation*

(3) $$X' = X + A, \quad X = X' - A$$

transforms each X to an $X' > A$ and conversely each $X' > A$ to an X, preserving rank ordering. A pantachie $\mathfrak{P} = \{X\}$ goes into a *similar ordered domain* $\mathfrak{P}' = \{X'\}$, which obviously has the following properties: all X' are $> A$; there is no element that would surpass all X'; for a cut of $\mathfrak{P}'$, there is no element that would fall between both pieces; there is no element that would fall between A and $\mathfrak{P}'$. Accordingly, $\{A, \mathfrak{P}'\}$ is an ordered domain that can only be expanded to a larger ordered domain or a pantachie by elements $< A$. If we complete it in any way to a pantachie $\mathfrak{D}$, then $\mathfrak{P}'$ is thereby represented as an *end segment* $\mathfrak{D}_A$ of a pantachie. Conversely, each pantachie end segment $\mathfrak{P}' = \mathfrak{D}_A$ goes into a pantachie $\mathfrak{P}$ by the transformation $X = X' - A$. *There exists for each pantachie a similar pantachie end segment and conversely.*

In a completely analogous way, the transformations

(4) $$X' = \frac{BX}{1+X}, \quad X = \frac{X'}{B - X'},$$

(5) $$X' = \frac{A + BX}{1 + X}, \quad X = \frac{X' - A}{B - X'} \quad (A < B)$$

map the totality of numerical sequences X onto a subset ($X' < B$, respectively, $A < X' < B$), whereby a pantachie $\mathfrak{P}$ goes into a similar pantachie initial segment $\mathfrak{D}^B$, respectively, pantachie middle segment $\mathfrak{D}_A^B$. *For each pantachie there is a similar pantachie segment (initial segment, middle segment, end segment) and conversely.*

Also mappings of pantachie initial segments onto middle segments, etc., are obtained by projections.

Note further that one can always meld arbitrary segments of (distinct or the same) pantachies into a pantachie type. For example, if an initial segment of $\mathfrak{P}$ is to be put together with an end segment of $\mathfrak{D}$, first bring both the elements in question into correspondence with a dilatation; then if

$$\mathfrak{P} = \mathfrak{P}^A + A + \mathfrak{P}_A, \quad \mathfrak{D} = \mathfrak{D}^A + A + \mathfrak{D}_A,$$

obviously $\mathfrak{P}^A + A + \mathfrak{D}_A$ is also a pantachie.

Summarizing what has been said, we know:

Each inverse pantachie type is again a pantachie type; each pantachie segment is a pantachie type; again, a pantachie type can always be put together from arbitrary pantachie types as segments.

So the formula

(6) $$\begin{cases} \Pi = \Pi_1 + 1 + \Pi_2, \\ \Pi = \Pi_1 + 1 + \Pi_2 + 1 + \Pi_3, \end{cases}$$
etc.

holds, where for an arbitrary pantachie type Π, the types $\Pi_1, \Pi_2, \ldots$ are also always pantachie types and for arbitrary pantachie types $\Pi_1, \Pi_2, \ldots$, the type Π is also always a pantachie type.

The simplicity of the preceding considerations depends essentially upon the *final* rank ordering; for the *infinitary* rank ordering, the sphere of *similarity* transformations is markedly limited (e.g., of the projections, only dilatation and involution are admissible), and I have not been able to find a mapping of an entire pantachie onto a pantachie piece, onto an end segment, onto a middle segment or anything of that kind. By a method analogous to the transition from a function to its inverse, one can assign to each numerical sequence that is infinitarily greater than

$$N = (1, 2, 3, \ldots, n, \ldots)$$

an inverse numerical sequence that lies between 1 and N. With this, the mapping of an end segment onto a middle segment seems to become a reality; but this correspondence, as has already been noted (§1), in no way has the property of uniquely inverting the infinitary rank ordering.

From formula (6), it can be concluded that if all pantachies have the same type Π, this type is certainly *homogeneous*, i.e., all its segments are similar; nevertheless, this question still remains open, and we only know that on the basis of Theorem IV in §3 the answer has to be positive if the continuum is of the second infinite cardinality. All the more noteworthy is the fact that, without any hypothesis, by an appropriate method one arrives at

the construction of a homogeneous pantachie,

whereby one has come at least a step closer to the ideal of regular generation of these remarkable types. This method will be explained next.

Obviously, a pantachie would certainly be homogeneous were it equal to a *field* [[*Körper*]] of numerical sequences, i.e., along with two elements, it would also always contain their sum, difference, product and quotient; it would permit infinitely many projective transformations into itself, by virtue of which the whole pantachie could easily be mapped similarly onto each of its segments. Still, the construction of ordered domains or pantachies that are supposed to be simultaneously fields presents certain problems, namely, in this respect: to ensure the *comparability* of its elements, from the outset not only the rational elements but also the algebraic elements must be included. For example, the totality of all positive rational numbers $r = (r, r, \ldots, r, \ldots)$ is a field domain [[*Körperbereich*]]; however, if one adjoins a new element A that is comparable with all the r, then even A^2 need no longer be comparable with all the r, and the totality of rational functions in A with rational coefficients is certainly a field but not an ordered domain.[a] Here one would be able to manage by first adjoining to $\{r\}$ all positive real algebraic numbers a and only then extending the field domain $\{a\}$ to a more comprehensive field domain by including some A that is comparable with all a; however, again one would have to include besides the rational also the algebraic functions of A, and this gives rise to complications, which in any case, though perhaps not insurmountable, destroys the simplicity of the method. Now the field property is not absolutely necessary for the construction of a homogeneous pantachie; restricting the feasibility of rational operations certainly suffices, and the following transparent method, by which the inclusion of algebraic elements is avoided, is based on this.

In this connection, for simplicity's sake we are going to temporarily suspend the restriction to *positive* numerical sequences that was assumed throughout. And we are going to consider numerical sequences

$$X = (x_1 \, x_2 \, \ldots \, x_n \, \ldots)$$

in which the x_n are arbitrary real numbers ($\gtreqless 0$); also let there be no further distinction between finally equal numerical sequences, so that each equation may have only the meaning of a *final* equation and in each X an arbitrary finite numerical set $x_1 \, x_2 \, \ldots \, x_n$ can be changed. (This mode of expression would be more precise, but at the same time more drawn out, were we to collect all the finally equal numerical sequences into a *class* and to carry out the rational operations on these classes instead of on their representatives.) Addition, subtraction and multiplication are now defined unrestrictedly; on the other hand, one must not divide by $0 = (0, 0, \ldots, 0, \ldots)$ nor by "divisors of zero," i.e., by such numerical sequences that contain infinitely many zeroes; whereas a numerical sequence with a finite number of zeroes can be changed to a finally equal sequence without zeroes and so can occupy the place of a divisor. Every division by elements of an ordered domain that contains the element 0 (otherwise only elements > 0 and < 0) can be carried out, with the exception of division by zero itself.

142

First, we now construct a well-ordered sequence of increasing elements in the following way. After the elements

$$1 = (1, 1, \ldots), \quad 2 = (2, 2, \ldots), \quad \ldots, \quad p = (p, p, \ldots), \quad \ldots$$

we let follow an element A_1 that surpasses each p, for example,

$$A_1 = (1, 2, 3, \ldots).$$

Then it is the case that $A_1 < A_1^2 < A_1^3 < \cdots$. Let A_2 follow this sequence of powers, for example,

$$A_2 = (1, 2^2, 3^3, \ldots).$$

Let A_3 follow the sequence of powers $A_2, A_2^2, A_2^3, \ldots$, etc. Then we let a finally greater element A_ω again follow the sequence $A_1, A_2, A_3, \ldots$, and we

---

[a] The restriction to positive r needs to be dropped for closure under differences. In [H 1909a, 312], A is taken to be a sequence converging to $\sqrt{2}$ from above and below. It is comparable to all the r, but $A^2 \parallel 2$.

let $A_{\omega+1}$ follow all the positive powers of A_ω. Since each countable set of elements can always be finally surpassed by a later element, this procedure can be continued up to each index of the second number class and possibly even beyond, and thus we obtain a sequence

$$\mathfrak{A} = \{A_\alpha\} = A_1, A_2, \ldots A_\omega, \ldots A_\alpha, \ldots$$

of type $\geqq \Omega$ with the property that there exists no element $> \mathfrak{A}$ ($\mathfrak{A}$ is a "transcendent sequence" [["transzendente Reihe"]] according to the terminology of §6) and with the property that

(7) $\qquad A_1 > p, \quad A_\beta > A_\alpha^p \quad (\beta > \alpha,\ p = 1, 2, 3, \ldots).$

143 We now call an ordered domain a *semifield domain* [[*Halbkörperbereich*]] if it has the following properties:

(α) *Sums and differences of two elements of the ordered domain likewise belong to the ordered domain.*

(β) *If X is an element of the ordered domain, then $X \cdot A_\alpha$ and $X : A_\alpha$ also belong to the ordered domain.*[b]

Accordingly, a pantachie with these properties is called a semifield pantachie. Our next aim is to prove the existence of one such. This proof is carried out in the following steps.

1st: *The set of all rational functions of one or more (finite in number) A_α with rational coefficients is a semifield domain.*

Let $A, B, C, \ldots, L$ be any of the A_α. If we form the integral function

$$F = \sum r A^a B^b C^c \cdots L^l,$$

where the r's are rational numbers, the exponents $a\ b\ c\ \ldots\ l$ are integral rationals that are not negative numbers, then it is claimed that the aggregate of elements $F : G$ is a semifield domain. Since this aggregate satisfies the requirements (α) and (β), it only remains to show that it is an ordered domain, thus that any two elements are comparable.

We imagine that the integral function F is arranged as follows: if $A > B > C > \cdots > L$, then we order the terms according to descending powers of the highest term A, then terms containing the same A^a according to descending powers of B, etc. If in this order of succession

$$F = rA^a B^b C^c \cdots L^l + r_1 A^{a_1} B^{b_1} C^{c_1} \cdots L^{l_1} + \cdots,$$

then one easily sees that $F \gtreqless 0$ according to whether the coefficient r of the leading term is $\gtreqless 0$. The simple proof of this relies on formula (7) or, more concisely stated, on the fact that A is "infinitely large" compared

---

[b] In [H 1909a, 310n], Hausdorff offers a "small correction" to this proof: he strengthens (β) to closure under multiplication by elements of $\mathfrak{H}_\circ$, the set of rational functions in the A_α, and he alters the proof of step 2 by offering $\{AX + B \mid A \in \mathfrak{H}_\circ, B \in \mathfrak{H}\}$ as a semifield domain (under the strengthened definition) containing the elements of $\mathfrak{H}$ and the element X.

with B, B is "infinitely large" compared with C, and so on. Namely, if $P = A^a B^b C^c \cdots L^l$ denotes the highest product of powers and $P_1 = A^{a_1} B^{b_1} C^{c_1} \cdots L^{l_1}$ denotes a lower one (thus whereby the first non-vanishing difference $a - a_1, b - b_1, \ldots$ is positive), then it follows that P surpasses P_1 as well as each rational multiple of P_1. Consequently,

$$P + s_1 P_1 > 0, \quad P + s_2 P_2 > 0, \quad \ldots,$$

and after addition and division by an integer,

$$P + s_1 P_1 + s_2 P_2 + \cdots > 0$$

for arbitrary rational coefficients s. Therefore it is the case that $\dfrac{F}{r} = P + \dfrac{r_1}{r} P_1 + \cdots > 0$, and F has the same sign as r.

Accordingly, each integral function that does not identically vanish, i.e., that does not have only coefficients $r = r_1 = \cdots = 0$, has a determined sign $\gtreqless 0$; the same holds for each rational function, and since the difference of two rational functions is again one, we always have $\dfrac{F}{G} \gtrsim\!\!\!\lessgtr \dfrac{F'}{G'}$, in which the case $\dfrac{F}{G} \sim \dfrac{F'}{G'}$ occurs only if $\dfrac{F}{G} = \dfrac{F'}{G'}$ identically (for undetermined variables A, B, C, ...).
Therefore our first assertion is proved.

2nd: *Each semifield domain that is not yet a pantachie can be extended to a more comprehensive semifield domain.*

Let $\mathfrak{H}$ be a semifield domain, and let X be an element not contained in it but comparable to each element of $\mathfrak{H}$. The aggregate of elements $AX + B$, where A and B run through all the elements of $\mathfrak{H}$, again forms a semifield domain. Indeed, requirements (α) and (β) are satisfied; moreover, $AX + B = A(X + \dfrac{B}{A})$, and thus it has a determined sign since A and $X + \dfrac{B}{A}$ have determined signs ($A \gtreqless 0$ is assumed since otherwise $AX + B$ reduces to B and so belongs to $\mathfrak{H}$); consequently, the difference of two such elements also has a determined sign, and all these elements are comparable.

3rd: *If $\mathfrak{H}_0\, \mathfrak{H}_1 \ldots \mathfrak{H}_\alpha \ldots$ is a well-ordered sequence of semifield domains each of which is contained in all the following ones, then the smallest multiple [[kleinste Vielfache]] $\mathfrak{H}$ of these sets, i.e., the collection of all elements that occur in at least one $\mathfrak{H}_\alpha$, is again a semifield domain.*

Either the set has a last term, which is then $= \mathfrak{H}$, or if not, then for any two elements X and Y of $\mathfrak{H}$ there is certainly a semifield domain $\mathfrak{H}_\alpha$ to which both belong. Then X and Y are certainly comparable, and also $X \pm Y$, $X \cdot A_\beta$, $X : A_\beta$ belong to $\mathfrak{H}_\alpha$ and to $\mathfrak{H}$ as well; so $\mathfrak{H}$ satisfies the requirements of an ordered domain and simultaneously those of a semifield domain.

4th: *There exists a semifield pantachie.*

From the 2nd and 3rd steps, it follows that if one were not to exist, then the construction of semifield domains could be continued up to each transfinite index, and the cardinality of the continuum would have to be greater than any aleph.

At this point, we can prove the crucial theorem:

Each semifield pantachie is isomeric; i.e., all its middle segments are similar.

Since $\mathfrak{P}$ permits addition and subtraction, first of all one can map each middle segment $\mathfrak{P}_Y^Z$ onto one $\mathfrak{P}_0^X$ ($X > 0$) beginning with 0, call its type ξ. Since the middle segment $\mathfrak{P}_{-X}^0$ is both directly and inversely similar to $\mathfrak{P}_0^X$, it follows that $\xi^* = \xi$, and since $\mathfrak{P}_0^{2X}$ is likewise similar to $\mathfrak{P}_0^X$, then $\xi = \xi + 1 + \xi$. Furthermore, let $0 < X < Y$; if one denotes the types of $\mathfrak{P}_0^X$, $\mathfrak{P}_0^Y$, and $\mathfrak{P}_X^Y$ by ξ, η, and ϱ, then

$$\eta = \xi + 1 + \varrho = (\xi + 1 + \xi) + 1 + \varrho = \xi + 1 + \eta,$$

and inverted

$$\eta^* = \eta^* + 1 + \xi^* = \eta + 1 + \xi = \eta;$$

thus

$$\eta = \xi + 1 + \eta = \eta + 1 + \xi.$$

Likewise, it follows that for $0 < X < Y < Z$

$$\zeta = \eta + 1 + \zeta = \zeta + 1 + \eta,$$

thus for $\xi = \zeta$, also $\xi = \eta$, i.e., if $\mathfrak{P}_0^X$ and $\mathfrak{P}_0^Z$ are similar, they are also similar to any middle segment $\mathfrak{P}_0^Y$ lying between. — Now, if X and Y are any two elements ($0 < X < Y$), then Y cannot surpass all the elements $X \cdot A_\alpha$ (for were $Y > X \cdot A_\alpha$, then it would be that $\dfrac{Y}{X} > A_\alpha$, i.e., contrary to hypothesis, $\mathfrak{A}$ is a sequence for which there exists a subsequent element). Consequently, since all $X \cdot A_\alpha$ belong to $\mathfrak{P}$ and are comparable with Y, there must be an element $Z = X \cdot A_\alpha > Y$. Then, however, since $\mathfrak{P}$ goes into itself through multiplication by A_α, $\mathfrak{P}_0^X$ is similar to $\mathfrak{P}_0^Z$ and also to $\mathfrak{P}_0^Y$ because $X < Y < Z$. Thus any two middle segments are similar, q.e.d.

It is also the case that all initial segments are similar to each other and likewise for all end segments ($\mathfrak{P}$ is "isotomic"), which arises from the possibility of addition within $\mathfrak{P}$; of course, the similarity of all segments does not yet follow. And if ω_ν was the type of the sequence $\mathfrak{A} = \{A_\alpha\}$ originally used, we can only say that each element of $\mathfrak{P}$ is an $\omega_\nu \omega_\nu^*$-limit and that each segment of $\mathfrak{P}$ is coextensive with $\omega_\nu^* + \omega_\nu$ since $\mathfrak{P}$ is cofinal with $\mathfrak{A}$ and coinitial with the sequence $\{-A_\alpha\}$, and each element X is the limit of the sequences $X(1 \pm \dfrac{1}{A_\alpha})$. But each middle segment of $\mathfrak{P}$ is a *homogeneous* type, and as we have seen earlier, it is indeed again a pantachie type; we can now once more restrict ourselves to positive numerical sequences. Hence by the transformation $X' = X : (A - X)$, the middle segment $\mathfrak{P}_0^A$ of our semifield pantachie goes into a homogeneous pantachie, which of course need no longer be a semifield pantachie.

§5
Separation and Mixture

Among the numerous remaining methods for transforming numerical sequences into other numerical sequences while preserving the final rank ordering, a procedure that can be designated as *separation* and *mixture* still deserves attention. For instance, from a numerical sequence

$$X = (x_1\, x_2\, x_3 \cdots x_n \cdots)$$

one can single out the x with odd indices and combine them into a new numerical sequence

$$(x_1\, x_3\, x_5 \cdots x_{2n-1} \cdots).$$

Generally, if

$$P = (p_1\, p_2\, p_3 \cdots p_n \cdots)$$

denotes a sequence of increasing integers, the subsequence

$$X_P = (x_{p_1}\, x_{p_2}\, x_{p_3} \cdots x_{p_n} \cdots)$$

can be separated from X.[1]) Thereby

$$X_P \sim Y_P, \quad X_P < Y_P, \quad X_P > Y_P$$

follows from

$$X \sim Y, \quad X < Y, \quad X > Y, \quad \text{respectively}$$

(while nothing about the relation between X_P and Y_P follows from $X \| Y$, nevertheless $X \| Y$ certainly follows from $X_P \| Y_P$). So for a fixed P a similar ordered domain $\mathfrak{B}_P$ arises from the ordered domain $\mathfrak{B}$; in general, only an ordered domain, which in special cases however can be a pantachie, arises from a pantachie. Conversely, if $\mathfrak{X} = \{X\}$ and $\mathfrak{Y} = \{Y\}$ are two *similar* ordered domains in which the element Y is assigned to the element X, then X and Y can, for instance, be *mixed*[2]) into a new numerical sequence according to the formula

$$T = (X, Y) = (x_1, y_1, x_2, y_2, \ldots, x_n, y_n, \ldots);$$

these new elements T obviously form an *ordered domain* $\mathfrak{T} = \{T\}$ similar to $\mathfrak{X}$ and $\mathfrak{Y}$. In order that this be a pantachie, it is in no way necessary that $\mathfrak{X}$ and $\mathfrak{Y}$ be pantachies, rather for that, it suffices that in $\mathfrak{X}$ and $\mathfrak{Y}$ new elements cannot simultaneously be inserted into *corresponding* cuts [[Schnittstestellen]]

$$\mathfrak{X} = \mathfrak{X}' + \mathfrak{X}'', \quad \mathfrak{Y} = \mathfrak{Y}' + \mathfrak{Y}''$$

---

[1]) This separation procedure was already used in §1 with the assignment of numerical sequences to functions.

[2]) [H 1907a, 147n1] The earlier declared abandonment of monotonicity is especially influential in the following considerations.

(including here the cuts [[Zerschneidungen]] $\mathfrak{X} = 0 + \mathfrak{X} = \mathfrak{X} + 0$, i.e., there should not be elements simultaneously $< \mathfrak{X}, < \mathfrak{Y}$, and there should not be elements simultaneously $> \mathfrak{X}, > \mathfrak{Y}$). Indeed: if $\mathfrak{T}$ is an extendible ordered domain for which the element E can be inserted in the position $\mathfrak{T}' + \mathfrak{T}''$, set

$$E = (A, B) = (a_1, b_1, a_2, b_2, \ldots, a_n, b_n, \ldots),$$

then $\mathfrak{T}' + E + \mathfrak{T}''$, $\mathfrak{X}' + A + \mathfrak{X}''$, $\mathfrak{Y}' + B + \mathfrak{Y}''$ are again ordered domains. Thus if no such pair of elements A, B exists, there also exists no element E, and $\mathfrak{T}$ is a pantachie. This mixing process can be altered in infinitely many ways by letting any varying numbers of the x and y alternate; one can also extend this to three or more numerical sequences.

In order to give an application of this, we consider the behavior of a pantachie $\mathfrak{P}$ with respect to an element F *disjoint* [[*fremd*]] from it; so F has the relation of incomparability with at least one element of $\mathfrak{P}$. $\mathfrak{P}$ divides into three classes of elements, namely, the class $\mathfrak{P}_1$ of elements $< F$, the class $\mathfrak{P}_2$ of elements $\parallel F$, and the class $\mathfrak{P}_3$ of elements $> F$, in which $\mathfrak{P}_1$ and $\mathfrak{P}_3$ can indeed even be empty, while $\mathfrak{P}_2$ certainly contains elements; one easily sees that these three classes are *pieces* that follow each other in the ordering,

$$\mathfrak{P} = \mathfrak{P}_1 + \mathfrak{P}_2 + \mathfrak{P}_3.$$

It can now easily be shown through the mixing process that $\mathfrak{P}_2$ can be both the entire pantachie as well as an arbitrary segment, with or without its boundary points. For the sake of simplicity, let $F = 1$. Moreover, let $\mathfrak{D}$ be a pantachie of type $\Pi + 1 + \Pi$, which (possibly after applying a dilatation) contains the numerical sequence 1 and is of the form

$$\mathfrak{D} = \{X\} + 1 + \{Y\},$$

thus where the initial segment $\{X\}$ and the final segment $\{Y\}$ are similar. The elements resulting from the mixing $T = (X, Y)$ then form a pantachie in which each element is incomparable with 1 since, owing to $X < 1$, $Y > 1$,

$$T = (X, Y) \parallel (1, 1) = 1.\text{[1]})$$

Moreover, let $\mathfrak{D}$ be a pantachie of type $\Pi' + 1 + \Pi + 1 + \Pi$; it is clearly of the form

$$\mathfrak{D} = \{U\} + A + \{X\} + 1 + \{Y\},$$

where once again $\{X\}$ and $\{Y\}$ are similar sets. The elements

$$(U, U), \quad (A, A), \quad (X, Y),$$

that arise from mixing when U runs through all its elements and X and Y run through all matching element pairs form a pantachie $\mathfrak{P}$ in which all the elements (X, Y) are an *end segment* incomparable with 1. If one replaces the one element (A, A) by $(A, 1)$ in $\mathfrak{P}$, then the pantachie property is preserved, and this segment is then, in addition, an end segment whose (left) boundary

---

[1]) Also the elements $T' = (Y, X)$ form a pantachie, and both pantachies $\{T\}$ and $\{T'\}$ are element for element incomparable $(T \parallel T')$.

point is still incomparable with 1. — The corresponding result holds for initial segments.

If $\mathfrak{D}$ is a pantachie of type $\Pi' + 1 + \Pi + 1 + \Pi + 1 + \Pi''$,
$$\mathfrak{D} = \{U\} + A + \{X\} + 1 + \{Y\} + B + \{V\},$$
then the mixed elements
$$(U, U), \quad (A, A), \quad (X, Y), \quad (B, B), \quad (V, V)$$
yield a pantachie $\mathfrak{P}$ in which all the elements (X, Y) are a *middle segment* incomparable with 1. If one replaces (A, A) by $(A, 1)$ or (B, B) by $(1, B)$, or does both, then a left boundary point or a right boundary point, or both, are added to this middle segment.

To investigate further possibilities with regard to the subset $\mathfrak{P}_2$ may be pointless. On this occasion, we have constructed for a given element F a pantachie $\mathfrak{P}$ with appropriate characteristics; conversely, it is not always possible to construct an element F for a given pantachie $\mathfrak{P}$ so that the subset $\mathfrak{P}_2 \| F$ becomes, for instance, the entire pantachie or an initial segment or an end segment, at least not when $\mathfrak{P}$ contains *scales* [[*Skalen*]] (see §6). Only if $\mathfrak{P}_2$ is an arbitrary middle segment *with* boundary points, thus
$$\mathfrak{P}_2 = A + \mathfrak{P}_A^B + B,$$
can one always accomplish it, for example, with
$$F = (a_1 \ b_2 \ a_3 \ b_4 \ a_5 \ b_6 \ \ldots).$$

These considerations show that a pantachie piece can completely change its character through separation. $\mathfrak{P}$ can be a pantachie and $\mathfrak{P}_P$ merely a pantachie piece or a disconnected subset of a pantachie; $\mathfrak{B}$ can be an end piece and $\mathfrak{B}_P$ a middle piece, etc. If $\{X\} + A + \{Y\} + B + \{Z\}$ is a pantachie, then
$$\{(X, X)\} + (A, A) + \{(Y, Y)\} + (B, B) + \{(Z, Z)\}$$
is again one, but
$$\{(X, X)\} + (A, B) + \{(Z, Z)\}$$
is also a pantachie; in the latter case, the boundary segments $\{(X, X)\}$, $\{(Z, Z)\}$ enclose a unique element, in the former case, they enclose a middle segment with two boundary points. The requirement of monotonicity would be an appropriate means of limiting these possibilities and of causing the ordered domains $\mathfrak{B}_P$ to be extendible to no other place than the ordered domain $\mathfrak{B}$ containing them.

The details sketched are also of influence on the nature of the *boundary between convergence and divergence* within a pantachie. If we decompose a pantachie $\mathfrak{P}$ into the subset of convergent sequences and the subset of divergent sequences, then $\mathfrak{P} = \mathfrak{P}_c + \mathfrak{P}_d$, i.e., $\mathfrak{P}_c$ is an initial piece and $\mathfrak{P}_d$ is an end piece. However, it can be that $\mathfrak{P}_c = 0$; thus the pantachie could consist of just divergent sequences, as the example of the pantachie consisting of the mixed elements (X, Y), where $X < 1, Y > 1$, shows; on the other hand, $\mathfrak{P}_d$ cannot be empty since each convergent sequence is < 1 and a domain that consists of only convergent sequences is always extendible by the element 1.[1]) Even if $\mathfrak{P}_c \neq 0$, the question is then what kind of cut is $\mathfrak{P}_c + \mathfrak{P}_d$; a priori, there is the possibility that $\mathfrak{P}_c$ could have a last element or $\mathfrak{P}_d$ could have a first element (both occurring at the same time is excluded by the everywhere denseness of $\mathfrak{P}$), or thirdly that $\mathfrak{P}_c$ could have no last element and $\mathfrak{P}_d$ could have no first element; in the last case, the character of the gap would give occasion for further classification. By the mixing procedure, one can easily show that each of these cases really occurs and that each gap-type appearing in a pantachie can be represented by $\mathfrak{P}_c + \mathfrak{P}_d$. For from a pantachie that contains the convergent element C and the divergent element D, one could form a pantachie

$$\{X\} + C + \{Y\} + D + \{Z\}$$

of type $\Pi + 1 + \Pi' + 1 + \Pi$, where thus $\{X\}$ and $\{Z\}$ are similar ordered domains; all the X are convergent and all the Z are divergent. Then if $\Pi = \Pi_1 + \Pi_2$ is any decomposition of Π and if $\{X\} = \{X_1\} + \{X_2\}$ and $\{Z\} = \{Z_1\} + \{Z_2\}$ are the corresponding decompositions, then the mixed elements

$$(X_1, X_1), \quad (X_2, Z_2)$$

yield a pantachie in which the convergent elements (X_1, X_1) have type Π_1 and the divergent elements (X_2, Z_2) have type Π_2. It is noteworthy that *inside a pantachie* there can very well exist a last convergent element or a first divergent element and, moreover, that the set of convergent elements can be cofinal with ω or the set of divergent ones can be coinitial with ω^*, whereas these cases are excluded in the domain of all numerical sequences by the theorems in §2.

§6
The Pantachie Problem

The investigation in §2 showed the membership of pantachie types in the class of H-types. The next problem would be to more precisely specify their place within this class. However, this problem poses incomparably greater difficulties, and for the time being, it is as unlikely to find a full solution as is the continuum problem with which it is intimately related. A pantachie certainly contains subsets of type Ω and of type Ω^* since otherwise it would have to be cofinal with ω and coinitial with ω^*. The most likely conjecture is that it would not contain well-ordered subsets (or their inverses) of higher than the second infinite cardinality. If this conjecture — which is the analogue of CANTOR's Continuum Hypothesis in the domain of order

---

[1]) According to A. PRINGSHEIM (Math. Ann. **35**, p. 356), $C < \varepsilon$, where ε is an arbitrary positive number, is the only necessary convergence condition, i.e., for each $D < \varepsilon$, it can still be that $C \parallel D$. For monotonic (never increasing) numerical sequences, $C < \varepsilon H$, where $H = (1, \frac{1}{2}, \frac{1}{3}, \ldots, \frac{1}{n}, \ldots)$, is in the same sense the only necessary convergence condition.

types — is true, then each pantachie is cofinal with Ω and coinitial with Ω^* and each of its elements is an $\Omega\Omega^*$-limit; thus it belongs (in the wider sense of §4 of the previous article, thus apart from homogeneity, which, by the way, can be enforced according to §4 of the present treatise) to group IV of the unbounded everywhere dense types. Since it has no $\omega\omega^*$-gaps, it must belong to one of the two species IV 15, 16 and in any case, it must contain $\omega\Omega^*$-gaps and $\Omega\omega^*$-gaps. If it also contains $\Omega\Omega^*$-gaps, then it must be assigned to species IV 16; if it contains none, then to the species IV 15, but in this last case, it would have to be (according to §3, III) that $2^{\aleph_0} = 2^{\aleph_1} > \aleph_1$. Finally, if the continuum is of the second infinite cardinality, then the pantachie question is completely solved: all pantachies have one and the same homogeneous type H of species IV 16.

These results are admittedly in hypothetical form, yet for all that, they may well represent (especially the last one) a clarification compared to the present obscurity of the pantachie question. For the time being, there seems to be essentially no prospect of getting beyond it. Therefore we have to confine ourselves to compiling some of the still unanswered questions.

(α) *Is there a pantachie without $\Omega\Omega^*$-gaps?*

If yes, then the continuum is of cardinality $2^{\aleph_1}(>\aleph_1)$.

(β) *Does there exist a non-homogeneous pantachie type?* or which is the same thing, *are there two distinct pantachie types?*

If yes, then the continuum is not of the second infinite cardinality.

(γ) *What are the sequences of largest cardinality that occur in a pantachie?*

That is, we allow the possibility that a pantachie could contain well-ordered subsets of types $\omega, \omega_1, \omega_2, \ldots$ (and well-ordered subsets of types $\omega^*, \omega_1^*, \omega_2^*, \ldots$ — their inverses), and we ask if among them there is a maximum ω_μ (the cardinality of the continuum is then $\geqq \aleph_\mu$) and what is this maximum.

(δ) *Are all pantachies cofinal with the same ω_ν?*

In this case, all pantachies and pantachie segments are cofinal with ω_ν and coinitial with ω_ν^*, and each element in them is an $\omega_\nu\omega_\nu^*$-limit. This ω_ν need not agree with the maximum ω_μ indicated in (γ); the sequences differing from ω_ν, may they be $< \omega_\nu$ (like the fundamental sequences ω) or $> \omega_\nu$, then represent the lefthand boundary of *gaps*. As long as question (δ) has not been answered, we can only conclude: if there exists a pantachie that is cofinal with ω_ν, then there also exists a pantachie that is coinitial with ω_ν^* or both at the same time (coextensive with $\omega_\nu^* + \omega_\nu$), and there are pantachies with ω_ν-limits, with ω_ν^*-limits, and with $\omega_\nu\omega_\nu^*$-limits. In particular, the question would be with which ω_ν are the *homogeneous* pantachies constructed in §4 cofinal.

(ε) *Is there an ω_ν-sequence whose elements finally surpass any arbitrary numerical sequence?*

In closing, about this question, which we are going to call

the Scale Problem,

there are still a few things that need to be noted.

Let us consider any ω_ν-sequence of elements
$$\mathfrak{A} = \{A^\alpha\} = A^0, A^1, A^2, \ldots, A^\omega, \ldots, A^\alpha, \ldots,$$
where thus without exception $A^\alpha < A^\beta$ for $\alpha < \beta$. To abbreviate a bit, we are going to call such sequences *immanent* if there exists an element X that *finally surpasses all the A^α* ($X > A^\alpha$ for each α, or more concisely $X > \mathfrak{A}$); if there is no such element, call the sequence $\mathfrak{A}$ *transcendent*. The corresponding would hold for ω_ν^*-sequences, where the existence of an element $X < \mathfrak{A}$ enters into consideration. If $\mathfrak{A}$ is a transcendent ω_ν-sequence, then there is a pantachie that is cofinal with $\mathfrak{A}$ and consequently with ω_ν; conversely, if there is a pantachie that is cofinal with $\mathfrak{A}$, then $\mathfrak{A}$ is a transcendent ω_ν-sequence. Fundamental sequences are therefore always immanent; but there certainly also exist immanent Ω-sequences and Ω^*-sequences (those that lie inside an initial segment, respectively, final segment of a pantachie), and there certainly exist immanent ω_ν-sequences and ω_ν^*-sequences if there are transcendent ones.

We are going to further divide the transcendent sequences into *scales and quasiscales*. If a transcendent ω_ν-sequence $\mathfrak{A}$ has the property that it eventually surpasses *any arbitrary element*[1] X, that is, if there exists for any arbitrary X an index α so that $X < A^\alpha$ (and of course also $X < A^{\alpha+1}$, $X < A^{\alpha+2}$, ...), then $\mathfrak{A}$ is said to be an ω_ν-*scale*; if this is not the case, then $\mathfrak{A}$ is said to be an ω_ν-*quasiscale*. The corresponding would hold for ω_ν^*-sequences.

The necessity for this distinction is clear from the circumstance that one can (§5) construct a pantachie $\mathfrak{P}$ that is incomparable throughout with a given element, e.g., 1. Thus if $\mathfrak{P}$ is cofinal with $\mathfrak{A}$, then of course $\mathfrak{A}$ is a transcendent sequence, but it is not a scale since for the element 1 no $A^\alpha > 1$ is available, rather it is always the case that $A^\alpha \| 1$. Hence it follows from our mixing process that if there are transcendent ω_ν-sequences at all, then there certainly exist among them ω_ν-quasiscales, whereas the existence of ω_ν-scales is definitely not evident from this.[2]

---

[1]) In this case, one could also call $\mathfrak{A}$ *absolutely* transcendent and designate a quasiscale as *relatively* transcendent since the latter finally surpasses only the elements of a pantachie in which it is embedded. With respect to an ω_ν-quasiscale, all elements fall into two classes $\{X\}$ and $\{Y\}$; from a certain index α on, for the one it is the case that $X < A^\alpha$ and for the other $Y \| A^\alpha$.

[2]) In regard to §5, one could still make the following classification of transcendent sequences: if, besides $\mathfrak{A}$, each sequence $\mathfrak{A}_P$ arising from it through separation is also transcendent, then call $\mathfrak{A}$ *essentially* transcendent, otherwise call it inessentially transcendent. For example, an ω_ν-sequence formed from monotonically increasing numerical sequences

The possible existence of scales is now of the greatest importance for the pantachie question in that it permits a positive answer to question (δ), i.e., the following theorem holds:

If there exists a single ω_ν-scale, then all scales and quasiscales are ω_ν-sequences or ω_ν^*-sequences; each pantachie is then coextensive with $\omega_\nu^* + \omega_\nu$ and contains $\omega_\nu \omega_\nu^*$-limits exclusively.

Indeed, let $\mathfrak{A}$ be an ω_ν-scale and let $\mathfrak{P} = \{X\}$ be any pantachie. To each element X assign as $\alpha = \varphi(X)$ the *first* index α for which $X < A^\alpha$. It follows immediately from this that $\varphi(X) \leq \varphi(Y)$ for $X < Y$. In the sequence of transfinite numbers of type ω_ν

$$0\ 1\ 2\ \ldots\ \omega\ \ldots\ \beta\ \ldots,$$

there cannot be any number β that would not be surpassed by some $\varphi(X)$; were it the case that $\beta \geq \varphi(X)$ for each X, then $X < A^\beta$ would be true for each X, and $\mathfrak{P}$ would not be a pantachie, but rather a mere ordered domain that would be extendible to $(\mathfrak{P}, A^\beta)$. So $\varphi(X)$ runs through a sequence of values

$$\alpha_0\ \alpha_1\ \alpha_2\ \ldots\ \alpha_\omega\ \alpha_\beta\ \ldots$$

that lies in no initial segment [[Abschnitt]] of ω_ν, but rather itself has type ω_ν. If X^β denotes one of the elements for which $\varphi(X^\beta) = \alpha_\beta$, then $\mathfrak{P}$ contains a subset

$$X^0\ X^1\ X^2\ \ldots\ X^\omega\ \ldots\ X^\beta\ \ldots$$

of type ω_ν with which it is *cofinal* since for each element X of $\mathfrak{P}$ there exists an $\alpha_\beta > \varphi(X)$, and so $X^\beta > X$. So each pantachie is cofinal with ω_ν, i.e, case (δ) occurs along with the consequences given there.

The scale question has an intimate connection with the continuum problem, which the following theorem reveals:

If the continuum is of the second infinite cardinality, then there surely exists an Ω-scale.

Considering (C) of §2, the proof is extremely simple. Imagine that we have arranged all our numerical sequences in a set of type Ω

(1) $$X^0\ X^1\ X^2\ \ldots\ X^\omega\ \ldots.$$

If α is a number of the first two number classes, then the finite or countable set $X^0\ X^1\ \ldots\ X^\alpha$ is certainly finally surpassed by elements of the sequence (1). Let X^β be the first of these elements, where thus $\beta = \psi(\alpha) > \alpha$ is uniquely determined by α. We then form:

$$\alpha_1 = \psi(0), \quad \alpha_2 = \psi(\alpha_1), \quad \alpha_3 = \psi(\alpha_2), \quad \ldots,$$

A^α is always essentially transcendent. Since the examples of quasiscales known to me are only inessential transcendent sequences, it seems not excluded that both classifications coincide: thus essentially transcendent sequences are scales and inessentially transcendent sequences are quasiscales; however, a proof for this is yet to be furnished.

and if α_ω is the limit of the numbers $\alpha_1\ \alpha_2\ \alpha_3\ \ldots$, then
$$\alpha_{\omega+1} = \psi(\alpha_\omega), \quad \alpha_{\omega+2} = \psi(\alpha_{\omega+1}), \quad \ldots,$$
and so on in the usual way. Then
$$X^0\ X^{\alpha_1}\ X^{\alpha_2}\ \ldots\ X^{\alpha_\omega}\ \ldots$$
is an Ω-scale since for any index α there exists an $\alpha_\beta > \alpha$, and then according to the definition, $X^{\alpha_{\beta+1}} = X^{\psi(\alpha_\beta)}$ is $> X^\alpha$.

Certainly, proving[1]) the existence of an Ω-scale independently of the Continuum Hypothesis would thus be substantial progress as it would establish the first close connection between the continuum and the second number class; meanwhile, it seems that the difficulty of the matter is in keeping with its significance. Obviously, for the construction of a scale one can restrict oneself to *integer* elements
$$X = (x_1\ x_2\ x_3\ \ldots) \qquad (x_n = 1, 2, 3, \ldots).$$
According to Theorem (C) of §2, the construction of an Ω-sequence of such elements is not difficult; for instance, define $A^{\alpha+1} = A^\alpha + 1$, while for a limit ordinal λ, which might be the limit of $\alpha\ \beta\ \gamma\ \ldots$, A^λ can be constructed as prescribed there so that it finally surpasses $A^\alpha\ A^\beta\ A^\gamma\ \ldots$ (and thereby also all A^ξ with $\xi < \lambda$).[2]) So one would, say, obtain the following beginning of an Ω-sequence (we write $A(\alpha)$ instead of A^α):

$$A(1) = (1\ 1\ 1\ 1\ \ldots)$$
$$A(2) = (2\ 2\ 2\ 2\ \ldots)$$
$$A(3) = (3\ 3\ 3\ 3\ \ldots)$$
$$\cdot\ \cdot\ \cdot\ \cdot\ \cdot\ \cdot\ \cdot$$
$$A(\omega) = (1\ 2\ 3\ 4\ \ldots)$$
$$A(\omega+1) = (2\ 3\ 4\ 5\ \ldots)$$
$$A(\omega+2) = (3\ 4\ 5\ 6\ \ldots)$$
$$\cdot\ \cdot\ \cdot\ \cdot\ \cdot\ \cdot\ \cdot$$

---

[1]) E. BOREL declares it an *axiom* (Théorie des Fonctions p. 117). We of course cannot be content with that. That there it is a question of the infinitary ordering of increasing functions and here it is a question of the final ordering of numerical sequences is without essential significance. However, it may be explicitly emphasized that in the case of the (infinitary or final) ordering of *arbitrary functions* $f(x)$ there surely does not exist an Ω-scale since we have proved (p. 114) that for each set of functions of cardinality $\leqq \aleph$ there exists another function that is incomparable with all functions in the set.

[2]) G. H. HARDY (A theorem concerning the infinite cardinal numbers, Quarterly Journ. **35** (1903), 87–94) has proposed a similar method, but without regard to rank ordering, in order to choose from the continuum a subset with $\aleph_1$ elements. E. W. HOBSON (On the general theory of transfinite numbers and order types, Proc. Lond. Math. Soc. (2) **3** (1905), 170–188) has wrongly objected to this method. It is clear that the requirement that *for each n* it should be the case that $a_n^\alpha < a_n^\beta$ ($\alpha < \beta$) is unrealizable, but it is also unnecessary for ensuring the rank relation $A^\alpha < A^\beta$.

$$A(\omega\, 2) = (2\ 4\ 6\ 8\ \ldots)$$

$$A(\omega\, 3) = (3\ 6\ 9\ 12\ \ldots)$$

$$A(\omega\, 4) = (4\ 8\ 12\ 16\ \ldots)$$

$$A(\omega^2) = (1\ 4\ 9\ 16\ \ldots)$$

$$A(\omega^3) = (1\ 8\ 27\ 64\ \ldots)$$

$$A(\omega^\omega) = (1\ 2^2\ 3^3\ 4^4\ \ldots)$$

$$A(\omega^{\omega^\omega}) = (1\ 2^{2^2}\ 3^{3^3}\ 4^{4^4}\ \ldots)$$

$$A(\varepsilon_0) = (1\ 2^2\ 3^{3^3}\ 4^{4^{4^4}}\ \ldots)[1)$$

The difficulty exists solely in this, to prove that with this or another construction one obtains a *transcendent* Ω-sequence and a *scale* in particular. It can immediately be seen that the mere rule $A^{\alpha+1} = A^\alpha + 1$ does not suffice for this, but that on the contrary, an immanent Ω-sequence can also satisfy this condition. For instance, let B be any element so that $B > n = (n, n, n, \ldots)$. Then (§4) the elements

$$B_1 = B - 1, \quad B_2 = B - 2, \quad B_3 = B - 3, \quad \ldots$$

are definable, and together with B, they form an ω^*-sequence $\mathfrak{B}$; for example, let

$$B\ = (1\ 2\ 3\ 4\ 5\ \ldots),$$
$$B_1 = (1\ 1\ 2\ 3\ 4\ \ldots),$$
$$B_2 = (1\ 1\ 1\ 2\ 3\ \ldots),$$

If $\mathfrak{A}$ is an at most countable set $< \mathfrak{B}$, then there exist elements $\mathfrak{A} < X < \mathfrak{B}$ and then also $X + 1 < \mathfrak{B}$; should X not yet be an integer sequence, replace it by an integer sequence whose components lie between x_n and $x_n + 1$. Consequently, one can construct an Ω-sequence of integer elements $A^\alpha < \mathfrak{B}$, thus an immanent sequence, so that $A^{\alpha+1} = A^\alpha + 1$. One sees immediately that rules such as

$$A^{\alpha+1} = A^\alpha \cdot M\ (M > 1), \quad A^{\alpha+1} = (A^\alpha)^2,$$

and the like would not have any other outcome. Thus in general for the construction of an Ω-scale, there must also appear, besides the first construction

---

[1]) By ε_0 we understand the smallest "ε-number"; cf. G. CANTOR, Beiträge zur Begründung der transfiniten Mengenlehre, Math. Ann. **59** (1897), 207–246; §20.

principle that relates $A^{\alpha+1}$ to A^α, necessarily yet a second construction principle for the A^λ with limit number index that guarantees sufficiently swift growth of A; and it is quite doubtful whether the simple "diagonal procedure," which up to now has been given (by DU BOIS-REYMOND, BOREL, HARDY and others) for surpassing an ω-sequence, is enough for this task.

The following remarks may also be appropriate to illustrate the difficulty of the problem. We are going to say that an infinite (countable or not countable) set of numerical sequences $A, B, C, \ldots$ *remains in the finite* if there exists a numerical sequence X and a fixed index n_0 after which

$$x_n > a_n, b_n, c_n, \ldots \quad \text{(for } n \geqq n_0\text{)}.$$

One can also express this by saying that the numerical sequence X, which obviously finally surpasses all $A, B, C, \ldots$, *surpasses them uniformly*. If an Ω-sequence $\{A^\alpha\}$ remains in the finite, then it is certainly immanent since it is uniformly surpassed by a specific X. On the other hand, an immanent Ω-sequence $\{A^\alpha\}$, of course, does not have to remain in the finite; however it certainly contains a subset $\{B^\beta\}$ of type Ω that remains in the finite. For if $X > \{A^\alpha\}$ and if n_α is the least index after which $x_n > a_n^\alpha$ without exception, then among the values $1, 2, 3, \ldots$ for n_α at least one is represented $\aleph_1$ times; let us call it n_0; the associated A^α (for which $n_\alpha = n_0$) form an uncountable set that is uniformly surpassed by X. — According to this, to construct a transcendent Ω-sequence $\{A^\alpha\}$, one must prevent the occurrence of an uncountable subset $\{B^\beta\}$ that remains in the finite; and already it would seem reasonable to see whether the occurrence of a *countable* subset $B^0\ B^1\ B^2\ \ldots$ that remains in the finite can be prevented. But this is impossible. For an argument analogous to the one above shows that among the values for a_1^α of the first place there is at least one, let us say b_1, that must occur $\aleph_1$ times. Let U be the first element starting with b_1; among the remaining $\aleph_1$ there must again be $\aleph_1$ elements that have the second place b_2 in common, and let V be the first among these that begins with the elements $b_1\ b_2$. Continuing like this, we see the necessary existence of an ω-sequence of numerical sequences of the form

$$U = (b_1\ u_2\ u_3\ u_4\ \ldots),$$
$$V = (b_1\ b_2\ v_3 v v_4\ \ldots),$$
$$W = (b_1\ b_2\ b_3\ w_4\ \ldots),$$
$$\cdot\ \cdot\ \cdot\ \cdot\ \cdot\ \cdot\ \cdot$$

and thus of a sequence that remains in the finite and that is uniformly surpassed by a numerical sequence, such as

$$(b_1, u_2 + b_2, u_3 + v_3 + b_3, \ldots).$$

We would like to conclude with the following reflection. The difficulties of the pantachie problem, as the analogue of the continuum problem, can be stated concisely: one is carried forth beyond the countable, but one does not know how far since the starting point still lies in the countable and no immediate relation with the next higher level, the second infinite cardinality, reveals itself. We have seen how one can gain not only (like G. Cantor) new cardinal numbers from covering sets but also new order types, be it by rank ordering according to first differences or by the final rank ordering (the power concept and the pantachie concept); but as long as the sets, the base and the exponent, used for the cover adhere to the countable, the result, be it in the positive or in the contradictory negative sense, also seems not to be freed from this category and to permit no sharper specification of the uncountable. On the other hand, one of course obtains a relation with the second number class if one has from the beginning assumed such a relation in the elements under consideration. If we form, for example, a pantachie $\mathfrak{P}(\mu, \omega)$ whose base μ is a homogeneous type of group IV^1) (with $\Omega\Omega^*$-limits), then our questions (α)–(ε) can be partially answered: in this case there are Ω-scales, each pantachie is coextensive with $\Omega^* + \Omega$ and contains only $\Omega\Omega^*$-limits, etc.; the question of whether sequences of higher than the second infinite cardinality are present even now remains undecided. But one owes the apparently more transparent structure of these pantachies to the circumstance that the base already contained sequences of the second infinite cardinality and thus embodied a definite outward manifestation of the uncountable.

[Declared ready to print 5/5/1907.]

---

[1]) The notation according to §3, towards the end. For instance, one could choose $\mu = H$.

Introduction to "About Dense Order Types"

Felix Hausdorff spoke on dense order types at a meeting of the **DMV** in Dresden on September 18, 1907.[1] *About Dense Order Types* [[*Über dichte Ordnungstypen*]] is the published version of that talk. In it, he announces results whose "detailed explanation" are to be given elsewhere. In fact, the results were restated and proved in the monumental *Mathematische Annalen* article [H 1908].

About Dense Order Types concentrates on the definitions and theorems that allow Hausdorff to describe his classification scheme for unbounded dense sets and to state his major existence theorem for "complete character sets," which is proved in [H 1908, 474-484].[2] There are no proofs in [H 1907b], but as a preview of things to come, it is a highly readable and undemanding introduction to the classification machinery of [H 1908].

This brief announcement is also of historic significance. For the first time, Hausdorff uses the concept of cofinality, which he introduced in [H 1906b, 124], to identify the important class of *regular* initial ordinals. He also states the pivotal result that every ordered set is cofinal with a unique regular initial ordinal; this lies behind his classification of unbounded dense sets by the limit nature of their elements and gaps. In stating his basic existence theorem for complete character sets, he informs us apologetically that he needs to restrict the regular limit numbers ω_π, ω_ρ that bound his character tableau by assuming that π and ρ are not limit ordinals. He then observes:

> [W]hether there are regular initial numbers with limit indices is very problematic; in any case, the smallest among them is already of such an exorbitant magnitude that all sets considered up until now and all sets still to be taken into consideration are probably exceeded. [H 1907b, 546]

Of course, Hausdorff is talking about the *weakly inaccessible cardinals* that we now know (assuming ZFC is consistent) cannot be proved to exist in ZFC and whose non-existence can be consistently added to ZFC.[3] He is the discoverer of the first class of "large" cardinals.

Notes

1. The handwritten notes for this talk are in Hausdorff's *Nachlaß* [Kapsel 26a; Fasz. 81]. They are said to be almost identical to the published paper [Pu 1995, 97].

2. A limited version of this classification scheme for unbounded dense sets with sequences of cardinality $\leq \aleph_1$ appeared at the end of *Homogeneous Types of the Cardinality of the Continuum* ([H 1907a, 99–101]).

   Non-trivial ordered sets without consecutive elements were called *everywhere dense* [[*überalldicht*]] by Cantor. This was the term also used by Hausdorff in his previous articles; in [H 1907b], he shortens it to *dense* [[*dicht*]]. Also in [H 1907b] the parenthetical accompaniment of "in Dedekind's sense" is dropped from the term *continuous* [[*stetig*]].

3. According to [SiT 1930, 292], C. Kuratowksi proposed the term "inaccessible" for regular $\aleph_\alpha$ where α is a limit ordinal. The qualifier "weakly" is a later addition. The "inaccessibility" defined in Sierpiński and Tarski is now called *strong inaccessibility*. The two kinds of inaccessibility are equivalent under GCH.

About Dense Order Types

By F. Hausdorff in Leipzig

In my last works on order types[1], I classified the dense types having sequences of the first and second infinite cardinality, and I constructed actual representatives for a part of the classes enumerated, namely, for the so-called homogeneous types. In the following, I extend this classification to all dense types having sequences of arbitrary cardinality, and I sketch a general existence proof for the types that are a priori possible; the detailed explanation will be given in another place.

The ordered set M is called *cofinal* with its subset A if no element of M follows A (i.e., follows all elements of A); M is called coinitial with A if no element of M precedes A. For example, the set of rational numbers is cofinal with the set of positive integers and coinitial with the set of negative integers, or expressed in types, η is cofinal with ω and coinitial with ω^*.

A decomposition $M = A + B$ presents four possibilities that are designated as *jump, cut, or gap* [[*Sprung, Schnitt, Lücke*]]:

jump: A has a last element, B has a first element.

cut: $\begin{cases} A \text{ has a last element,} & B \text{ has no first element.} \\ A \text{ has no last element,} & B \text{ has a first element.} \end{cases}$

gap: A has no last element, B has no first element.

A set with no jumps, thus with no pair of consecutive elements, is called *everywhere dense* [[*überalldicht*]] (G. Cantor) or, briefly, *dense* [[*dicht*]]; a dense set without gaps is called *continuous* [[*stetig*]]. A subset A of M is called *dense in M* (dense relative to M) if between any two elements of M there lies at least one element of A; obviously, then M and A are themselves dense. For example, the set of rational numbers is dense in the set of real numbers.

If by *middle segment* [[*Mittelstrecke*]] of M we understand any subset included between two elements of M, then we can say: A is dense in M when each middle segment of M contains at least one element of A.

Theorem I. *A dense set M can be split into two subsets each of which is dense in M.*

---

[1] Untersuchungen über Ordnungstypen, Berichte der K. Sächs. Gesellschaft d. Wissenschaften **58** (1906), 106–169 and **59** (1907), 84–159

542 This important theorem's proof, omitted here, depends on the possibility of well-ordering the set of all subsets of M. For example, the set of real (or the algebraic) numbers can be split into the rational and irrational numbers.

As usual, we designate the cardinality of a well-ordered set by $\aleph_\alpha$ where the *index* α is an ordinal number, namely, the type of the set of all smaller alephs. The smallest ordinal number of cardinality $\aleph_\alpha$ is named ω_α, and it is called an *initial number* [[*Anfangszahl*]].

Theorem II. *Each set of cardinality $\aleph_\alpha$ is cofinal with an ordinal number $\beta \leq \omega_\alpha$.*

Theorem III. *Each set of cardinality $\aleph_\alpha$ contains at least one subset of type $\beta + \beta^*$ ($\beta \leq \omega_\alpha$) between whose symmetric halves lies no further element or only a single element.*

Both theorems follow from well-ordering the elements and element pairs of M.

According to II, an ordinal number that is not an initial number is certainly cofinal with a smaller ordinal number. An initial number can be cofinal with smaller numbers as well, e.g., $\omega_\omega = \omega_0 + \omega_1 + \omega_2 + \cdots$ with $\omega_0 = \omega$; such initial numbers may be called *singular* initial numbers, while we call those initial numbers that are not cofinal with a smaller number *regular* initial numbers. Thus ω_0 and each initial number $\omega_{\alpha+1}$ with predecessor (ω_α) are surely regular. Sets of type an initial number or its inverse may be called *sequences*, and, correspondingly, they can be divided into singular and regular sequences. Finally, it then follows:

Theorem IV. *Each set without a last element is cofinal with one and only one regular initial number.*

Now let M be a dense set without a first or last element. If $M = A + m + B$ is the decomposition produced by an element m, then A has no last element, and by IV, it is cofinal with a definite regular initial number ω_α. Likewise, B is coinitial with the inverse ω_β^* of such an initial number. Then we call m an $\omega_\alpha \omega_\beta^*$-element or, briefly, a $c_{\alpha\beta}$-element; $c_{\alpha\beta}$ is called the *character* of the element. Moreover, if $M = C + D$ is a gap, again C is cofinal with a definite regular ω_γ and D is coinitial with ω_δ^*; we then speak of an $\omega_\gamma \omega_\delta$-gap or of a $c_{\gamma\delta}$-gap and of the character $c_{\gamma\delta}$ of this gap. Now we divide the dense types into *species* according to the element characters and gap characters occurring in them, i.e., we form the sets

$$U = \{c_{\alpha\beta}\}, \quad V = \{c_{\gamma\delta}\},$$

543 and we classify two types with the same U, V to be in the same species, which we briefly denote as the species (U, V). As a result, $V = o$ signifies a species of *continuous* types.

For example, η belongs to the species (c_{00}, c_{00}) and the linear continuum λ belongs to the species (c_{00}, o). The infinitary and final *pantachies* considered by me (loc. cit.), with which I tried to salvage a failed speculation of P. Dubois-Reymond, probably belong to the species $(c_{11}, c_{01} c_{10} c_{11})$.

If each middle segment of M belongs to the same species as M itself, then call M *irreducible*; otherwise, call it reducible. So λ and η are irreducible, while the type $\lambda + \lambda$ is reducible.

In addition, we combine species into *genera* [[*Geschlechtern*]] by forming the union $W = \mathfrak{M}(U, V)$ of all characters and by classifying two sets with the same W to be of the same genus. Each genus contains a unique species of continuous types, namely, (W, o).

If we imagine that a new element $\overline{m}$ is placed in each gap of a dense set M of the species (U, V) and if we make the obvious arrangement with respect to the ordering, the new elements form a set $\overline{M}$ (in general not dense); the old and new elements together form a continuous set $[M]$ of species (W, o). Call $\overline{M}$ the *completion* [[*Ergänzung*]] of M and $[M]$ the *filling* [[*Ausfüllung*]] of M. Each $c_{\alpha\beta}$-element of M is simultaneously a $c_{\alpha\beta}$-element of $[M]$, and each $c_{\gamma\delta}$-gap of M corresponds to a $c_{\gamma\delta}$-element of $[M]$. In particular, if M is *everywhere discontinuous* [[*überall unstetig*]], i.e., each middle segment of M is discontinuous, then $\overline{M}$ is dense, everywhere discontinuous, and of species (V, U); then M is also the completion of $\overline{M}$. For example, the sets of rational and irrational numbers are completions of each other and the set of real numbers is their common filling. For each individual character, if $M_{\alpha\beta}$ denotes the subset of all $c_{\alpha\beta}$-elements of M and likewise $\overline{M}_{\gamma\delta}$ denotes the set of elements of $\overline{M}$ corresponding to the $c_{\gamma\delta}$-gaps of M, then it is necessary and sufficient for the irreducibility of M that all the sets $M_{\alpha\beta}$ and $\overline{M}_{\gamma\delta}$ be dense in $[M]$. If M is itself already continuous ($\overline{M} = o$), the condition says that each $M_{\alpha\beta}$ must be dense in M.

Now turning to the question of which genera and species actually exist, we notice immediately that the set W cannot be prescribed completely arbitrarily. First, from III comes the fact that each dense set must contain *symmetric* elements or gaps, i.e., $\omega_\sigma \omega_\sigma^*$-elements or $\omega_\sigma \omega_\sigma^*$-gaps. Thus at least one symmetric character $c_{\sigma\sigma}$ must be present in W. Furthermore, let ω_π be the first regular initial number with the property that no middle segment of M contains an ω_π-sequence, briefly, the first regular initial number not contained *within* M (so that ω_π can still be contained *in* M, namely, M can be cofinal with ω_π). Then also there is no regular initial number $\geq \omega_\pi$ within M; however, each regular initial number $\omega_\alpha < \omega_\pi$ is contained within M, and after any ω_α-sequence, there follows either an element or a gap, i.e., for each of these indices α at least one character $c_{\alpha\beta}$ must be present in W. The corresponding fact holds for the first regular sequence type ω_ϱ^* not present within M and for the regular sequences ω_β^* ($\beta < \varrho$). So if we form the tableau

$$\Omega: \begin{array}{cccc} c_{00} & c_{01} & \cdots & c_{0\beta} & \cdots \\ c_{10} & c_{11} & \cdots & c_{1\beta} & \cdots \\ \cdot & \cdot & & \cdot & \\ c_{\alpha 0} & c_{\alpha 1} & \cdots & c_{\alpha\beta} & \cdots \\ \cdot & \cdot & \cdot & \cdot & \cdot & \cdot & \cdot \end{array} \qquad \begin{array}{c} \alpha < \pi, \ \beta < \varrho \\ \\ \omega_\alpha, \omega_\beta, \omega_\pi, \omega_\varrho \ \text{regular} \end{array}$$

W must contain at least one element from each row and each column and from the main diagonal. We are then going to call W a *complete* character set.

This simple condition is necessary, but also — and here is the crux of the present consideration — sufficient for the existence of the species (W, o) and all species of the same genus, sufficient even with the strengthening that irreducible representatives of these species are required. That is, the following holds:

Theorem V. *For each complete character set W, there exists irreducible types of species (W, o) and of all species of the same genus.*

The importance of this theorem, which can very well be called the fundamental theorem of the theory of dense types, will become clear through an enumeration of the simplest cases. Should only countable sequences (of types ω and ω^*) occur within M, then $W = c_{00}$ is the only character set at hand, and there exists one genus with the two species (c_{00}, o) and (c_{00}, c_{00}), represented by the known types λ and η. However, by admitting sequences of the first and second infinite cardinality, there are already 9 genera with 210 species, and by admitting sequences through the third infinite cardinality, there are 302 genera and 243376 species. If one of the indices π, ϱ is transfinite, then the number of species is at least equal to the cardinality of the continuum.

The proof of Theorem V splits into two parts: (1) an existence proof for an irreducible continuous type of species (W, o); (2) an existence proof for the remaining species of the same genus.

The second part is easier. It depends on the concept of relative density and on Theorem I. Let M be an irreducible set of species (W, o), and let $M_{\alpha\beta}$ be the set of $c_{\alpha\beta}$-elements contained in M. In addition, let $W = \mathfrak{M}(U, V)$ be a prescribed composition of W out of two subsets that can contain common characters; emphasizing this, we put

$$U = (W_0, W_1), \quad V = (W_0, W_2), \quad W = (W_0, W_1, W_2),$$

in which one of the three sets W_0, W_1, W_2 can be empty, or even both W_1, W_2 can be empty. If we distinguish the characters in these three components by corresponding indices,

$$W_i = \{c_{\alpha_i \beta_i}\}, \quad i = 0, 1, 2,$$

then, according to I, we now split each set $M_{\alpha_0 \beta_0}$ into two, relative to it, dense subsets $P_{\alpha_0 \beta_0}$ and $Q_{\alpha_0 \beta_0}$ and thereby M itself into the two sets

$$P = \{P_{\alpha_0 \beta_0}, M_{\alpha_1 \beta_1}\}, \quad Q = \{Q_{\alpha_0 \beta_0}, M_{\alpha_2 \beta_2}\}.$$

Then P and Q are irreducible sets of species (U, V) and (V, U), and, by the way, they are completions of one another and have M as a filling.

I base the harder first part of the proof, namely, the construction of an irreducible type of species (W, o), on the same method with which I accomplished the extension of the product and power concepts to arbitrary, even transfinite sets of factors (loc. cit.): this method is the rank ordering of covering sets [[Belegungsmengen]] and combination sets [[Verbindungsmengen]] *by first differences*. Let A be a well-ordered set, whose elements we denote by 0 1 2 ... ξ ...; an ordered set M_ξ corresponds to each place ξ. We form the element combinations [[Elementverbindungen]]

$$x = (x_0\, x_1\, x_2\, \ldots\, x_\xi\, \ldots),$$

where each x_ξ runs through the set M_ξ, and we order them so that x and y have the rank relations [[Rangbeziehungen]] of their first differing elements, i.e.,

$$x \lesseqgtr y \quad \text{for} \quad x_\xi \lesseqgtr y_\xi, \quad x_\eta = y_\eta \quad (\eta < \xi).$$

The set $\{x\}$, so ordered, has the character of a product $\ldots M_\xi \ldots M_2\, M_1\, M_0$ (with this sequence of factors, not $M_0\, M_1\, M_2 \ldots M_\xi \ldots$), and, in the case of equal factors, it has the character of a power M^{A^*} (not M^A). From here, one reaches the general concepts of power and product (also encompassing the case of an arbitrary ordered set A) by stipulations that the current problem, however, does not require. Rather, here we are going to generalize the concept of element combinations from another point of view: A remains well-ordered, but the set M_ξ, through which x_ξ runs, is even allowed to depend on earlier elements x_η. Then if $W = \{c_{\alpha\beta}\}$ is a prescribed complete character set and $c_{\sigma\sigma}$ is its smallest symmetric character, we have to form element combinations

$$x = (x_0\, x_1\, x_2\, \ldots\, x_\xi\, \ldots), \quad \xi < \omega_\sigma$$

of type ω_σ in which the sets M_ξ are additive combinations of initial numbers and their inverses and, as said, are dependent not only on the place ξ but also on the previous elements; e.g., they can even be reduced to single elements; in general, a type like

$$\sum (\omega_\alpha + 1 + \omega_\beta^*)$$

plays a main role, where, for instance, the sum, taken in the appropriate sequence, extends over all the desired element characters $c_{\alpha\beta}$. Through appropriate rules, whose fairly complicated details cannot be gone into here, it can be gotten that an irreducible type of species (W, o) arises through rank ordering of the set $\{x\}$ by first differences. But I have not yet been able to free myself from a restriction: I must assume that the regular initial numbers $\omega_\pi, \omega_\varrho$ that outwardly bound the tableau Ω have predecessors $(\omega_{\pi-1}, \omega_{\varrho-1})$; thus π and ϱ are not to be limit numbers. In general, whether there are regular initial numbers with limit indices is very problematic; in any case, the smallest among them is already of such an exorbitant magnitude that all sets considered up until now and all sets still to be taken into consideration are probably exceeded.

The cardinality of our irreducible set of species (W, o) is $\aleph_{\pi-1}^{\aleph_\sigma}$ or $\aleph_{\varrho-1}^{\aleph_\sigma}$, depending on whether $\pi \geqq \varrho$ or $\varrho \geqq \pi$. This is the smallest cardinality that such a set can have, whereas the cardinality of types of the same genus can be lowered by the occurrence of $c_{\sigma\sigma}$-gaps.

For all species, the existence proof can still be strengthened by showing their representation by infinitely many distinct types. For instance, besides λ, the species (c_{00}, o) is represented by at least $\aleph_1$ types of the same cardinality that have almost all the properties of λ in common with the exception that only the linear continuum contains a countable subset that is dense in it.

Introduction to "The Fundamentals of a Theory of Ordered Sets"

The book-length second installment of Schoenflies's report to the **DMV**, *Die Entwickelung der Lehre von den Punktmannigfaltigkeiten* ([Sch 1908]), is a testament to the influential position that Hausdorff had achieved with his work on ordered sets. In the first paragraph of his forward, Schoenflies mentions his report's more detailed discussion of the abstract problems of set theory, "especially Hausdorff's newer works on linear order types." The thirty-two page second chapter on ordered sets starts with Hausdorff's generalization of well-ordered sets, the *gestufte* sets, introduced in [H 1901b]. And in the very first sentence of the next paragraph Schoenflies writes:

> We also owe the other advances in ordered sets that we possess to Hausdorff. [Sch 1908, 40]

Schoenflies does mention the work of others, e.g., the non-Archimedean continuum of Veronese and Du Bois-Reymond's infinitary pantachie and his ideal boundary between convergence and divergence, but his second chapter is dominated by the results of *Investigations into Order Types* [H 1906b; 1907a]. Even in discussing Veronese's continuum, Schoenflies relates it to Hausdorff's powers of order types; Du Bois-Reymond's infinitary pantachie and ideal boundary between convergence and divergence are viewed in light of Hausdorff's criticisms in [H 1907a, 105–159], and Hausdorff's own take on these concepts is featured prominently.

In the short span from 1901–1907, Hausdorff had attained recognition as a leader in the second generation of Cantorians. Yet with the turmoil of the 1904 Heidelberg International Congress and the continued frenzied attacks on Zermelo's 1904 choice-assisted proof of the Well-Ordering Theorem, Cantor and his followers, Hausdorff among them, could not but feel besieged.[1]

Hausdorff's own feelings of defensiveness are revealed in a letter that he wrote to Hilbert on July 15, 1907.[2] Hausdorff had met with Cantor some two weeks before in Kösen at a gathering of faculty from Leipzig, Jena, and Halle.[3] In this letter, which is really a tactful initiation of a negotiation between an author and an editor with whom the author has an established friendship, Hausdorff reports Cantor's urgings that he produce a short synopsis of *Investigations into Order Types* and that he submit it to the *Mathematische Annalen*; in Cantor's opinion, it was desirable that Hausdorff's work be given exposure to a wider readership than that of the Reports of Leipzig's *Königlich-Sächsischen Gesellschaft der Wissenschaften*, and Cantor assumed that a short, systematic account limited to the essentials would be more apt to achieve this.

Making sure that Hilbert understands Cantor's role in this, Hausdorff seeks to ascertain his opinion of such an undertaking:

> You, dear Geheimrat, will easily be able to judge by glancing at the works (Leipz. Berichte 1906 and 1907), which I had the honor of sending you as reprints, whether Cantor's suggestion, to which I am expressly authorized to refer, deserves consideration. Thus I take the liberty of asking whether you are inclined *in principle* to receive an article for the Annals in the range of 2-3 sheets [[Bogen]] entitled, say, "Theory of Order Types."[4]

Assuring Hilbert that he is only interested in a statement of principle and not expecting a blanket acceptance that would usurp his right to judge the article once submitted, Hausdorff explains:

> I would like only to spare myself the trouble in case, perhaps, the editors of the Annals should be disposed from the start to exclude the field of set theory, which is nowadays so often challenged. (And with such medieval weapons!)

Hausdorff closes with an appeal to Hilbert's loyalty to Cantor's ideas:

> In the hope that you, dear Geheimrat, still consider "Cantorism," which Poincaré declared dead, as somewhat alive, and that a work that adds something new to set theory with regard to contents is not denied your interest, I am faithfully yours ...

Hilbert's reply must have been very encouraging indeed because by November of 1907 he had accepted Hausdorff's memoir-length manuscript, entitled *The Fundamentals of a Theory of Ordered Sets* [[*Grundzüge einer Theorie der geordneten Mengen*]].[5] It appeared in 1908 and was an auspicious debut for Hausdorff in the pages of the *Mathematische Annalen*. This major work, complete with a table of contents and a short index listing the first appearances of "fixed terms," combines the tutorial aspects of a primer on the basics with the demands of an advanced research monograph.

Though as an introductory source for ordered sets [H 1908] was to be overshadowed by Hausdorff's classic 1914 text, *Grundzüge der Mengenlehre*, it still can be read as a lexicon for the subject. It also serves as an introduction to the tools previously created by Hausdorff for the representation of ordered sets (infinite sums, infinite products, and powers), and it is the only place that one can find his power-thinning technique, multiplication with variable factors. [H 1908] is widely credited for the first statement of the generalized continuum hypothesis. It also contains worthy results that are either absent from or only mentioned briefly in [H 1914a], such as: the decomposition of a dense set M into subsets that are dense in it (absent), type rings (absent), the existence of continuous, dense subsets corresponding to a given complete character set (mentioned), and sharp lower bounds on the cardinalities of irreducible, continuous $c_{\sigma\sigma}$-sets (absent). The article's opening is also particularly revealing about Hausdorff's views on the controversies embroiling set theorists. Up to now Hausdorff had maintained a public posture of aloofness. In [H 1906b, 107], he had declared his noncombatant status in the wars over foundations and the paradoxes. His imperturbability was remarked upon by Hessenberg in a September 20, 1907 postcard to the philosopher Leonard Nelson. Hessenberg, who had just heard Hausdorff speak to the **DMV** on what was to become [H 1907b], writes of Hausdorff:

> He is the only productive set theorist who is so little irritated by the paradoxes, ... [Pu 2002, 14]

As for the controversial Axiom of Choice, he still had not voiced an opinion; in [H 1907a, 117] he just started using well-orderings with a footnote citing Zermelo's 1904 proof of the Well-Ordering Theorem, but without further comment.

In his memorable introduction to [H 1908], particularly the second paragraph, Hausdorff abandons his air of aloofness and enters the fray. Expecting his own refusal to discuss the principles of set theory to give offense "in the places where presently a somewhat misplaced degree of ingenuity is squandered in these discussions," Hausdorff, speaking impersonally as "an observer, who in the face of skepticism, is not wanting in skepticism," offers a succinct categorization of the " 'finitistic' objections against set theory." There are three such categories: the first consists of those objections that "reveal a serious need for a, perhaps axiomatic, sharpening of the set concept," with this first group an understanding will be arrived at "sooner or later"; the second consists of those objections that "would affect the whole of mathematics together with set theory," and these one can "safely let rest"; the last category contains the objections that are "the simple absurdities of a scholasticism that clings onto words and letters," and accordingly, this category "deserves the sharpest and clearest disapproval." (Perhaps these anonymous scholastics were the wielders of the "medieval weapons" mentioned in Hausdorff's earlier cited letter to Hilbert.)

Hausdorff continues by stating the doctrinal underpinnings of his article. "The totality of 'all' ordinal numbers or cardinal numbers does not exist free from contradiction either as a set or as a subset of a set." (Hessenberg took essentiallly the same position in [Hes 1906, 550, 552].) In fact, early in his exposition, Hausdorff makes the working assumption that "a sufficiently large ordinal number Δ that exceeds in cardinality all the sets that we consider" has been chosen, and that W, the set of all ordinals $< \Delta$, and its subsets are the subject of his article [H 1908, 441]—an assumption that would warm the heart of any model theorist. He also "accepts Cantor's Well-Ordering Theorem in its formulation and proof by Herr Zermelo." He will make "use of the 'choice postulate,' even without mentioning it"—so looking for explicit vs. implicit uses of AC in [H 1908] is a fruitless pursuit. In a rebuff to nameless critics, Hausdorff "places no value upon the statement that a part of its [AC's] conclusions could also be derived independently of it." However, he does concede a point to those who abjure making transfinite sequences of successive, dependent choices; he will avoid doing that. Instead, he will opt for well-orderings or, equivalently, simultaneous, independent choices.[6]

Given the rhetorical power of this part of Hausdorff's introduction, it is easy to overlook the opening paragraph, in which Hausdorff describes the article at hand as the first attempt at "a sustained introduction to the still practically unknown field of simply ordered sets." In this field created by Cantor, Hausdorff says that "only the well-ordered sets and sets of reals have actually experienced a detailed treatment." Interestingly, these words echo a judgment that he expressed in the opening lines of [H 1901b], his first article on ordered sets:

> In the field of order types that was developed by G. Cantor, it is really only the special realm of ordinal numbers about which we are somewhat well-informed; extremely little is known about general types, the types of non-well-ordered sets. [H 1901b, 460]

But what of Hausdorff's own contributions in *Investigations into Order Types*, the ones soon to be highly praised in [Sch 1908]? Hausdorff sees this earlier work as pursuing a "special direction," namely the study of types having a particularly regular structure, with all segments similar (homogeneity) and with all well-ordered subsets (sequences) of cardinality $\leq \aleph_1$. He then states that in the present article these special types will "fall back into the role of occasional illustrative examples." The representation of "complex structures" [arbitrary types] through the application of certain operations of generation to "primitive atoms" will be at the fore.

Hausdorff continues with a brief outline of the article's results and methods. The primitive atoms are the regular initial numbers and their inverses. The regular initial numbers (and the corresponding regular alephs) were first defined in [H 1907b], using the relation of cofinality that first appeared in [H 1906b]. Under generalized addition, these initial numbers and their inverses yield what Hausdorff calls the "scattered sets," which he later identifies as those ordered sets that do not embed the rationals. Hausdorff then announces a remarkable result: arbitrary types are either scattered or the sum of scattered sets over a dense index set. (The usually modest Hausdorff can not help but label this result as "fundamental.") The role of dense sets in this representation makes knowledge of their structure imperative. Hausdorff ends this opening paragraph promising a classification of dense sets in terms of the "characters" of their elements and gaps; these characters turn out to be pairs of regular initial numbers and their inverses. The existence of dense sets with certain admissible character sets will be proved using the operations of multiplication and raising to powers, first studied in some generality in [H 1906b], and a new operation that Hausdorff calls *multiplication with variable factors*.

The body of the article itself is divided into three main sections: *Foundations*; *Sums, Products, and Powers of Order Types*; and *Dense Sets*. In [H 1901b], Hausdorff introduced some new terminology for segments of ordered sets (pieces) in a first attempt to add more expressiveness to the existing, but meager, set of terms introduced by Cantor; in [H 1906b, 123–125], he greatly increased the segment nomenclature. In *Foundations*, Hausdorff begins with a recapitulation of his previous vocabulary for segments, of the Cantorian operation of ordered addition, and of the important relation of cofinality that he first introduced in [H 1906b].

What was missing in the [H 1906b] introduction of cofinality was the concept of an initial number (ordinal) that is not cofinal with any smaller ordinal, a *regular initial number*.[7] Hausdorff shows that these ordinals are indeed abundant; he sketches a proof that each initial number $\omega_{\alpha+1}$ is regular (as claimed in [H 1907b]). He is immediately led to the question of the existence of regular ω_α where α is a limit ordinal, and he says it "must remain undecided here." However, he does engage in brief speculation on the properties of such a number. First, it must be contained in the set of initial numbers that are equal to their own index, those ω_ζ for which $\omega_\zeta = \zeta$. Naming these ζ-numbers after Hessenberg, Hausdorff notes that they can be indexed in increasing order by the ordinals, and he again concludes that a regular initial number with limit index must be contained among the fixed points of this enumeration of the ζ-numbers.[8] Naming these new fixed points η-numbers and noting that they also can be indexed in increasing order by the ordinals, he observes that the illusive regular initial number with limit index occurs among the fixed points of the enumeration of the η-numbers.[9] And he indicates how this analysis can be iterated into the transfinite, making the existence of such a regular initial number "appear at least problematic." Later ([H 1908, 478]), he uses these observations to conclude more strongly that "the existence of of regular initial numbers with limit index is altogether questionable; ... the smallest among them would have to be of an exorbitant cardinality, one that likely exceeds [the cardinalities of] all known sets."[10]

The regular initial numbers, whatever the provenance of their indices, are truly the "building blocks" for arbitrary types in [H 1908]. Using an external well-ordering, Hausdorff proves that an ordered set without a last element is cofinal with a unique regular initial number. This has immediate ramifications. For an unbounded (open) dense set M, the sets A and B in the decompostion $M = A + m + B$ induced by an element m have no last or first elements, respectively. Thus A is cofinal with a unique regular ω_α

and B^* is cofinal with a unique regular ω_β, which means B is coinitial with the unique ω_β^*. The element m is said to be an $\omega_\alpha \omega_\beta^*$-element or to have character $c_{\alpha\beta}$. A gap in M is a decomposition $M = C + D$ where C has no last element and D has no first element; reasoning as before, C and D^* are cofinal with unique regular ω_γ and ω_δ, respectively. The gap is then said to be an $\omega_\gamma \omega_\delta^*$-gap or to have character $c_{\gamma\delta}$. For an unbounded dense set, the existence of symmetric $\omega_\alpha \omega_\alpha^*$-elements or gaps for some regular ω_α follows from applying an external well-ordering to its intervals and using transfinite induction to define a descending, nested sequence of intervals that intersect in a unique point or have empty intersection. The least index of such a symmetric element or gap turns out (surprisingly) to play a significant role in the representation theory of dense sets. *Foundations* ends with a clever use of well-ordering to decompose a given dense set M into two complementary subsets that are dense in M, i.e., each open interval of the original set contains points of the two subsets.[11] Hausdorff shows how to extend this to a decomposition into countably many subsets dense in M. The decomposition into two subsets dense in M plays an important role in the Existence Theorem in §21.

The second main division of [H 1908], *Sums, Products, and Powers of Order Types*, presents the operations of generation that Hausdorff will need to represent ordered sets in terms of the "primitive atoms"—the regular initial numbers and their inverses. It begins with an easy generalization of Cantor's addition of ordered sets. For A an arbitrary ordered set and M_a a family of pairwise disjoint ordered sets indexed by the elements of A, one obtains the general or transfinite sum of the M_a over the "generator" [index set] A by inserting each M_a into A in place of a. When all the M_a are identical, say B, M is Cantor's product of ordered sets BA—the insertion of B into A.

The importance of general addition becomes clear through Hausdorff's introduction of the concept of a *type ring*. A type ring is a set of types closed under ordinary addition and under general addition when the generator and the summands belong to the given set of types. (So in particular a type ring is also closed under ordinary multiplication). For any set of types A, if the type β does not belong to some type ring containing A, Hausdorff says "β is independent of A," and if it belongs to every type ring containing A, "β is dependent on A." He uses [A] to denote the set of types dependent on A. Hausdorff notes that the non-empty intersection of any collection of type rings is again a type ring and hence that [A] is the smallest type ring containing A. Moreover, if no type of A is dependent on the remaining types in A, Hausdorff calls A a basis for the ring [A]. Our understanding of type rings of the form [A] is enhanced by the fact that any property of types that is preserved by finite and general addition and that is shared by all the types in A is inherited by the types in [A]. For example, being cofinal with a type in A is such a property; a type that is not cofinal with any type in A is independent of A [H 1908, 456].

Type rings are ubiquitous, for as Hausdorff remarks, "almost every naturally presented set of types is a type ring." For regular ω_α, Hausdorff makes spectacular use of the type ring [A], where A consists of the type 1 together with the regular $\omega_\xi < \omega_\alpha$ and their inverses ω_ξ^*. The types in [A] have a particularly nice characterization. Hausdorff calls an ordered set that does not contain a dense subset (does not embed the rationals) *scattered* [[*zerstreut*]]. The ring [A] consists of all the types of scattered sets of cardinality $< \aleph_\alpha$ and A is a basis for this ring.

For elements a and b of an arbitrary ordered set M of cardinality $< \aleph_\alpha$, Hausdorff defines a and b to be *coherent* if the type of the interval $[a, b]$ is in [A]. Although he does not use the language of equivalence relations and equivalence classes, he essentially shows that *coherence* is such a relation and that the elements with which a given a is coherent (the equivalence class of a) form a piece of M whose type belongs to [A]; elements that are not coherent belong to disjoint pieces; furthermore, if there is more than one piece, the pieces are densely ordered by the ordering that M induces on them. So, in general, an ordered set is either scattered or the infinite sum of scattered sets with respect to a densely ordered index set—the "fundamental conclusion" mentioned in the introduction.[12] Specializing to the at most countable case, Hausdorff combines his theorem with Cantor's order characterization of the rationals to conclude that the ring of order types of cardinality $\leq \aleph_0$ has a finite basis, namely: $1, \omega, \omega^*, \eta$. Hausdorff notes that this finite-basis result is no doubt exceptional, since, as he later (§25) shows, the basis for the ring of all types of cardinality $\leq 2^{\aleph_0}$ has at least $\aleph_1$ types.

In the article's final part, *Dense Sets*, Hausdorff pursues the representation of dense sets in terms of the scattered sets. Since the latter are closed under general addition, he will need some new kind of operation. Fortunately, he is well prepared. He makes strategic use of the operation of raising to powers, which he began to develop in [H 1904a] and which he employed successfully in his representation theory for homogeneous dense sets in both [H 1906b] and [H 1907a]. In [H 1906b, 108–122], he introduced power formation for arbitrary ordered sets and also the more general operation of multiplying a family of ordered sets indexed by an ordered set. Hausdorff based his definitions of power and product on the concept of *covering set* that Cantor used to define cardinal exponentiation [Cantor 1895, 486–488]. As with all these operations, there are corresponding operations on order types.

Hausdorff ends *Sums, Products, and Powers of Order Types* with a reworking of his [H 1906b] discussion of products and powers for ordered sets. In distinction to [H 1906b], he begins with products rather than powers, a pedagogically effective choice since powers are a special case and the laws of exponents follow from the general associative law for products. For A an ordered set and M_a a family of (non-empty) ordered sets indexed by A, an "element combination" is an A-sequence whose entries in slot a come from the set M_a. (*A-sequence* is our terminology not Hausdorff's.) The collection of all A-sequences is the "combination set of all the M_a." Hausdorff's idea, dating back to [H 1904a], is to order the combination set by the principle of first differences. In general, the result is not a totally ordered set; there may be incomparable A-sequences. However, if A is well-ordered, the result is a totally ordered set that Hausdorff calls the "complete product of the factors M_a with argument A." When all the $M_a = M$, one has the "complete power with base M and argument A."

As in [H 1906b], Hausdorff solves the problem of A-sequences that are incomparable by choosing "principal elements" p_a in each M_a; the A-sequence P consisting of the p_a is called "the principal covering." Hausdorff then considers the set T consisting of all A-sequences X for which $\{a \mid x_a \neq p_a\}$ is well-ordered. The set T, which depends on the M_a and P, is totally ordered by the principle of first differences. Hausdorff calls it a "maximal product." No explanation is given here for this terminology, but in [H 1914a, 153] Hausdorff shows that the maximal linearly ordered subsets of the partially ordered (by first differences) combination set of all the M_a are the sets T. For ω_π a regular initial number, T^π is the set of all X where $\{a \mid x_a \neq p_a\}$ is well-ordered and $< \omega_\pi$; the set T^π is called the product of the $(1+\pi)$th class. Of course, T and T^π are identical for $\overline{\overline{A}} < \aleph_\pi$. The T^π with $\pi < \omega$ were the only products considered in [H 1906b]. When all the $M_a = M$ and all the $p_a = p$, T becomes the "maximal power" and the T^π become the "powers of the $(1+\pi)$th class of base M with argument A and principal element p."

After proving the general associative law for products of the $(1+\pi)$th class, Hausdorff derives the laws of exponents for the corresponding powers. He also derives expressions for the initial and final segments of these products. Specializing to the case where A^* and M are well-ordered and p is the first element of M, Hausdorff uses his segment calculations to show that the resulting powers are really those that Cantor had defined by transfinite induction for the (countable) ordinals [Cantor 1897, 231–235].[13] Hausdorff ends this section with a crucial theorem about ω_ν-sequences of coverings in complete powers. Such sequences are cofinal with either an an argument-like sequence (an ω_ν-sequence associated with the argument A) or with a base-like sequence (an ω_ν-sequence associated with the base M). He first stated and outlined the proof of such a theorem in [H 1907a, 89–90], and it plays an important role in *Dense Sets* where sequences from complete powers (and certain of their subsets) are studied intensively.

Dense Sets begins with a classification scheme for unbounded dense sets M. As noted before, each element and gap of M can assigned a unique character $c_{\alpha\beta}$ ($\omega_\alpha, \omega_\beta$ regular). The set M is said to be of *species* (U, V), where U is the set of characters belonging to the elements of M and V is the set of characters belonging to the gaps of M. M is called *irreducible* if each middle segment of M is also of species (U, V). In addition, M is said to be of *genus* $W = U \cup V$. For a continuous set M, $V = 0$ (the empty set).

Using his theorem on decomposing a dense set into subsets that are dense in it, Hausdorff shows:

> If an irreducible continuous set of species $(W, 0)$ exists then there also exist irreducible sets of any species of the same genus. [H 1908, 476]

This focuses the question of the existence of sets of genus W on the continuous, unbounded dense sets. For such an M, W must satisfy certain necessary conditions. Let ω_κ and ω_λ be the first regular initial numbers that cannot be embedded in a middle segment of M and M^*, respectively. The indices κ and λ are called the "boundary numbers" of W. (Hausdorff always assumes that $\kappa \geq \lambda$; he notes that if that were not the case, he could reverse the roles of M and M^*.) If one thinks of the characters $c_{\alpha\beta}$, $\alpha < \kappa, \beta < \lambda$ as an array, then W must contain at least one character from each row and one character from each column. Furthermore, the existence of symmetric elements implies that W must contain at least one principal diagonal element $c_{\alpha\alpha}$. A subset W of $\{c_{\alpha\beta} \mid \alpha < \kappa, \beta < \lambda\}$ satisfying these row, column, and diagonal membership conditions is called a *complete* character set with boundary numbers κ, λ.

These necessary conditions turn out to be (almost) sufficient conditions. Hausdorff shows that for each complete character set W whose boundary numbers κ, λ are non-limit ordinals there exists an irreducible continuous set of species $(W, 0)$. Given his earlier comments on regular initial numbers with limit indices, Hausdorff views the restrictions on κ and λ as "no important loss in generality" [H 1908, 478].

Hausdorff starts his existence proof by well-ordering the $c_{\alpha\beta}$ in W according to the well-ordering of the index pairs (α, β) by first differences; using the assumption that κ is a non-limit ordinal, he concludes that the type of this ordering is $< \omega_\kappa$. He then forms the sum Φ over W of scattered types of the form $\omega_\alpha + 1 + \omega_\beta^*$, one for each $c_{\alpha\beta}$ in W; this places $\omega_\alpha \omega_\beta^*$-elements at his disposal. In general, $\mu(\gamma, \delta)$ denotes the scattered (gap-free) type $1 + \omega_\gamma^* + \Phi + \omega_\delta + 1$. For $\delta < \kappa, \gamma < \lambda$, each $\mu(\gamma, \delta)$ can be considered an interval of $\mu = \mu(\lambda - 1, \kappa - 1)$.

Hausdorff then introduces the innovation that he announced in the introduction, *multiplication with variable factors*. He proceeds to inductively define a gap-free set N of element combinations that live within the complete power with base $\mu = \mu(\lambda - 1, \kappa - 1)$ and argument ω_σ, where $c_{\sigma\sigma}$ is the symmetric character of least index in W. (The choice of argument seems mysterious at this point.) "Variable factor" means that in an admissible ω_σ-sequence X the range of $x_{\varrho+1}$ depends on properties of x_ϱ, and for ϱ a limit ordinal, the range of x_ϱ depends on properties of $\{x_\nu \mid \nu < \varrho\}$ and (possibly) the cofinality of ϱ. The range of an x_ϱ is either a fixed singleton $\{p\}$ or a set of the form $\mu(\gamma, \delta)$, where in the latter case γ and δ are intimately connected to W. Once an $x_\varrho = p$, all subsequent $x_\xi = p$. For example, x_0 always has range μ and the range of x_1 depends on properties of x_0 within μ: if x_0 is a boundary element or two-sided limit in μ, then $x_\xi = p$ for all $\xi \geq 1$, and if x_0 has an immediate predecessor and an immediate successor in μ, then the range of x_1 is $\mu(\sigma, \sigma)$, etc.

The definition of N is crafted to maintain control of the characters of its elements. Because N lies in a complete power, the limit behavior of sequences from N is determined by that of its argument-like and base-like sequences. As Hausdorff proceeds with his analysis of such sequences, it becomes clear why the range restrictions in the definition of N were chosen and why ω_σ is the smallest well-ordered argument for which this construction works. The set N, though gap-free, can have consecutive elements. By construction, the left member of such a pair is an ω_σ-element, while the right member is an ω_σ^*-element. In his final step, Hausdorff removes N's boundary elements and collapses pairs of consecutive elements, which then become single elements of character $c_{\sigma\sigma}$, to obtain the continuous, unbounded dense set N' whose set of element characters (and that of any middle segment of N') is precisely W. The set N' has cardinality $\aleph_{\kappa-1}^{\aleph_\sigma}$. Later, in a series of remarkable results on $c_{\sigma\sigma}$-sets Hausdorff shows that this is not accidental.

After a brief foray into combinatorics where he calculates the number of genera and species for κ, λ finite and small, Hausdorff introduces certain parameters (indices of regular initial numbers) $\sigma, \varrho, \mu, \nu, \pi$ for each unbounded dense set M of species (U, V): σ is the least index of the symmetric characters in the genus $W = U \cup V$, ϱ is determined by the species in a way soon to be described, μ and ν are the indices of the cofinalities ω_μ, ω_ν of M and M^*, respectively, and $\pi = \min(\mu, \nu, \varrho)$. The set M itself is then called a $c_{\sigma\sigma}$-set, a d_ϱ-set, and an e_π-set.

The definition of ϱ is best understood by changing the order of Hausdorff's presentation and first considering his definition of η_τ-set. An unbounded dense set M is an η_τ-set "if it is not cofinal or coinitial with any set of cardinality $< \aleph_\tau$ and if it contains no pair of adjacent subsets each of cardinality $< \aleph_\tau$"; the latter condition is equivalent to saying that in each pair of adjacent sets at least one set must have cardinality $\geq \aleph_\tau$.[14] The cardinal $\aleph_\tau$ is not required to be regular, but Hausdorff notes that for singular $\aleph_\tau$ any η_τ-set is also an $\eta_{\tau+1}$-set. (The η_0-sets, just called the η-sets by Hausdorff, are the entire class of unbounded dense sets.)

Suppose M is an η_τ-set of species (U, V). What can one say about the indices of the characters in U and V and about the indices of the unique regular ω_μ and ω_ν with which M and M^* are cofinal, respectively? If $c_{\alpha\beta}$ is in U, then M has an element m that is the right limit of an ω_α-sequence and the left limit of an ω_β^*-sequence. If $c_{\gamma\delta}$ is in V, then M has a gap-decomposition as $A + B$ where A is cofinal with ω_γ and B is coinitial with ω_δ^*. Then by the cardinality condition on adjacent subsets, necessarily both α and β are $\geq \tau$, and at least one of γ and δ is $\geq \tau$, that is, $\max(\gamma, \delta) \geq \tau$. Both μ and ν must be $\geq \tau$ by the condition on the cardinality of cofinal and coinitial subsets.

In defining ϱ for a given M, Hausdorff reverses the above analysis. Say M is of species (U, V). For each $c_{\gamma\delta}$ in V, let $\varepsilon = \max(\gamma, \delta)$. Then ϱ is defined to be the minimum of all the $\alpha, \beta, \varepsilon$, where $c_{\alpha\beta}$ is in U. Clearly, ω_ϱ is regular and each middle segment of the d_ϱ-set M is an η_ϱ-set. Recall that $\pi = \min(\mu, \nu, \varrho)$, so ω_π is regular, and the e_π-set M is an η_τ-set for all $\tau \leq \pi$. (For N any η_τ-set, $\pi_N \geq \tau$.) By the definition of π, in an e_π-set at least one of the following must happen: a cofinal subset has cardinality $\aleph_\pi$, a coinitial subset has cardinality $\aleph_\pi$, or there is some pair of adjacent subsets in which one subset has cardinality $\aleph_\pi$, while the other has cardinality $\leq \aleph_\pi$. So an e_π-set is not an $\eta_{\pi+1}$-set.

Hausdorff proves that e_π-sets are universal for ordered sets of cardinality $\leq \aleph_\pi$ and that any two e_π-sets of cardinality $\aleph_\pi$ are similar. He remarks that both results hold for η_π-sets. Whether there are any e_π-sets of cardinality $\aleph_\pi$ is undecided at this point. In [H 1907a, 127–128], he had proved universality and similarity theorems for H-types (η_1-types) with respect to $\aleph_1$; there, for the similarity result, he invented the *back-and-forth* construction of an isomorphism. (See [Pl 1993] for more on the history of back-and-forth.) Here, to prove the similarity of any two e_π-sets of cardinality $\aleph_\pi$, he surprisingly reverts to Cantor's *forth* argument, with which Cantor proved that any two countable, unbounded dense sets are similar. This approach requires a careful inductive proof that the defined mapping is onto, which the more elegant back-and-forth avoids. However, in [H 1914a, 182–183], Hausdorff uses a back-and-forth argument to prove both Cantor's theorem and his own more general theorem that any two η_ξ-sets of (regular) cardinality $\aleph_\xi$ are similar. (Neither the forth argument nor the back-and-forth argument for similarity uses AC.) Of course, it was [H 1914a] that introduced back-and-forth to the wider mathematical world.

In the final two sections Hausdorff pursues the study of $c_{\sigma\sigma}$-sets. These pages contain some of the deepest results in the article and shed light on the role of the least index of a symmetric character in the existence proof of continuous sets with a given complete character set.

In a clever technical result (Theorem XX), Hausdorff singles out certain subsets $S(\omega_\sigma)$ and S of the complete power with base $N = \omega_\beta^* + \omega_\alpha$ and argument ω_σ and shows that any $c_{\sigma\sigma}$-set M, each of whose middle segments contains an ω_α-sequence and an ω_β^*-sequence, has a subset similar to $S(\omega_\sigma)$ and, in case M has no $c_{\sigma\sigma}$-gaps, a subset similar to S. A simple consequence of the fact that σ is the least index of the symmetric characters of M is that each middle segment of M contains both ω_σ and ω_σ^*-sequences. So with $N = \omega_\sigma^* + \omega_\sigma$, the important question becomes: what are the cardinalities of $S(\omega_\sigma)$ and S? These are easy to calculate: the cardinality of $S(\omega_\sigma)$ is

$$\aleph_\sigma + \aleph_\sigma^2 + \cdots + \aleph_\sigma^{\aleph_0} + \aleph_\sigma^{\aleph_1} + \cdots + \aleph_\sigma^{\aleph_\xi} + \cdots (\xi < \sigma),$$

which Hausdorff denotes by $(\aleph_\sigma)_\sigma$, and the cardinality of S is $\aleph_\sigma^{\aleph_\sigma}$. Thus for M a $c_{\sigma\sigma}$-set, in general, $\overline{\overline{M}} \geq (\aleph_\sigma)_\sigma$, and if M has no $c_{\sigma\sigma}$-gaps, $\overline{\overline{M}} \geq \aleph_\sigma^{\aleph_\sigma}$.

For irreducible sets with boundary numbers the non-limit ordinals κ and λ, the lower bounds become $(\aleph_{\kappa-1})_\sigma$ and $\aleph_{\kappa-1}^{\aleph_\sigma}$, respectively, assuming $\kappa \geq \lambda$. So the smallest possible cardinality for an irreducible, continuous $c_{\sigma\sigma}$-set is $\aleph_{\kappa-1}^{\aleph_\sigma}$, exactly what Hausdorff achieved in the proof of his existence theorem.

Hausdorff notes that for a $c_{\sigma\sigma}$-set of cardinality $\aleph_\sigma$ to exist it is necessary that $(\aleph_\sigma)_\sigma = \aleph_\sigma$. He calls this equality for regular $\aleph_\sigma$ (and he insists that he is only speaking of regular cardinals) *Cantor's Aleph Hypothesis*, because for $\sigma = 1$ it becomes $\aleph_1 = \aleph_1^{\aleph_0}$, the continuum hypothesis.[15] He also notes that for $\sigma = (\sigma - 1) + 1$ the Aleph Hypothesis becomes $\aleph_\sigma = \aleph_\sigma^{\aleph_{\sigma-1}} = 2^{\aleph_{\sigma-1}}$. Since $\aleph_{\sigma+1}$ is always regular, the assertion of the Aleph Hypothesis for $\aleph_{\sigma+1}$ is equivalent to an instance of the Generalized Continuum Hypothesis.[16] Because of this, Hausdorff is often credited as the first to state GCH. (For example, see [Ka 1997, 16], [Koe 1996, 78], and [Mo 1989, 111]). Hausdorff uses the Aleph Hypothesis for regular $\aleph_\sigma$ to show that there exist $c_{\sigma\sigma}$-types of cardinality $\aleph_\sigma$ and that there is a unique e_σ-type of cardinality $\aleph_\sigma$.

Using Theorem XX, the technical result alluded to above, specialized to the base $N = \omega^* + \omega$, Hausdorff shows that each $c_{\sigma\sigma}$-set contains a subset of type η_σ, where η_σ is the power of class $(1 + \sigma)$ with base $3 = \{q_1 < p < p_1\}$, argument ω_σ, and principal element p. Furthermore, η_σ is of species (U, V) where $U = \{c_{\sigma\sigma}\}$ and $V = \{c_{\sigma\tau}, c_{\tau\sigma} \mid \tau < \sigma\} \cup \{c_{\sigma\sigma}\}$. Thus a set of type η_σ is both an e_σ-set and a $c_{\sigma\sigma}$-set of smallest cardinality, and that cardinality happens to be $(\aleph_\sigma)_\sigma$. So if $(\aleph_\sigma)_\sigma = \aleph_\sigma$, there exists only one e_σ-type of cardinality $\aleph_\sigma$, namely η_σ. Hausdorff points out that by "filling" the $c_{\sigma\sigma}$-gaps of η_σ one obtains a $c_{\sigma\sigma}$-set and e_σ-set of species $(\{c_{\sigma\sigma}\}, \{c_{\sigma\tau}, c_{\tau\sigma} \mid \tau < \sigma\})$. This set has cardinality $2^{\aleph_\sigma}$ and is contained in every $c_{\sigma\sigma}$-set without $c_{\sigma\sigma}$-gaps.[17]

Hausdorff ends [H 1908] with a virtuoso performance. He strengthens his existence theorem by once more resorting to multiplication with variable factors. Two dense sets are said to be "essentially distinct" if no middle segment of one is similar to a middle segment of the other. Using multiplication with variable factors, he shows how to produce $\aleph_\lambda$ essentially distinct irreducible sets of species $(W, 0)$ from an irreducible $c_{\sigma\sigma}$-set M of species $(W, 0)$. Here, λ is the minimum of the two boundary numbers of W. Hausdorff operates within the complete power with base M and argument ϱ, where $\omega_\sigma \leq \varrho < \omega_\lambda$ and ϱ is cofinal with ω_σ. The variable factors are either a fixed singleton $\{p\}$ or intervals of M whose end points are both $c_{\alpha\beta}$-elements of M. Hausdorff arranges for M to have the additional property that any set of exclusive intervals (any two intervals in the set have at most one point in common) in M cannot have the same type as a $c_{\sigma\sigma}$-set without $c_{\sigma\sigma}$-gaps. This property and the use of principal ordinals (ordinals similar to all of their remainder segments) as arguments in multiplication with variable factors lead to the desired distinctness results. This is a most satisfying ending, since conditions on exclusive interval sets and the use of principal ordinal numbers led to his earlier constructions of $\aleph_1$ distinct continua and $\aleph_1$ distinct ultracontinua [H 1906b, 143 and 169].

Notes

1. The tenor of the times is succinctly described by Walter Purkert:
   > The central target was the choice axiom, and the antinomies hung over everything like the sword of Damocles. [Pu 2002, 14]

   G. H. Moore's excellent book, *Zermelo's Axiom of Choice* [Mo 1982], provides a thorough account of the attacks launched by the critics of AC. Zermelo's own reply to his critics ([Ze 1908a]), which is a masterpiece of polemics, is also a good source.

2. This handwritten letter is in Hilbert's Nachlaß at the Niedersächsische Staats- und Universitätsbibliotek in Göttingen (Hilbert 136). Excerpts from the letter are reprinted in [Pu 2002, 13].

3. The location of the meeting and its purpose comes from a handwritten letter from Cantor to Hilbert on August 8, 1907. Cantor takes the opportunity in replying to news of Zermelo's progress to put in a good word for Hausdorff's work and to establish himself as the instigator of Hausdorff's finite basis result for the finite and countable types. (Hausdorff does not mention Cantor in his presentation of type rings in [H 1908].) The finite basis result is a corollary of his fundamental discovery that an arbitrary ordered set is either scattered or the dense sum of scattered sets.

   In his letter, Cantor writes:
   > I am glad to hear that Herr Zermelo is successfully working on set theory. Give him my best. I also consider the work of Hausdorff in the theory of types to be useful, thorough, and promising. For this reason, I arranged to meet with him at the last meeting of the three universities, Leipzig, Jena, and Halle, in Kösen on June 28, and I suggested an investigation of a question that he appears to have completed successfully. It concerns the type class of the cardinal $\aleph_0$, i.e., the countable types; I was always of the opinion that these types could be reduced to the three ur-types ω^*, ω, and η and the finite ur-types 1 and 2 through composition. This is confirmed by Hausdorff in the following form, as he wrote to me on July 19 ...

   Cantor then details Hausdorff's basic results on type rings.

   This letter is in Hilbert's Nachlaß at the Niedersächsische Staats- und Universitätsbibliotek in Göttingen (Hilbert 54). The excerpt I have cited is reprinted in [PuI 1987, 227–228].

4. In printing, a *Bogen* is a large sheet of paper which is folded (like a quarto) to produce eight pages. Hausdorff is probably offering an article of sixteen to twenty-four pages.

5. The publication manuscript is found in Hausdorff's *Nachlaß* [Kapsel 26b; Fasz. 89]. The first page has Hilbert's handwritten notation: "Accepted Nov. 1907 Hilbert" [Pu 1995, 101].

6. The imbroglio over the status of the totality of all ordinal numbers W was an embarrassment. For example, both Bernstein and Jourdain published papers in volume 60 of the *Mathematische Annalen* insisting that W was a set. Bernstein concluded that the set of all subsets of W was the

simplest example of a non-well-orderable set [Berns 1905b], while Jourdain concluded that every set could be well-ordered [Jo 1905].

As for the Axiom of Choice, Bernstein said that it could be dispensed with: it was replaceable by his concept of *multiple valued equivalence* [[*vielwertigen Äquivalenz*]][Berns 1905b]. Jourdain insisted that the Well-Ordering Theorem could be proved without choice, and he continued to attempt such proofs until his dying day. (See [Mo 1982, 188–192] for the poignant story.)

There was apparent agreement that making infinitely many successive dependent choices had to be ruled out because of temporal considerations. (See Hadamard's 1905 letter to Borel, which is translated in [Mo 1982, 311–312].) Hausdorff seems to have shared these views. However, he (and Hadamard) saw no such problem with Zermelo's choice functions, because the selection of elements from the non-empty subsets of a given set was deemed "simultaneous and independent." In [H 1914a, 134], Hausdorff presents the case against successive dependent choices; he praises Zermelo for having had the "fortunate idea" of making all choices at the outset.

Much later, dependent choice was recognized as a principle in its own right. Bernays formulated such an axiom and considered it useful for analysis [Bern 1942, 86]. Tarski independently formulated dependent choice [Ta 1948, 96]. Though weaker than AC, dependent choice is known to be stronger than the the axiom of choice for countable collections of non-empty sets [Je 1966].

7. The terms *singular initial number* and *regular initial number* were introduced in [H 1907b], *singular* for those cofinal with an ordinal of smaller cardinality and *regular* for those that are not. (For more on cofinality, see the *Introduction to "Investigations into Order Types I, II, III,"* Note 7, p. 42.)

König had used the adjective singular [[singulären]] in quotes to refer to well-ordered sets of cardinality $\aleph_\omega$ and $\aleph_{\mu+\omega}$. He noted that for such sets one could not assume that a countable subset was always contained in a proper initial segment [Ko 1905, 180].

8. In [Hes 1906, 121], Hessenberg states that the limit of the sequence $\omega_1 = \Omega_\omega$, $\omega_2 = \Omega_{\omega_1}$, $\omega_3 = \Omega_{\omega_2}$, ..., which is the initial number Ω_μ, is equal to its own index. He credits his knowledge of Ω_μ to an oral communication from Zermelo. Hessenberg names an initial number that is equal to its own index a ζ-number, apparently because ζ comes after ε and there already are ε-numbers. He also shows how to obtain (as the limit of a sequence) the first ζ-number greater than a given α, and he notes that the limit of a sequence of ζ-numbers is again a ζ-number.

Jourdain erroneously claimed that $\omega_\gamma > \gamma$ holds for every initial ordinal ω_γ [Jo 1905, 466]. In a September 6, 1905, letter to Bertrand Russell, the amateur set theorist and paradox connoisseur G. G. Berry, citing Jourdain's claim, described how to obtain the first ω_γ with $\omega_\gamma = \gamma$ and how to obtain the subsequent initial ordinals with this property. Berry's letter is reprinted in [Ga 1992, 171–173].

9. Changing the direction of Hausdorff's analysis, Mahlo postulated the existence of fixed points in certain enumerations of regular initial numbers

to obtain hierarchies of (weakly) inaccessible cardinals [Ma 1911; 1912; 1913]. For more on Mahlo, see the *Introduction to "Investigations into Order Types I, II, III,"* Note 14, p. 43.

10. This is Hausdorff's second pronouncement on the "exorbitant" size of regular initial numbers with limit index, the numbers now known as weakly inaccessible. For his first remarks on the size of regular initial numbers with limit index see [H 1907b, 546]. His most quoted comment appears in [H 1914a, 131]:

> Thus if there exist regular initial numbers with limit index (so far no contradiction has been discovered in this assumption), then the smallest among them is of such an exorbitant magnitude that it will hardly ever come into consideration for the usual purposes of set theory.

Hausdorff was eventually proved wrong. Now set theorists are quite at home with a variety of inaccessibles and the assumption, when convenient, that a weakly inaccessible exists is considered harmless. (See [Ka 2003].)

11. In [H 1907b] where he announced his decomposition theorem for dense M, Hausdorff indicated its proof would use a well-ordering of the set of all subsets of M. Here his interesting proof just employs a well-ordering of M.

In [Sk 1920], Skolem used a *back-and-forth* argument to show that any two countable, unbounded dense sets, each of which is decomposed into two disjoint subsets that are dense in their parent sets, are isomorphic under a mapping that takes the components of the respective decompositions to one another (reprinted in [Sk 1970, 103-136]). He also extended his result to two countable, unbounded dense sets that are decomposed into countably many pairwise disjoint subsets that are dense in them. He remarked that analogous results must hold for Hausdorff's η_ξ-sets in [H 1914a]. Skolem references both [H 1908] and [H 1914a], but only for terminolgy and not for the decomposition theorem or for back-and-forth.

12. Type rings do not appear in [H 1914a]. The identification of the regular initial numbers and their inverses as a "basis" for the scattered sets is omitted and the "fundamental conclusion" is proved without reference to type rings. (See [H 1914a, 95–97].) The basis result for the case of countable scattered sets was rediscovered by Erdős and Hajnal [ErH 1962]. They cited [H 1914a] in their paper, but did not know of [H 1908].

13. Hausdorff defined raising to powers for arbitrary order types as bases and ordinals as exponents in [H 1904a] and then for arbitrary order types as both bases and exponents in [H 1906]. In each case, he showed that his definition agreed with Cantor's for ordinals both as bases and as exponents. (See [H 1904a, 570–571; 1906b, 115–117].) His previous proofs were more in the category of proof-by-example. The inductive proof given here is completely rigorous.

14. This definition of an η_τ-set is the same one that Hausdorff gave in [H 1907a, 132] with only a slight change in wording and with the removal of the restriction to $\tau < \omega$. The earlier definition was an easy generalization of the concept of an η_1-type; η_1-types were reformulations of the H-types that were abstracted by Hausdorff from his study of the maximal linearly

ordered subsets (pantachies) of the set of sequences of positive reals under the final rank ordering (a partial ordering) [H 1907a, 121].

It is fair to say that widespread knowledge of η_τ-sets came from their appearance in [H 1914a, 180-185]. See [Fe 2002] for the post-1914 history of η_τ-sets and their influence outside of set theory.

15. In a footnote to his presentation of Cantor's Aleph Hypothesis ([H1908, 494n]), Hausdorff conjectures a version of the Aleph Hypothesis for singular $\aleph_\beta$, where ω_β has cofinality ω_α ($<\omega_\beta$). First he notes that an "obvious generalization" of König's lemma ([König 1905]) yields $\aleph_\beta^{\aleph_\alpha} > \aleph_\beta$ (showing that he possessed such a generalization). He then states that "at best" the Aleph Hypothesis for the singular $\aleph_\beta$ is $(\aleph_\beta)_\alpha = \aleph_\beta$.

In modern notation, both the regular and singular versions can be combined into the hypothesis: $\aleph_\sigma^{<cf(\aleph_\sigma)} = \aleph_\sigma$.

16. Hausdorff first flirted with GCH in [H 1907a, 133]. After he defined the η_ν-types for $\nu < \omega$, he wrote:

Should CANTOR's hypothesis $2^{\aleph_0} = \aleph_1$ turn out to be true in an extended range, that is, $2^{\aleph_\nu} = \aleph_{\nu+1}$ also holds for each finite ν, then there exists one and only one η_ν-type of cardinality $\aleph_\nu$ ...

This seems to be the first statement of an extension of Cantor's continuum hypothesis.

The phrase *generalized continuum hypothesis* [[*hypothèse généralisée du continu*]] for the assertion $(\forall \alpha)\, 2^{\aleph_\alpha} = \aleph_{\alpha+1}$ is due to Tarski [Ta 1925, 10].

17. The existence of a power with a three element base that can be embedded in every $c_{\sigma\sigma}$-set generalizes a similar result (without the genus information) that was proved for H-types and stated for η_ν-types with $\nu < \omega$ in [H 1907a, 129–133]. See also [H 1914a, 183–185] where it is proved that every η_ξ-set contains a subset similar to such a power.

Sierpiński improved the results in [H 1914a] by using a two element base [Si 1949].

The Fundamentals of a Theory of Ordered Sets

By

F. Hausdorff in Leipzig

In what follows, a sustained introduction that is systematic and as general as possible to the still practically unknown field of simply ordered sets, a field developed by Herr G. Cantor, is attempted for the first time. Up to now, only the well-ordered sets and sets of reals have actually experienced a detailed treatment. On the whole, even my own earlier studies[*], of which nothing more is assumed here, pursue a special direction; their principal subject matter is certain types that are distinguished by especially regular structure (homogeneity) and that have sequences up to the second infinite cardinality, and as far as generalizations are strived for, they are restricted to the nearest levels, those that correspond to the alephs with finite index. In the present article, lest it swell into a book, these special types definitely had to fall back into the role of occasional illustrative examples and general methods had to occupy the foreground. For instance, the ideal had in mind was control of the world of types, in the sense that complex structures should appear to be built from elementary ones by operations of generation; however, certain difficulties, namely, the unsolved cardinality problems, still compel one to resignation. Of course, the building material cannot be in doubt: it is the well-ordered sets and their inverses, whose cardinalities certainly must no longer be limited to the alephs with finite index; especially, the *regular initial numbers* [[*regulären Anfangszahlen*]] and their inverses are to be regarded as the ultimate building blocks and primitive atoms [[Uratome]] of the world of types. But the operations of construction are not so simple and obvious as one would wish. Generalized *addition* allows just the "scattered sets" to be

---

[*] *Untersuchungen über Ordnungstypen*, Ber. d. K. Sächs. Ges. d. Wiss. 58 (1906), 106–169, and 59 (1907), 84–159. The individual articles are entitled:
    I. The Powers of Order Types
    II. The Higher Continua
    III. Homogeneous Types of the Second Infinite Cardinality
    IV. Homogeneous Types of the Cardinality of the Continuum
    V. On Pantachie Types

produced from the basic elements and it allows the deduction of the certainly fundamental conclusion that each set is either scattered [[zerstreut]] or a sum of scattered sets over a "dense" generator [[Erzeuger]]. New construction methods are needed for building dense sets; *multiplication* and *raising to powers* [[*Potenzierung*]] are developed in their most general form, and since even these operations have limited scope, as a last resort, *multiplication with variable factors* has to be dragged in. It is still an open question whether with this the analysis, mentioned above, of arbitrary types down to the primitive types [[Urtypen]] will succeed; but the synthetic side of the problem can now be tackled with a satisfactory result. The dense sets can be classified on the basis of the "characters" of their elements and gaps, characters that are derived again from regular initial numbers. And with the help of generating operations of the most general kind, it now follows that all a priori possible "species" of dense sets actually exist and are represented not by just one, but by uncountably many, essentially different types.

With this, the program and main content of the present article is outlined. That an investigation such as this, which endeavors to add, by an increment even if only modest, to the positive inventory of the still so new theory of sets in the spirit of its creator, cannot prae limine spend any time to enter into a discussion of the principles of set theory will probably give offense in the places where presently a somewhat misplaced degree of ingenuity is squandered in these discussions. To an observer, who in the face of skepticism is not wanting in skepticism, the "finitistic" objections against set theory might roughly fall into three categories: into one such fall those that reveal a serious need for a, perhaps axiomatic, sharpening of the set concept; into another fall those that would affect the whole of mathematics together with set theory; finally, there are those that fall into the category of the simple absurdities of a scholasticism that clings onto words and letters. One shall be able to come to an understanding with the first group sooner or later; one may safely let the second rest; the third deserves the sharpest and clearest disapproval. In the present work, these three responses are tacitly enforced. It takes the point of view that the totality of "all" ordinal numbers or cardinal numbers does not exist free from contradiction either as a set or as a subset of a set; it accepts Cantor's Well-Ordering Theorem in its formulation and proof by Herr Zermelo, and it places no value upon the statement that a part of its conclusions could also be derived independently of it; it makes use of the "choice postulate," even without mentioning it; and it makes concessions to only one of the, *perhaps* justified, objections of the first group with an occasional reference to the fact that a transfinite sequence of successive choices each of which presumes the earlier ones can be avoided by reason of a well-ordering or, which is the same, can be replaced by a simultaneous set of independent choices. The terminology could not simply follow what currently prevails, but rather a definite decision had to be made in cases of wavering usage (e.g., between segment and interval), and still more often neologisms had to be resorted to. An index for the recurring terms with fixed meaning is added for ease of usage.[a]

Table of Contents
Foundations

			Page
§	1.	Pieces and Segments	439
§	2.	Well-Ordered Sets	441
§	3.	Regular Initial Numbers	443
§	4.	Intervals	445
§	5.	Relatively Dense Subsets	447
§	6.	Completion and Filling	448
§	7.	Splitting into Relatively Dense Subsets	450

Sums, Products, and Powers of Order Types

§	8.	Sums	451
§	9.	Finite products and Powers	453
§	10.	Type Rings	454
§	11.	The Structure of Arbitrary Types	457
§	12.	Combination Sets and Covering Sets	458
§	13.	General Products and Powers	461
§	14.	The Associative Law	464
§	15.	Segments of Product Types	467
§	16.	Cantor's Powers and Products	468
§	17.	The Case of Well-Ordered Arguments	470

Dense Sets

§	18.	Species and Genera	474
§	19.	The Derivation of Discontinuous Irreducible Types	476
§	20.	Existence Conditions	477

---

[a] We have alphabetized the translated terms.

§	21.	The Existence Proof for Complete Character Sets.............	478
§	22.	The Number of Genera and Species...........................	484
§	23.	e_π-Sets..	487
§	24.	$c_{\sigma\sigma}$-Sets......................................	490
§	25.	A Strengthening of the Existence Proof.......................	498

Index

(The numbers give the section where the term is defined.)

adjacent 1
argument 12
argument sequence 17
associative law
— of addition 8
— of multiplication 14
base of a power 12
base sequence 17
basis of a ring 10
boundary element 1
boundary numbers 20
bounded 1
Cantor's powers 16
— products 16
character 18
class of a power 13
— of a product 18
coextensive 1
cofinal 1
coinitial 1
combination 9, 12
combination set 9, 12
complete character
   set 20
complete power 12
complete product 12
completion 6
consecutive 1
continuous 1
covering 12
covering set 12
cut 1
dense 1

— in a set 5
differences, rank ordering
   by first — 9, 12
element (ω_α-, ω_β^*-, $\omega_\alpha\omega_\beta^*$-) 3
enclosing intervals 4
end piece 1
end segment 1
essentially distinct 25
everywhere dense 1
— discontinuous 6
exclusive 4
exponent 12
factors 12
filling 6
gap 1
   (ω_α-, ω_β^*-, $\omega_\alpha\omega_\beta^*$-) 3
generator 6
genus 18
higher species 18
independent 10
index of an aleph 2
— of an initial number 2
initial number 2
initial piece 1
initial segment 1
inner elements or
   subsets 1
interval 1
irreducible 18
jump 1
limit 1
limit element 1
limit number 2

lower species 18
maximal power 13
maximal product 13
piece 1
place 12
power 13
principal combination 13
principal covering 13
principal element 13
principal number 16
product 13
— with variable
   factors 17
rank ordering by first
   differences 9, 12
reducible 18
regular 2
relatively dense 5
ring 10
scattered 11
segment 1
separated subsets 1
sequence 2
set (η-)
— (e_π-, d_ρ-, $c_{\sigma\sigma}$-, η_τ-) 23
singular 2
species 18
summand 18
symmetric 4
type ring 10
type, see set
unbounded 1
variable factors 17

Foundations
§1
Pieces and Segments

The subsets $A, B, \ldots$ of an ordered set M are also ordered.

Two subsets are called *separated* [[*getrennt*]] if each element of one precedes each element of the other; we say that *one subset precedes the other*, and we write $A < B$. This holds even if one of the sets is a single element.

Two ordered sets M and N without common elements combine, with the stipulation that $M < N$, into an overall set that is designated by $M + N$; this *addition* can be extended to arbitrary, even transfinite sets of summands (§8).

If
$$M = A + C \quad \text{or} \quad M = A + B + C,$$
A is called an *initial piece* [[*Anfangsstück*]], B a *middle piece* [[*Mittelstück*]], and C an *end piece* [[*Endstück*]] of M. A piece can also be defined as a subset P of M from which each remaining element is separated ($\gtrless P$).

The aggregate M^A of all elements that precede a specific subset A is an initial piece of M; the aggregate M_A of all elements that follow A is an end piece of M; the aggregate M_A^B of all elements that lie between two separated sets ($A < B$) is a middle piece.

In particular, the aggregate M^a of elements that precede a specific element a is called an *initial segment* [[*Anfangsstrecke*]] of M; the aggregate M_a of elements that follow a is called an *end segment* [[*Endstrecke*]]; the aggregate M_a^b of elements that lie between two elements ($a < b$) is called a *middle segment* [[*Mittelstrecke*]]. Thus an element or element pair decomposes M into segments:
$$\begin{aligned}M &= M^a + a + M_a = M^b + b + M_b \\ &= M^a + a + M_a^b + b + M_b.\end{aligned}$$

The segments together with their enclosing elements
$$M^a + a, \quad a + M_a^b + b, \quad b + M_b$$
are called *intervals* (initial, middle, or end).

Initial segments and end intervals were termed segments [[*Abschnitte*]] and remainders [[*Reste*]] by Herr G. Cantor.

In special cases, the defined subsets can be empty; i.e., they contain no elements.

If $M^a = 0$, then a is the first element of M.

If $M_a = 0$, then a is the last element of M.

If $M_a^b = 0$, then the elements a and b are called *adjacent* [[*benachbart*]] or *consecutive* [[*konsekutiv*]].

The first and last elements together are called *boundary elements* [[*Randelemente*]]; all others are *inner* elements. A set with two boundary elements is called *bounded* [[*begrenzt*]]; one without boundary elements is called *unbounded*. A set without pairs of consecutive elements is called *everywhere dense* [[*überall dicht*]] or, briefly, *dense* [[*dicht*]].

If $M^A = 0$, M is called *coinitial* with A.
If $M_A = 0$, M is called *cofinal* with A.
If $M_A^B = 0$, the subsets A and B are called *adjacent* or *consecutive*.

So M is cofinal (coinitial) with A if no element of M follows A (precedes A). If both are the case, M is called *coextensive* with A; if neither one is the case, then A is called an *inner* subset of M.

We say that the type μ is cofinal (coinitial, coextensive) with the type α if a set M of type μ contains at least one subset A of type α with which it is cofinal (coinitial, coextensive).

Examples: the set of rationals, as well as the set of reals, in its natural ordering, is cofinal with the set of natural numbers; i.e., the types η and λ (the unbounded linear continuum) are cofinal with ω. Both are also coinitial with ω^* and coextensive with $\omega^*+\omega$; furthermore, they are cofinal with each limit number of the second number class.

The following easily seen theorem, in which the word cofinal can be replaced by coinitial or coextensive, holds.

Theorem I. *If M is cofinal with P and P is cofinal with Q, then M is also cofinal with Q. If Q is a subset of P and P is a subset of M and M is cofinal with Q, then M is also cofinal with P and P is cofinal with Q.*

If A is a subset without a last element that is adjacent to a later element b $(A < b)$, then b is called the *upper limit* [[*obere Limes*]] (the upper limit element [[obere Grenzelement]]) of A. Then it is the case that b is the first element of the end piece M_A and the initial segment M^b is cofinal with A. Each element b without an immediate predecessor, if it is not the first element of M, is the upper limit of the initial segment M^b and of all subsets with which M^b is cofinal. The *lower limit* is defined correspondingly.

A decomposition $M = A + B$ is designated as a

Jump [[*Sprung*]]	if	(α)	A has a last, B has a first
Cut [[*Schnitt*]]	if	(β)	A has a last, B has no first
	or	(γ)	A has no last, B has a first
Gap [[*Lücke*]]	if	(δ)	A has no last, B has no first

element.

A dense set is free of jumps. A set that is free of jumps and gaps, or a gapless dense set, is called *continuous* [[*stetig*]].

§2
Well-Ordered Sets

A set is called well-ordered (its type an ordinal number) if it and each of its end pieces has a first element. We assume that the theory of well-ordered sets is known.*) To start with, we imagine that we have chosen a

---

*) G. Cantor, Beiträge zur Begründung der transfiniten Mengenlehre, Math. Ann. 46 (1895) and 49 (1897).

A. Schoenflies, Die Entwickelung der Lehre von den Punktmannigfaltigkeiten, Jahresber. d. D. M.-V. 8 (1900).

G. Hessenberg, Grundbegriffe der Mengenlehre, Abh. d. Friesschen Schule, neue Folge 4 (1906).

sufficiently large ordinal number Δ that exceeds in cardinality all the sets that we consider, and by W we understand the set of all ordinal numbers $< \Delta$; in what follows, the subject is only the elements and subsets of W. By including the number 0, we get that each number α is, at the same time, the type of its initial segment W^α. If W^α has no last element, α is called a *limit number*. If A is a set of ordinal numbers without a last element and W is not cofinal with A, the first number following A, thus the first element of W_A, is the (upper) limit of A and is designated by $\lim A$. The cardinality of a transfinite well-ordered set is called an aleph and is designated by $\aleph_\alpha$; the *index* α is an ordinal number and clearly the type of all the preceding alephs. The smallest ordinal number of cardinality $\aleph_\alpha$ is called ω_α and termed an *initial number*. One writes $\omega_0 = \omega$, $\omega_1 = \Omega$ for the two smallest initial numbers; they are the smallest numbers of the second and third number classes (G. Cantor); likewise, ω_n with finite index is the smallest number of the $(n+2)$th number class.

Each set contains well-ordered subsets, or, as we are going to say, each set contains ordinal numbers. If α is the first ordinal number not contained in M, then each number $< \alpha$ and no number $\geq \alpha$ is contained in M; e.g., $\omega, \omega+1, \Omega$ are the first ordinal numbers not contained in ω^*, ω, η. The corresponding holds of the ordinal numbers contained *inside* M (i.e., in middle segments of M).

Theorem II. *Each set of cardinality $\aleph_\alpha$ is cofinal with an ordinal number $\leqq \omega_\alpha$.*

Proof. Let M be of cardinality $\aleph_\alpha$; thus its elements can be written in the form of a well-ordered set of type ω_α

$$m_0\, m_1\, m_2\, \cdots\, m_\omega\, \cdots\, m_\xi\, \cdots \qquad (\xi < \omega_\alpha).$$

Let us consider the set P of those elements m_ξ in M that follow all elements of lower index ($m_\xi > m_\eta$ for $\xi > \eta$); here m_0 is counted as in it. The set P is similar to the sequence of its element indices; thus it is well-ordered and of type $\leqq \omega_\alpha$. M is certainly cofinal with P. For if Q is the complementary set, thus the set of all elements m_ξ that are succeeded by at least one element of lower index, then $q = m_\xi$ is followed by an m_η with smaller index, which, if it belongs to Q, is followed by an m_ζ with smaller index, etc., and since each decreasing sequence $\xi > \eta > \zeta > \cdots$ of ordinal numbers breaks off, we surely arrive at an element $p > q$ after a finite number of steps. Thus there is no element $q > P$.

The set P can also be defined by induction; it contains $p_0 = m_0$ and after that $p_1 = m_{\xi_1}$, the element in M of lowest index ξ_1 that follows p_0; generally, $p_\varrho = m_{\xi_\varrho}$ is the element in M of lowest index that follows the totality P^{p_ϱ} of all elements of P already constructed.

From II, it follows that an ordinal number that is not an initial number is cofinal with smaller numbers. Even among the initial numbers, there are those that are cofinal with smaller numbers. And we call these *singular* initial numbers, while each initial number that is not cofinal with any earlier ordinal number shall be called *regular*. In particular, since the limit of a set of initial numbers is again an initial number, for α a limit number, ω_α is the limit of the set of all smaller initial numbers and W^{ω_α} is cofinal with this set; thus ω_α is cofinal with α: each initial number with limit index is cofinal with this index.

For example, $\omega_\omega = \omega_0 + \omega_1 + \omega_2 + \cdots$ is cofinal with ω ($< \omega_\omega$); each initial number ω_α whose index is a limit number of the second number class is cofinal with ω; ω_{ω_1} is cofinal with ω_1; the initial number $\omega_{\omega_\omega} = \omega_{\omega_0} + \omega_{\omega_1} + \omega_{\omega_2} + \cdots$ is cofinal with ω_ω and therefore again cofinal with ω, etc.

Let an aleph $\aleph_\alpha$ be called regular or singular just in case the initial number ω_α is regular or singular. We will call sets of type or inverse type a regular initial number *sequences*.

§3
Regular Initial Numbers

Theorem III. *Each initial number whose index is not a limit number is regular.*

That $\omega_0 = \omega$ is regular, i.e., not cofinal with any smaller (in this case finite) number, is evident. Thus the theorem needs to be proved only for initial numbers of the form $\omega_{\alpha+1}$. The quite simple proof, which can be omitted, depends on the aleph equation $\aleph_\alpha^2 = \aleph_\alpha < \aleph_{\alpha+1}$ and on the fact that each number $< \omega_{\alpha+1}$ has at most cardinality $\aleph_\alpha$.

The question whether Theorem III is reversible or whether there are regular initial numbers with limit index must remain undecided here. Of course, the following can be noted. An initial number with limit index is singular by §2 if its index is smaller than it itself; consequently, a regular initial number with limit index can only be contained among those initial numbers that are equal to their own index ($\omega_\zeta = \zeta$), which we temporarily are going to call (along with Herr Hessenberg) ζ-numbers. If α is an arbitrary ordinal number, but not a ζ-number, then the first ζ-number that follows α is the limit of the sequence

$$\alpha_1 = \omega_\alpha, \quad \alpha_2 = \omega_{\alpha_1}, \quad \alpha_3 = \omega_{\alpha_2}, \quad \ldots;$$

moreover, the limit of a set of ζ-numbers is again a ζ-number. Consequently, if we write the ζ-numbers, as we wrote the initial numbers, in the form

$$\zeta_0 \, \zeta_1 \, \zeta_2 \, \cdots \, \zeta_\omega \, \cdots \, \zeta_\alpha \, \cdots,$$

so that the index α represents the type of the set of all ζ-numbers $< \zeta_\alpha$, then each ζ-number whose index is not a limit number is the limit of an ω-sequence and thus cofinal with ω. And each ζ-number with limit index is cofinal with this index. Thus a regular initial number with limit index can again only be contained among those ζ-numbers that are equal to their own index ($\zeta_\eta = \eta$), which we designate as η-numbers

$$\eta_0 \, \eta_1 \, \eta_2 \, \cdots \, \eta_\omega \, \cdots \, \eta_\alpha \, \cdots.$$

Certainly, the same holds for these and again we would have to restrict ourselves to the η-numbers that are equal to their own indices. The continuation of this reasoning leads to the following abstraction: let M_1 be the set of all ζ-numbers in W, M_2 the set of all η-numbers, M_3 the the set of all η-numbers that are equal to their own indices, etc.; let M_ω be the set of all numbers occurring simultaneously in $M_1\, M_2\, M_3 \cdots$; then again, let $M_{\omega+1}$ be the set of numbers in M_ω that, ordered one after the other, are equal to their own indices. In general, let $M_{\alpha+1}$ be the set of all numbers in M_α that are equal to their own indices and let M_α with limit index be the common part of all earlier sets. The number $\zeta_{\alpha\beta}$ that bears the index β in the set M_α is then either cofinal with α or with β or with ω. A regular initial number with limit index $\omega_\xi = \xi$ would have to occur in all sets M_α ($\alpha < \xi$) and would have to bear the index ξ in them; thus only in M_ξ would it have to receive the index 0. According to this, the existence of one such number ξ appears at least problematic; however, it must be considered as a possibility in what follows.

Now let M be a set without a last element (a set with a last element is cofinal with 1). For its part, the smallest ordinal number that M is cofinal with must not be cofinal with a smaller number, otherwise (by Theorem I) M would also be cofinal with this smaller number; thus it is a regular initial number. On the other hand, M cannot simultaneously be cofinal with two regular initial numbers, or more generally, M cannot simultaneously be cofinal with a regular initial number and a smaller number. In particular, if M is simultaneously cofinal with A and B, then (Theorem I) the union $C = \mathfrak{M}(A, B)$ is cofinal with A and B. If A and B are well-ordered, then so is C; if A is of type ω_α and B of smaller type, thus of cardinality $\aleph_\beta < \aleph_\alpha$, then C is not only of cardinality $\aleph_\alpha$ but also precisely of type ω_α (not $> \omega_\alpha$) since each initial segment C^c is of cardinality $< \aleph_\alpha$. However, C cannot then be cofinal with B if ω_α is a regular initial number.

With this, the following theorem is proved:

Theorem IV. *Each set without a last element is cofinal with one and only one regular initial number.*

Each subset with which M is cofinal is also cofinal with the same regular initial number with which M is cofinal.

Each set without a first element is coinitial with the inverse of one and only one regular initial number.

If $M = A + B$ and A has no last element, so A is cofinal with a regular initial number ω_α, then either B has a first element b ($A = M^b$), which as an upper limit of an ω_α-sequence may be called an ω_α-limit or ω_α-element, or B has no first element, in which case we speak of an ω_α-gap. The terms

$$\omega_\beta^*\text{-limit} \quad \text{or} \quad \omega_\beta^*\text{-element}, \quad \omega_\beta^*\text{-gap},$$
$$\omega_\alpha\omega_\beta^*\text{-limit} \quad \text{or} \quad \omega_\alpha\omega_\beta^*\text{-element}, \quad \omega_\alpha\omega_\beta^*\text{-gap}$$

are explained similarly. The regular initial numbers that belong to a limit element, respectively, to the left and right of a gap are thus uniquely determined; a classification of ordered sets, which we will carry out for dense sets (§18), can be based on them.

§4

Intervals

Simultaneously with the ordered set M is given the set (not ordered) of its *element pairs* [[*Elementpaare*]]. To each element pair (a,b), where $a < b$, there uniquely corresponds a middle segment M_a^b and the accompanying interval (more precisely: middle interval)

$$[a,b] = a + M_a^b + b.$$

If $a < c < d < b$, thus c,d belong to the middle segment M_a^b, we say that the element pair (a,b) *encloses* the element pair (c,d) or that the middle segment M_a^b encloses the middle segment M_c^d or that the interval $[a,b]$ encloses the interval $[c,d]$. To any set of intervals that pairwise enclose each other (expressed more precisely: where for any two intervals, one is an encloser, the other enclosed), we assign the order type of their left boundary elements.

On the other hand, we are going to say the intervals $[a,b]$ and $[c,d]$ *exclude* one another or are *exclusive* if the middle segments M_a^b, M_c^d are separated (§1); thus for instance,

$$M_a^b < M_c^d, \quad a < b \leqq c < d.$$

So exclusive intervals are either separated or have a single element in common. We also assign the order type of their left boundary elements to a set of pairwise exclusive intervals (§25).

The following theorem states for element pairs or intervals what Theorem II (§2) states for elements.

Theorem V. *Each set of cardinality $\aleph_\alpha$ contains a set of intervals of type $\leqq \omega_\alpha$ that pairwise enclose each other and that enclose no further interval.*

By separating the left and right boundary points of the intervals, we can also express this as follows:

Each set of cardinality $\aleph_\alpha$ contains two separated subsets of types β, β^* ($\beta \leqq \omega_\alpha$) that are either adjacent or enclose a single element.

For the proof, we consider the set of intervals n of M; this set $N = \{n\}$ is of cardinality $\aleph_\alpha^2 = \aleph_\alpha$ and thus can be put into the form of a well-ordered set of type ω_α

$$n_0\, n_1\, n_2\, \cdots\, n_\omega\, \cdots\, n_\xi\, \cdots \quad (\xi < \omega_\alpha).$$

From this, by induction, which is certainly more convenient at present, we construct a set P of intervals that pairwise enclose each other: let $p_0 = n_0$, then let $p_1 = n_{\xi_1}$ be the interval of lowest index that is enclosed by p_0; in general, let $p_\varrho = n_{\xi_\varrho}$ be the element of lowest index that is enclosed by each element of P^{p_ϱ}, the totality of all already found intervals. No further interval is enclosed by each element of $P = \{p_\varrho\}$, the totality of all these intervals, since certainly the construction of P would not yet otherwise be ended. If

p_σ is enclosed by p_ϱ, then $\xi_\varrho < \xi_\sigma$ because both p_ϱ, p_σ are enclosed by the intervals of P^{p_ϱ} and p_ϱ has the smallest index ξ_ϱ of all such intervals. Thus in the above sense, the set of intervals P is well-ordered and of type $\leqq \omega_\alpha$; and since these intervals enclose no further interval, they either enclose a single element or no element at all.

If one wants to replace the successive generation of P by a simultaneous definition, here and in similar cases, one can proceed according to the model of the proof that Herr Zermelo has given for Cantor's Well-Ordering Theorem.*) Let $R = \{n_\varrho\}$ be a subset of intervals in which n_0 occurs and, further, in which each n_ϱ is enclosed by all the intervals of smaller index belonging to R (call their aggregate R^{n_ϱ}) and in which n_ϱ has the smallest index among all those intervals of N enclosed by the intervals of R^{n_ϱ}. P is the union of all sets R so characterized. The proof that P is identical with the set constructed by induction above or the direct proof that the defined set P has type $\leqq \omega_\alpha$ and encloses no further interval offers no difficulty.

For a non-dense set, there already is a particular interval (a pair of consecutive elements) that encloses no further interval; thus the type of the interval set in question can be $=1$. For a dense set, the interval set in question is transfinite and cofinal with a regular initial number (that naturally need not be the same for each such interval set). By this, it follows:

Theorem VI. *In each dense set, there are symmetric limits or symmetric gaps, i.e., $\omega_\alpha \omega_\alpha^*$-limits or $\omega_\alpha \omega_\alpha^*$-gaps, where ω_α is a regular initial number.*

§5
Relatively Dense Subsets

The subset A of M is called *dense in M* (dense relative to M) if each middle segment of M contains at least one element of A.**)

In this case, M and A are themselves dense in the absolute sense of the word (§1). If A has a first (last) element, it is also the first (last) element of M, while on the other hand, any boundary element of M need not belong to A. Thus if M has no first (last) element, then A has none and M is coinitial (cofinal) with A. As is easily seen, the analogue of Theorem I (§1) holds.

Theorem VII. *If Q is dense in P and P is dense in M, then Q is also dense in M. If Q is a subset of P and P is a subset of M and Q is dense in M, then Q is dense in P and P is dense in M.*

Moreover: if M_1 is dense in M, A is a piece of M, and A_1 is the set of elements simultaneously belonging to A and M_1, then A_1 is dense in A. For each middle segment of A contains elements of M_1 that, since A is a piece, at the same time belong to A and thus belong to A_1. From this, the result is the following relationship between the elements and gaps of M and M_1, where we take both sets to be without boundary elements. If we

---

*) E. Zermelo, Beweis, daß jede Menge wohlgeordnet werden kann, Math. Ann. 59 (1904). Neuer Beweis für die Möglichkeit einer Wohlordnung, Math. Ann. 65 (1908).

**) [[H 1908, 447n*]] If each end segment (initial segment) of M contains at least one element of A, then M and A are sets without last (first) elements and M is cofinal (coinitial) with A.

consider (α) an element m_1 of M_1, (β) an element m of M that does not belong to M_1, and (γ) a gap of M, we obtain the following corresponding decompositions:

(α) $\quad M = A + m_1 + B, \quad M_1 = A_1 + m_1 + B_1;$
(β) $\quad M = A + m + B, \quad M_1 = A_1 + B_1;$
(γ) $\quad M = A + B, \quad M_1 = A_1 + B_1.$

In all three cases, A_1 is dense in A and B_1 is dense in B; since A has no last element and B has no first element, A is cofinal with A_1 and B is coinitial with B_1. Thus if A is cofinal with ω_α and B is coinitial with ω_β (ω_α, ω_β regular initial numbers), then the same holds for A_1 and B_1, and we have in the case

(α) an element that is an $\omega_\alpha \omega_\beta^*$-element in both sets,

(β) an $\omega_\alpha \omega_\beta^*$-element of M that corresponds to an $\omega_\alpha \omega_\beta^*$-gap of M_1,

(γ) a gap that is an $\omega_\alpha \omega_\beta^*$-gap in both sets.

448 If M_1 is continuous, then cases (β), (γ) are excluded, i.e., M_1 is identical to M. Thus a continuous set is not dense in any set other than itself.

§6
Completion and Filling

The initial pieces A or the decompositions $M = A + B$ of an ordered set can themselves be simply ordered by considering of two initial pieces the one initial piece that is contained in the other as the predecessor. The set of *all* initial pieces so ordered always contains pairs of consecutive elements since the initial interval $M^a + a$ immediately follows after the initial segment M^a.

Let M be an unbounded, dense set; we consider only the initial pieces without last elements and the corresponding decompositions $M = A + B$ ($A \neq 0$, $B \neq 0$). Such a decomposition is either a cut or a gap, depending on whether B has or does not have a first element; the cuts are in one to one correspondence with the elements of M. Now, if ordered in the above way, the cuts taken together form a set $\mathfrak{M}$ that is similar to M, and the gaps taken together form a new set $\overline{\mathfrak{M}}$, all the decompositions together form a third set

$$[\mathfrak{M}] = (\mathfrak{M}, \overline{\mathfrak{M}}),$$

the union of the first two. We call $\overline{\mathfrak{M}}$ the *completion* [[*Ergänzung*]] and $[\mathfrak{M}]$ the *filling* [[*Ausfüllung*]] of $\mathfrak{M}$.

Since the individual nature and designation of the elements of ordered sets are altogether of no concern to us, we can also say: a new element $\overline{m}$ is placed into each gap $M = A + B$, and it is ordered with respect to the elements of M according to the rule $A < \overline{m} < B$. The new elements, which are also hereby ordered among themselves, form a set $\overline{M} = \{\overline{m}\}$, the old and new elements together form a set $[M] = (M, \overline{M})$; $\overline{M}$ is called the completion of M and $[M]$ is called the filling of M.

Example: If M is the set of rational numbers, then $\overline{M}$ is the irrationals and $[M]$ is the set of reals; thus $\overline{\eta} = \iota$ (the type of the irrationals) and $[\eta] = \lambda$ (the unbounded linear continuum).

It now follows by quite simple considerations that $\mathfrak{M}$ is dense in $[\mathfrak{M}]$ and that each subset without a last element has an upper limit in $[\mathfrak{M}]$, unless $[\mathfrak{M}]$ is cofinal with it; for if $\mathfrak{A} = \{A\}$ is such a subset, the union of all the elements of M occurring in A, unless it is M itself, is again an initial piece without a last element and thus an element of $[M]$, and clearly it immediately follows after $\mathfrak{A}$. Thus the continuity of $[\mathfrak{M}]$ follows; furthermore, $[\mathfrak{M}]$ is unbounded since the improper decompositions $0 + M, M + 0$ are excluded. Thus:

Theorem VIII. *Each unbounded dense set is dense in its filling, and its filling is an unbounded continuous set.*

If M itself is continuous, then it is identical to $[M]$ and $\overline{M} = 0$. If M is not continuous, then, for example, $\overline{M}$ can consist of a single element and thus $\overline{M}$ need not be dense, let alone dense in $[M]$. Should this be the case (that $\overline{M}$ is dense in $[M]$), then each middle segment of M must enclose a gap; that is, it must be discontinuous; we then say M is *everywhere discontinuous* [[*überall unstetig*]]. This condition is also sufficient; if M is everywhere discontinuous, then $\overline{M}$ is dense in $[M]$. According to §5, the following decompositions belong to an element of $[M]$, depending on whether it is an element of M or of $\overline{M}$:

$(m) \quad [M] = [A] + m + [B], \quad M = A + m + B, \quad \overline{M} = \overline{A} + \overline{B};$

$(\overline{m}) \quad [M] = [A] + \overline{m} + [B], \quad M = A + B, \qquad \overline{M} = \overline{A} + \overline{m} + \overline{B}$

(in which A comprises the common elements of $[A]$ and M, likewise, $\overline{A}$ the common elements of $[A]$ and $\overline{M}$); by this, $[A]$ is cofinal with A and $\overline{A}$, and $[B]$ is coinitial with B and $\overline{B}$. Thus the gaps of M are in one to one correspondence with the elements of $\overline{M}$, and the gaps of $\overline{M}$ are in one to one correspondence with the elements of M; i.e., for its part, M is the completion of $\overline{M}$. From this immediately follows:

Theorem IX. *If M is unbounded, dense, and everywhere discontinuous, then the same holds for its completion $\overline{M}$; both sets are dense in their common filling $[M]$ and each is the completion of the other.*

If we again attend to the regular initial numbers that belong to $[A]$ and $[B]$, in the general case we get: each $\omega_\alpha \omega_\beta^*$-element of the dense, unbounded set M is also an $\omega_\alpha \omega_\beta^*$-element of $[M]$, and to each $\omega_\alpha \omega_\beta^*$-gap of M, there corresponds an $\omega_\alpha \omega_\beta^*$-element of $[M]$; in the special case where M is everywhere discontinuous: to each $\omega_\alpha \omega_\beta^*$-gap of M, there corresponds an $\omega_\alpha \omega_\beta^*$-element in $\overline{M}$ and $[M]$, and to each $\omega_\alpha \omega_\beta^*$-gap in $\overline{M}$, there corresponds an $\omega_\alpha \omega_\beta^*$-element in M and $[M]$.

If M is unbounded and continuous and P is dense in M, then $[P] = M$. For since M has no gaps, an element of M corresponds to each element and gap of P.

If M is unbounded and P is dense in M, then (by Theorems VII, VIII) P is also dense in $[M]$; so by the remark made above, $[P] = [M]$ and both sets have the same filling.

§7
Splitting into Relatively Dense Subsets

To conclude these basic considerations, we prove the following:

Theorem X. *Each dense set M can be split into two subsets each of which is dense in M.*

Besides the ordering of the elements, we imagine that an arbitrary well-ordering

$$m_0\, m_1\, m_2 \,\cdots\, m_\xi \,\cdots$$

is given to M, and we designate the set of elements with index smaller than ξ by

$$M_\xi = \{m_\eta\}, \quad \eta < \xi.$$

According to the ordering in M, these sets are separated into two summands by m_ξ: $M_\xi = P_\xi + Q_\xi$, $P_\xi < m_\xi < Q_\xi$, one of which can even be empty (for $\xi = 0$, $M_0 = P_0 = Q_0 = 0$). We designate a possible last element of P_ξ by p_ξ and a possible first element of Q_ξ by q_ξ. After this, we define the following partition of the set M into two subsets A and B by induction: if m_ξ has two neighboring elements p_ξ, q_ξ in M_ξ and both of these belong to the set A, then let m_ξ belong to the set B; in all other cases, let m_ξ belong to the set A. Thus the sets A and B are completely determined by the underlying well-ordering. We assert that both sets are dense in M, thus that each middle segment N of M contains elements from both sets. On the basis of the presumed density of M, N certainly contains (infinitely many) elements; let m_ξ, m_η be the two elements of lowest index that lie in N. For these two elements, if one belongs to A and the other to B, then the assertion is correct. However, if they both belong to the same set, let m_ζ be element of lowest index that falls into the middle segment enclosed by m_ξ and m_η; then it is the case that $\zeta > \xi, \eta$, and with the possession of m_ζ, the elements m_ξ, m_η have to be regarded as the neighboring elements p_ζ, q_ζ; so that m_ζ belongs to the set B or to the set A, depending on whether m_ξ, m_η belong to the set A or to the set B. Thus also in this case N contains elements of both subsets. And with this the theorem is proved.

For example, the set of real numbers (or also the set of real algebraic numbers) can be split into the sets of rational and irrational numbers each of which is dense in the overall set.

The theorem can be immediately extended to an arbitrary finite or countable set of subsets. For B, as a dense set, again allows a splitting $B = (B_1, C_1)$, where B_1 and C_1 are dense in B, thus also dense in M (Theorem VII); then $M = (A, B_1, C_1)$, where the three components are dense in M. If one proceeds similarly with $C_1 = (C_2, D_2)$, then $M = (A, B_1, C_2, D_2)$, and so forth. If one adheres to the underlying well-ordering also for subsets of M, then all the subsets $A, B; B_1, C_1; C_2, D_2; \ldots$ are completely determined and non-empty. If L is then the intersection of the sets $B, C_1, D_2, \ldots, L$ can be empty, and in any case, it need not be dense in M. However, the set $A_0 = (A, L)$ is dense in M (Theorem VII), and one also sees that a splitting

$$M = (A_0, B_1, C_2, D_3, \ldots)$$

into a countable set of subsets each of which is dense in M is possible and is unambiguously determined by the well-ordering mentioned above.

Sums, Products, and Powers of Order Types

§8
Sums

As everyone knows, the addition of two or finitely many ordered sets, which we have already used on occasion, is defined as follows: if $M_1 M_2 \cdots M_n$ are ordered sets without common elements, $M = (M_1, M_2, \ldots, M_n)$, their union, can be simply ordered by the stipulations

$$\begin{aligned} m_i &< m_i' \quad \text{in } M \quad \text{if } m_i < m_i' \text{ in } M_i, \\ m_i &< m_k \quad \text{in } M \quad \text{if } i < k \end{aligned}$$

or, briefly, by the stipulation that $M_i < M_k$ ($i < k$), understood in the sense of §1. And the sum is then denoted by

$$M = M_1 + M_2 + \cdots + M_n,$$

and its type is denoted by

$$\mu = \mu_1 + \mu_2 + \cdots + \mu_n.$$

M_i is a piece of M. The commutative law does not hold, but the associative law does:

$$M_1 + M_2 + \cdots + M_n$$
$$= (M_1 + M_2 + \cdots M_h) + (M_{h+1} + \cdots + M_i) + \cdots + (M_{k+1} + \cdots + M_n).$$

For an arbitrary, even transfinite, set of summands the following explanation of addition results. Let $A = \{a\}$ be an arbitrary ordered set, and to each element a let there correspond an ordered set M_a that has no elements in common with the others. The union $M = \{M_a\}$ is simply ordered by the stipulation that $M_a < M_b$ (for $a < b$ in A); the sum and its type shall be denoted by

$$M = \sum_{a}^{A} M_a, \quad \mu = \sum_{a}^{\alpha} \mu_a.$$

The M_a are called *summands* of the sum M, and A is called the *generator* [[*Erzeuger*]] of the sum M; the operation is designated as general or transfinite addition (also as the *insertion* of the summands into the generator), and for a finite generator, it turns into ordinary or finite addition. The summands are pieces of the sum.

In the above manner of writing, the notation for summation elements (standing under $\sum$) is naturally of no concern if only they run through all the elements of the generator (standing above $\sum$); we can use different summation elements with the same generator, or we can use the same summation elements with different generators. This is to be taken note of in what follows. As a consequence, the notation can also be applied to the generator itself, i.e., to each ordered set

$$A = \sum_a^A a, \quad \alpha = \sum_a^\alpha 1,$$

which hereby appears as a simple element sum with itself as generator, while to begin with a type sum can be read as a double element sum

$$M = \sum_a^A M_a = \sum_a^A \left(\sum_b^{M_a} b \right) = \sum_a^A \sum_b^{M_a} b$$

that then reduces to the simple $\sum_b^M b$. Now if we form the type sum $N = \sum_b^M N_b$ with this type sum M as new generator, simple reflection shows that we can collect the summands as groups $P_a = \sum_b^{M_a} N_b$ and with this obtain $N = \sum_a^A P_a$; these formulas give the *associative* law for generalized addition, from which the three-fold element sum

$$N = \sum_a^A \sum_b^{M_a} \sum_c^{N_b} c$$

obviously comes to be read one time as $\sum_a \sum_{bc}$ and another time as $\sum_{ab} \sum_c$.

453   If M is cofinal with the *well-ordered* set A, then M is a type sum with generator A; namely, $M = \sum_a^A B_a$, where B_a is the set of all elements of M that lie between a and all the earlier elements of A with a itself added (in the notation of §1, $B_a = M_{A^a}^a + a$). *In general*, if M is cofinal with A or A is another subset of M, M need in no way be a sum with generator A. On the other hand, if M is a sum with generator A and A has no last element, then μ is cofinal with α.

§9
Finite Products and Powers

For n sets $M_1 M_2 \cdots M_n$, we form the *element combinations* ⟦*Elementverbindungen*⟧

$$X = (x_1 \, x_2 \, x_3 \, \cdots \, x_n),$$

where each x_i runs through all the elements of M_i. The set $\{X\}$ is called the *combination set* ⟦*Verbindungsmenge*⟧ for the sets M_i, and it can be simply ordered by the following stipulations:

$$\begin{aligned}
(x_1 \, x_2 \, x_3 \, \cdots \, x_n) &< (y_1 \, y_2 \, y_3 \, \cdots \, y_n) \quad \text{if} \quad x_1 < y_1 \quad \text{in} \quad M_1, \\
(x_1 \, x_2 \, x_3 \, \cdots \, x_n) &< (x_1 \, y_2 \, y_3 \, \cdots \, y_n) \quad \text{if} \quad x_2 < y_2 \quad \text{in} \quad M_2, \\
(x_1 \, x_2 \, x_3 \, \cdots \, x_n) &< (x_1 \, x_2 \, y_3 \, \cdots \, y_n) \quad \text{if} \quad x_3 < y_3 \quad \text{in} \quad M_3,
\end{aligned}$$

and so forth; so the rank ordering of X and Y is determined by the rank ordering of the *first differing pair of elements* x_i, y_i. Briefly, we call this *rank ordering by first differences.* The combination set, so ordered, is denoted by $M_n \cdots M_3 M_2 M_1$ (not $M_1 M_2 M_3 \cdots M_n$), and its type is denoted by $\mu_n \cdots \mu_3 \mu_2 \mu_1$. With this, the product and, in the case of equal factors, the power M^n, μ^n is accounted for, but only for a finite number of factors. This multiplication is associative, but it is not commutative; it is one-sidedly distributive in relation to addition, namely, with respect to the *last factor*, i.e.,

$$\gamma(\alpha + \beta) = \gamma\alpha + \gamma\beta, \quad \gamma \sum_a^\alpha \beta_a = \sum_a^\alpha \gamma\beta_a.$$

In particular, for $\beta_a = 1$, $\gamma\alpha = \sum_a^\alpha \gamma$; so the product $\gamma\alpha$ is a type sum with generator α and all equal summands γ (the insertion of γ into α).

§10
Type Rings

General addition can be viewed as the fundamental operation for the generation of types from other types; its great importance, though not unlimited, will become clear with the following abstraction.

A system of types is called a type ring, provided that both the sum of two types of the system and the sum formed from infinitely many types of the system with a type of the system as generator again belong to the system.

Thus together with α, β and together with α, β_a both $\alpha + \beta$ and $\sum_a^\alpha \beta_a$ should belong to the system; in particular, the product of two types from the system again belongs to the system. In the case that the type 2 belongs to the system, the call for closure under ordinary addition is superfluous.

The set of types that are common to two or arbitrarily (even infinitely) many type rings is again a type ring — in case it is not empty.

Let $\mathsf{A} = \{\alpha\}$ be an arbitrary type set. The type β is said to be *independent* of A if there is a type ring which contains A, but not β; in the contrary case β is said to be dependent on A. The set $[\mathsf{A}]$ of all types dependent on A*), to which the α's themselves belong, is itself a ring and is contained in every ring that contains A, thus the smallest ring that contains A. If no type in A is dependent on the remaining types of A, A is called a *basis* for the ring $[\mathsf{A}]$. Obviously, the ring $[\mathsf{A}]$ consists of all types that come from the types α by a finite, even if unbounded, number of ordinary additions and by transfinite additions; i.e., each type ξ belonging to $[\mathsf{A}]$ is capable of a representation of the following form (in which the respective generators are distinguished by the index 0):

$$\xi = \sum_a^{\xi_0} \xi_a, \quad \xi_a = \sum_b^{\xi_{a,0}} \xi_{a,b}, \quad \xi_{a,b} = \sum_c^{\xi_{a,b,0}} \xi_{a,b,c}, \quad \cdots$$

that ends in nothing but types $\xi_{a,b,c,\cdots,k}$ that belong to A; only the generators,

---

*) This set is a subset of the ring of all types $\leq \aleph_\nu$ (see below) if $\aleph_\nu$ exceeds the cardinality of all the types α; thus this set is surely not contradictory or "inconsistent."

if they are finite, need not belong to **A**. For example, the type ω^ω belongs to the ring $[\omega]^*)$ since

$$\omega^\omega = \omega + \omega^2 + \omega^3 + \cdots = \sum_n^\omega \omega^n$$

and ω^n arises by $(n-1)$-fold addition with the generator ω.

From this immediately follows:

Theorem XI. *If a property is preserved by ordinary and transfinite addition* (i.e., the sum has the property as soon as the summands have it, respectively, as soon as the summands and the generator have it) *and all the types in the set* **A** *have the property, then all the types in the ring* [**A**] *have it.*

Examples of type rings are very numerous since almost every naturally presented set of types is a type ring. We mention the following cases:

(a) Each *type class* (G. Cantor), i.e., each set of all types of a particular cardinality $\aleph_\alpha$.

The ring property depends on the aleph relation

$$\aleph_\alpha + \aleph_\alpha = \aleph_\alpha \cdot \aleph_\alpha = \aleph_\alpha.$$

(b) The set of all types whose cardinality is smaller than a *regular* aleph $\aleph_\alpha$.

For from $\mathfrak{b} < \aleph_\alpha, \mathfrak{c} < \aleph_\alpha, \mathfrak{c}_b < \aleph_\alpha$, it follows that

$$\mathfrak{b} + \mathfrak{c} < \aleph_\alpha, \quad \sum_\mathfrak{b}^\mathfrak{b} \mathfrak{c}_b < \aleph_\alpha,$$

while for a singular $\aleph_\alpha$ it is only the case that $\sum_\mathfrak{b}^\mathfrak{b} \mathfrak{c}_b \leq \aleph_\alpha$, thus the mentioned type set does not form a ring. By comparison, for arbitrary $\aleph_\alpha$ the set of all types $\leq \aleph_\alpha$ is a ring since $\aleph_{\alpha+1}$ is regular (§3, Theorem III).

Since well-ordered sets remain well-ordered under addition, the following sets are type rings:

(c) each *number class*,

(d) the set of all ordinal numbers that are smaller than a *regular* initial number ω_α.

A basis**) can be immediately given for this ring. Let us mention beforehand: if **A** is an arbitrary type set, the property of being cofinal with a type in the set **A** is preserved by addition; so all the types of the type ring [**A**] have this property, and thus a type that is not cofinal with any type in **A** is independent of **A**. Especially if **A** is a set of regular initial numbers (to which we also going to add, for the moment, the number 1), then each regular initial number not in **A** is independent of **A**, and **A** always forms a

---

*) Confusion with the symbol [ ] for the filling (§6) is probably not to be feared.

**) There is only one basis for a ring of ordinal numbers, from which it easily follows that an ordinal number can never be dependent on greater ordinal numbers

basis. In particular, if
$$\mathsf{A} = 1, \omega_0, \omega_1, \ldots, \omega_\xi, \ldots \qquad (\xi < \alpha)$$
is the set of all regular initial numbers $< \omega_\alpha$ ($\omega_\xi, \omega_\alpha$ regular), then all ordinal numbers $< \omega_\alpha$ are contained in the ring [A] belonging to this basis, as is easily proved by induction. For if $\eta < \omega_\alpha$, then either $\eta = (\eta-1)+1$ or η is cofinal with a number ω_ξ; in this latter case (§8), η is representable as a sum with generator ω_ξ and with summands $< \eta$, and therefore η belongs to [A] in case all the numbers $< \eta$ belong to [A]. Since the numbers $< \omega_\alpha$ form a ring and since [A] is the smallest ring with basis A, both rings are identical; consequently, A is the basis of the ring (d).

For example, the ring [1] contains all finite numbers, [1, ω] contains all numbers of the first two number classes*), and [1, ω, Ω] contains all numbers of the first three number classes. The ring [ω] contains all limit numbers of the second number class; the ring [1, $\omega + 1$] contains all finite numbers and all non-limit numbers of the second number class.

Properties invariant with respect to addition separate out certain subsets from each type ring, which, if they are not empty, are again rings. So within a type ring (e.g., within the ring of all types $\leqq \aleph_\alpha$), the following sets (and others) form a ring:

(e) the unbounded types,

(f) the bounded types,

(g) the types with a last element (e.g., the non-limit numbers of a ring of ordinal numbers),

(h) the types that are cofinal with a fixed type or with a fixed set of types,

(i) the dense, unbounded types; the simplest example of a ring of dense, unbounded, countable types consists of the single type η,

(k) the dense types without a last element, e.g., the ring that consists of the two types η and $1 + \eta$.

§11
The Structure of Arbitrary Types

Let A and B be two arbitrary type sets. The property of a type expressed by the following sentence "β and all its subsets are dependent on A" is preserved by ordinary and transfinite addition. Thus all the types of the ring [B] have it if all the types of B have it. Thus all the types of the ring [A] have it if all the types of A have it, i.e., in this case, the ring [A] also contains all subsets of the types belonging to it.

Now if ω_α is a regular initial number and
$$\mathsf{A} = 1, \omega_0, \omega_1, \ldots, \omega_\zeta, \ldots, \omega_0^*, \omega_1^*, \ldots, \omega_\xi^*, \ldots \qquad (\xi < \alpha)$$
is the set of all regular initial numbers $< \omega_\alpha$ and their inverses, together with the number 1, then the case mentioned above is realized, and the ring [A] also contains all the subsets of the types belonging to it.

---

*) There is no finite basis for the second number class alone.

Now let M be an arbitrary set whose cardinality is $< \aleph_\alpha$. For the moment, we call two elements a and b *coherent* if the type of their interval $[a, b]$ belongs to the ring [A]. Accordingly, we collect all the elements coherent with a and which lie to the left and right of a, and we combine them along with a into the subset

$$M(a) = M'(a) + a + M''(a).$$

Here the following holds:

If (for $a < b < c$) a, b and b, c are coherent, then a, c are coherent too; conversely, if a, c are coherent, then a, b and b, c are coherent too.

For if $[a, b]$ and $[b, c]$ belong to the ring [A], then so does $M_b^c + c$ and the sum $[a, b] + M_b^c + c = [a, c]$; and if $[a, c]$ belongs to the ring [A], so do its subsets $[a, b]$ and $[b, c]$.

Thus if two elements are coherent with a third element, they are coherent with each other; i.e., if a, b are coherent, then $M(a) = M(b)$. If a, b are not coherent and $a < b$, then $M(a) < M(b)$; the subsets $M(a)$ are therefore *pieces* of M. They themselves of course belong to the ring [A]. Namely, if $M''(a)$ has a last element b, then, as part of $[a, b]$, $M''(a)$ belongs to [A]. If it has no last element, it is cofinal with a number ω_ξ from A, and thus (§8) it is a sum with generator ω_ξ and with summands that, as parts of intervals $[a, b]$, belong to our ring; consequently, even in this case, $M''(a)$ belongs to [A]; so $M''(a), M'(a)$, and $M(a)$ belong to the ring [A].

From this comes the fact that M is the sum of all these pieces,

$$M = \sum_a^A M(a),$$

in which the generator A designates a subset of M such that each element m is coherent with one and only one element a (such a set can be immediately defined in the well-known way from any well-ordering of M). If M itself belongs to the ring [A], then A has type 1; if not, then A is a *dense* set since for two consecutive elements a, b the corresponding pieces $M(a) + M(b)$ would have to contract to a single one.

If we say that a set is *scattered* [[*zerstreut*]] *if it contains no dense subset*, then all types of the set A have the property of scatteredness. As is easily seen, this property is preserved by addition, and so all the types of [A] have the property. However among the sets $< \aleph_\alpha$, only these types have the property since a set that does not belong to [A] is a type sum with a dense generator. The ring [A] can thus be defined as the *totality of scattered types* of cardinality $< \aleph_\alpha$, and we have:

Theorem XII. *Each set is either scattered or a sum of scattered sets with a dense generator. The scattered sets of cardinality $< \aleph_\alpha$ ($\aleph_\alpha$ regular) form a ring whose basis consists of all the regular initial numbers $< \omega_\alpha$ (including the number 1) and their inverses.*

For example, each countable type is either scattered, that is, belongs to the ring $[1, \omega, \omega^*]$, or it results from the summation of scattered types over a countable, dense generator. Since there are only four countable, dense types $(\eta, 1+\eta, \eta+1, 1+\eta+1)$, all types of the first two type classes belong to the ring $[1, \omega, \omega^*, \eta]$, which has a *finite basis*. However, in all likelihood this noteworthy result stands here isolated because there exist infinitely many independent dense types of the cardinality of the continuum on up (§25).

§12
Combination Sets and Covering Sets

By the invariance of certain properties under ordinary and transfinite addition, the range of these operations is seriously limited; thus, for example, dense types are unable to be constructed from non-dense types. Hence we must go on, and we must try to advance in a direction in which we also extend *multiplication* and the *raising to powers* that is connected with it to an arbitrary, even transfinite, set of factors.

Let $A = \sum a$ be an arbitrary ordered set (type α, cardinality $\mathfrak{a}$), and to each element a, let there correspond an arbitrary ordered set M_a (type μ_a, cardinality $\mathfrak{m}_a$). The (unordered) *combination set* of all the sets M_a can then be defined as the set of all element combinations

$$X = \sum_{a}^{A} x_a,$$

where each x_a runs through all the elements of M_a, and the cardinal number of the set $\{X\}$ can be defined to be the product of all the cardinalities $\mathfrak{m}_a$, and in the case where all these cardinalities are equal, it can be defined to be the power $\mathfrak{m}^{\mathfrak{a}}$. As is well known, this definition for power of cardinal numbers was given by Herr G. Cantor, and the one for product was given by Herr A. N. Whitehead. However, up until now there is no known method that in each case *orders* the *entire* combination set by a simple law, and we must content ourselves by selecting certain *subsets* of the combination set that are capable of a simple ordering that satisfies the demands for multiplication or raising to powers.

We call the sets M_a or their types μ_a the *factors*, and we call the set A or its type α the *argument* of the combination set (so the factors correspond to the summands and the argument to the generator of a type sum); the elements a of the argument may be designated as *places* [[Stellen]]. If all the factors are similar to the set M, $\mu_a = \mu$, then we also call an element combination a *covering* [[Belegung]] of the set A by M or of α by μ, and we call the combination set the *covering set* [[Belegungsmenge]]; the set M or its type μ is designated as the *base* [[Basis]].

In a special case that we handle to begin with, an ordering of the whole combination set immediately presents itself, namely, if the *argument* is *well-ordered*, if α is an ordinal number. Here, the set of places in which two combinations

$$X = \sum_{a} x_a, \quad Y = \sum_{a} y_a$$

differ, i.e., where $x_a \neq y_a$ in M_a, as a subset of A, is likewise well-ordered and has a first element; i.e., there is a place a where $x_a \lessgtr y_a$, while for all earlier places ($b < a$) $x_b = y_b$. It is obvious to stipulate the relation $X \lessgtr Y$, thus to give the combinations X and Y the rank ordering of first differing elements and with this to carry over the *rank ordering by first differences* (§9) to this case. As we shall show, the set $\{X\}$, so ordered, has the formal properties of a product in which, however, as in the case of finite arguments, the sequence of factors is the *reverse* of the sequence of places in the argument. Thus if we designate the places by

$$0 \ 1 \ 2 \cdots \xi \cdots \quad (\xi < \alpha),$$

the product is designated by $\cdots M_\xi \cdots M_2 \, M_1 \, M_0$, and in the case of equal factors, the power is designated by M^{A^*} (not by $M_0 \, M_1 \, M_2 \cdots M_\xi \cdots$, respectively, M^A).

This is a consequence of Cantor's newer product notation[*]), with which we of course had to comply, and of rank ordering by first differences, which, as less sacrosanct, could certainly be changed to rank ordering by last differences by inverting the argument; however, this would be altogether unsuitable since the very types of interest are more easily represented in our notation. Hence we will have to distinguish the *argument* A and the *exponent* A^* in powers.

We denote the above covering sets, respectively, their types by

$$\left(\!\!\left(\prod_{a}^{A} M_a\right)\!\!\right) = ((\cdots M_\xi \cdots M_2 \, M_1 \, M_0)),$$

$$\left(\!\!\left(\prod_{a}^{\alpha} \mu_a\right)\!\!\right) = ((\cdots \mu_\xi \cdots \mu_2 \, \mu_1 \, \mu_0)),$$

stipulating that in $\prod_a^A$ the factors of the place sequence are to be read from right to left, thus the reverse of how we read the summands in $\sum_a^A$. We call this the *complete product* [[*Vollprodukt*]] of the factors M_a with argument A. If all the factors are equal, we denote the covering set and its type by

$$M((A)), \quad \mu((\alpha)),$$

and we call them the *complete power* [[*Vollpotenz*]] with base M and argument A.

Examples. The complete power $\omega((\omega))$ is the type of the set of number sequences

$$X = (x_0 \, x_1 \, x_2 \cdots)$$

ordered by first differences, where each x_n runs through the set of natural numbers $1, 2, 3, \ldots$, taken in their natural rank ordering. If we think of

---

[*]) Herr G. Cantor has just reversed his old notation in consideration of a suitable way of writing powers of ordinal numbers (§16).

assigning to each such number sequence X a real number x through the dyadic expansion
$$x = (\tfrac{1}{2})^{x_0} + (\tfrac{1}{2})^{x_0+x_1} + (\tfrac{1}{2})^{x_0+x_1+x_2} + \cdots,$$
then $0 < x \leqq 1$, furthermore, $x > y$ for $X < Y$. Conversely, to each real number of the domain $0 < x \leqq 1$ there corresponds a unique (endless) dyadic expansion and a unique number sequence. Therefore $\omega((\omega)) = 1 + \lambda$, precisely the left bounded linear continuum.

The complete product $((\prod_n^\omega \mu_n)) = ((\cdots \mu_2 \mu_1 \mu_0))$, where $\mu_{2n} = \omega$ and $\mu_{2n+1} = \omega^*$, is the type of the set of number sequences
$$X = (x_0, -x_1, x_2, -x_3, x_4, \ldots)$$
ordered by first differences, where again $x_n = 1, 2, 3, \ldots$. If we assign to the number sequence X the continued fraction
$$x = x_0 + 1 : x_1 + 1 : x_2 + 1 : x_3 + \cdots,$$
then $x > 1$ and $x < y$ for $X < Y$. To each irrational number $x > 1$ there corresponds a distinct number sequence X. The above type, which can obviously be interpreted as the complete power $\varphi((\omega))$ where $\varphi = \omega^*\omega$, is thus identical to the type ι of the set of irrational numbers.

§13
General Products and Powers

Now let $A = \{a\}$ be an arbitrary argument, and let an arbitrary set M_a be assigned to each place a. We keep the rank ordering by first differences for the combinations $X = \sum_a^A x_a$ (x_a runs through M_a); i.e., for the two combinations X and Y, if the set of differing places where $x_a \neq y_a$ has a first element, and for this first place a, if the rank relation $x_a \lesseqgtr y_a$ exists in M_a, then $X \lesseqgtr Y$ shall also hold. In general, the combination set is not ordered by this, since the former place set need not even have a first element. However, in so far as the relations $X \lesseqgtr Y$ hold at all, they are asymmetric and transitive, i.e.,

if $X < Y$, then $Y > X$,

if $X < Y$, $Y < Z$, then $X < Z$.

Both assertions are evident, the second with the additional remark that if a_1 is the first differing place between X and Y and a_2 is the first differing place between Y and Z, then the smaller of the two (i.e., a_1 for $a_1 \leqq a_2$ or a_2 for $a_2 \leqq a_1$) is the first differing place between X and Z.

So in order to obtain a subset of the combination set that is ordered by the principle of first differences, we must see to it that the set of differing places for two combinations X and Y has to have a first element. To this end, we make the following arrangement: in each set M_a, let an arbitrary, but fixed, element p_a be chosen and designated as the *principal element* [[*Hauptelement*]], and let all the remaining elements be designated as secondary elements [[Nebenelemente]]. *Then we only allow such combinations*

X where the places covered by secondary elements form a well-ordered subset of *A*. Two such combinations can only differ in those places that in one or both are covered by secondary elements. Thus if *B* is the subset of *A* that bears the secondary elements of *X* and *C* is the same for *Y* and $D = \mathfrak{M}(B,C)$ is the union of both, then the set of differing places is a subset of *D* and, as *B*, *C*, and *D* are well-ordered, it thus has a first element.

The total set *T* of these special combinations (coverings) is, accordingly, simply ordered by the principle of first differences, and we will show that it has the character of a product (of a power). However, it has to be most strongly emphasized that this set *T* depends not only on the factors and the argument but also on the choice of all the principal elements p_a, whose aggregate, the *principal combination* (principal covering) $P = \sum_a^A p_a$ (in which certainly no secondary element appears), belongs to *T* in any case. This ambiguity of the product, in as much as it appears as a defect of the method, is unavoidable; on the other hand, it is an advantage since the domain of types that are representable by simpler types in product or power form is enormously extended.

However, the power concept includes even more: not only the total set *T* of these special combinations, but also other of its subsets have multiplicative character as well. Namely, if we understand by T^π the set of those combinations $X = \sum_a^A x_a$ in which the set of places covered by secondary elements is well-ordered and $< \omega_\pi$ (so by T^0 we understand those sets in which only a finite number of secondary elements occur), then we will show that T^π also has the character of a product $\prod_a^A M_a$ in case ω_π is a *regular* initial number. Thus these sets are

$$T^0 \, T^1 \, T^2 \, \cdots \, T^{\omega+1} \, \cdots .$$

Each of them is a subset of any of the following ones and of *T*; and the set T^π is identical with *T* from a certain index on, namely, as soon as ω_π has reached the first regular initial number not contained in the argument *A*. All these sets depend, aside from their index π, on the factors M_a, the argument *A*, and on the principal combination *P*, which belongs to each set T^π.

We now introduce the following names and notations. The set T^π is called the *product of the* $(1 + \pi)$*th class**) *of the factors* M_a *with argument A and with principal combination P*; it and its type are denoted by

$$\left(\prod_a^A M_a \right)_P^\pi, \quad \left(\prod_a^\alpha \mu_a \right)_P^\pi,$$

where in the product the factors are to be read from right to left; thus M_b stands before M_a if $a < b$ in *A*. The set *T* is called the *maximal product*

---

*) Thus for transfinite π, the πth class; for finite π, the notation follows that of Cantor's number classes.

and is denoted either by bringing in the appropriate index π or by

$$\left(\left(\prod_a^A M_a\right)\right)_P, \quad \left(\left(\prod_a^\alpha \mu_a\right)\right)_P.$$

In the case of a well-ordered argument A, where *all* combinations fulfill the stated conditions, the whole combination set is included in T and the principal combination has no effect; the maximal product (however, apart from this, even here the products of lower class that depend on P exist) turns into the earlier considered *complete product*

$$\left(\left(\prod_a^A M_a\right)\right), \quad \left(\left(\prod_a^\alpha \mu_a\right)\right).$$

If all the factors are equal, $M_a = M$, and the principal element is the same, $p_a = p$, we designate the set T^π as the *power of the $(1+\pi)$th class of base M with argument A and with principal element p*. And for this, we write

$$M(A)_p^\pi, \quad \mu(\alpha)_p^\pi,$$

and for the set T as *maximal power*, we write

$$M((A))_p, \quad \mu((\alpha))_p,$$

and for the case of well-ordered arguments, as the *complete power* we again write

$$M((A)), \quad \mu((\alpha)).$$

§14
The Associative Law

464

In order to justify the names introduced, the permanence of the laws of ordinary type multiplication has to be established, primarily that of the associative law, according to which the factors of a product can be collected groupwise in any fashion.

Thus let $(\prod_a^A M_a)_P^\pi$ be a product of the $(1+\pi)$th class with factors M_a, argument A, and principal combination $P = \sum_a^A p_a$, where p_a is the principal element of M_a. The product mentioned above is the set, ordered by first differences, of those combinations $X = \sum_a^A x_a$ where each x_a runs through the set M_a, but the set A^0 of places a^0 where the secondary elements ($x_a \neq p_a$) appear is well-ordered and smaller than the regular initial number ω_π.

Now let A be a type sum $A = \sum_b^B A_b$. According to the associative law for addition, each combination X decomposes through this into

$$X = \sum_a^A x_a = \sum_b^B \sum_a^{A_b} x_a = \sum_b^B X_b,$$

and in particular, the principal combination decomposes into

$$P = \sum_a^A p_a = \sum_b^B \sum_a^{A_b} p_a = \sum_b^B P_b,$$

where

$$X_b = \sum_a^{A_b} x_a, \quad P_b = \sum_a^{A_b} p_a;$$

i.e., X is now a combination with argument B of the elements X_b, which for their part are pieces of X and combinations of the elements x_a with argument A_b.

Here first of all, the rank ordering of the X by the first differences of the x_a is now identical to the rank ordering of the X by the first differences of the X_b if the X_b themselves are ordered by the first differences of the x_a. For if

$$X = \sum_a^A x_a = \sum_b^B X_b, \quad Y = \sum_a^A y_a = \sum_b^B Y_b$$

are two combinations and $X < Y$, then there exists a first differing place a with $x_a < y_a$ and $x_u = y_u$ for $u < a$; then of course, if a also belongs to A_b, a is the first differing place for X_b and Y_b, i.e., $X_b < Y_b$ and $X_v = Y_v$ for $v < b$.

Moreover, the set A^0 of the places a^0 in X covered by secondary elements decomposes into

$$A^0 = \sum_b^B A_b^0 = \sum_{b^0}^{B^0} A_{b^0}^0,$$

where we have denoted those places b of B where $A_b^0 \neq 0$ by b^0 and their set by B^0. For the remaining places b where $A_b^0 = 0$, A_b is free of places a^0 and thus $X_b = P_b$. Since it was assumed that $A^0 < \omega_\pi$, the subsets $B^0, A_{b^0}^0$ are also well-ordered and $< \omega_\pi$. Conversely, if B^0 and all $A_{b^0}^0$ are well-ordered and $< \omega_\pi$, then A^0, as a type sum, is likewise well-ordered and $< \omega_\pi$ since, along with the other sets, it belongs to the ring of all numbers $< \omega_\pi$ (§10, (d)); for this the hypothesis that ω_π is a *regular* initial number is essential. If ω_π were singular or generally not an initial number, the numbers $< \omega_\pi$ would not form a ring, and thus it could be that $B^0 < \omega_\pi, A_{b^0}^0 < \omega_\pi$, yet $A^0 \geqq \omega_\pi$.

As a result of these last considerations, the following two definitions define the same set $\{X\}$ with the same ordering:

(a) In $X = \sum_a^A x_a$ the set of places where $x_a \neq p_a$ is well-ordered and $< \omega_\pi$.

(b) In $X = \sum_b^B X_b$ the set of places where $X_b \neq P_b$ is well-ordered and $< \omega_\pi$, and conjointly, in each $X_b = \sum_a^{A_b} x_a$ the set of places where $x_a \neq p_a$ is well-ordered and $< \omega_\pi$.

In the first definition, each x_a runs through the set M_a, and the element combination X runs through the product $(\prod_a^A M_a)_P^\pi$ with principal combination $P = \sum_a^A p_a$; in the second definition, each element combination X_b runs through the product

$$N_b = \Big(\prod_a^{A_b} M_a\Big)_{P_b}^\pi$$

with principal combination $P_b = \sum_a^{A_b} p_a$, and X runs through the product $(\prod_b^B N_b)_P^\pi$ with principal combination $P = \sum_b^B P_b$. The equality of both products

$$\Big(\prod_a^A M_a\Big)_P^\pi = \Big(\prod_b^B N_b\Big)_P^\pi$$

is the content of the associative law that hereby is proved in a most general version. This says that by collecting factors into arbitrary groups a product of any class can be turned into a product of the same class of products of the same class, whereby only the choice of the new principal elements is not arbitrary, but rather is prescribed by the formula $P_b = \sum_a^{A_b} p_a$; i.e., if a product figures as a new factor, its principal combination figures as a new principal element.

If we add that the commutative law obviously does not hold in the multiplication we defined, but that the (one-sided) distributive law, which allows the resolution of the last factor into a sum, does hold, that 1 is the modulus of multiplication, i.e., in each product an arbitrary (even transfinite) set of factors $=1$ can be stricken or inserted, and that a product vanishes if and only if one of its factors vanishes, then with these (easily proved) propositions, it is shown that our products are admissible generalizations of products with a finite number of factors.

Among the numerous special cases of the associative law, we content ourselves to establish the two most important *power formulas* ($M_a = M, p_a = p$). One obtains for $A = A_1 + A_2$,

$$M(A_1 + A_2)_p^\pi = M(A_2)_p^\pi \cdot M(A_1)_p^\pi,$$

and for $A_b = C$, $A = CB$, $q = \sum_a^C p$,

$$M(CB)_p^\pi = N(B)_q^\pi, \quad N = M(C)_p^\pi,$$

where q (the earlier P_b) represents the principal covering of C by M and the new principal element of the base N as well. So two powers with the same base are multiplied by adding the arguments in reverse order; a power is raised to a power by multiplying the arguments; here of course we are talking about powers of the same class and with the same principal element (respectively, those defined by the above formulas). The first formula confirms that it is not the argument but rather the inverse of the argument that formally behaves as an exponent; for if one writes M^{A^*} for $M(A)_p^\pi$, then

$$M^{(A_1+A_2)^*} = M^{A_2^*+A_1^*} = M^{A_2^*} \cdot M^{A_1^*}$$

or

$$M^{B_2+B_1} = M^{B_2} \cdot M^{B_1}.$$

Thus two powers with the same base are multiplied by adding the exponents in the same order.

§15
Segments of Product Types

Let $N = (\prod_a^A M_a)_P^\pi$ be the product of the $(1+\pi)$th class of the factors M_a with argument A and principal combination $P = \sum_a^A p_a$. Moreover, let $X = \sum_a^A x_a$ be one of the allowable combinations, thus an element of N; we want to determine the accompanying initial segment N^X and end segment N_X.

To each place a of the argument, there corresponds the following decompositions

$$A = A^a + a + A_a,$$
$$X = X^a + x_a + X_a,$$
$$X^a = \sum_b^{A^a} x_b, \quad X_a = \sum_b^{A_a} x_b.$$

Now let $U(< X)$ be an element belonging to N^X, and let a be the first differing place between U and X, thus

$$U = X^a + u_a + U_a, \quad u_a < x_a.$$

With a fixed, if we let U vary in all possible ways, then u_a runs through the initial segment $(M_a)^{x_a}$; on the other hand, U_a runs through all combinations with argument A_a and with principal combination $P_a = \sum_b^{A_a} p_b$, under the restriction that all the places covered by secondary elements form a well-ordered set $< \omega_\pi$. Thus U_a runs through the product $(\prod_b^{A_a} M_b)_{P_a}^\pi$ and $u_a + U_a$ runs through the product that results from right multiplication of this last product by $(M_a)^{x_a}$. Furthermore, if U and U' are two combinations whose differing places with X are the distinct a and a', then $U < U'$ if $a < a'$ in A. Therefore N^X turns out to be the type sum:

$$N^X = \sum_a^A \left(\prod_b^{A_a} M_b \right)_{P_a}^\pi \cdot (M_a)^{x_a}.$$

The analogous result holds for combinations $U > X$, only here u_a runs through the end segment $(M_a)_{x_a}$ and $U > U'$ for $a < a'$, so that N_X is a type sum with generator A^*:

$$N_X = \sum_a^{A^*} \left(\prod_b^{A_a} M_b \right)_{P_a}^\pi \cdot (M_a)_{x_a}.$$

We omit a rather complicated formula for middle segments N_X^Y.

For powers of base M with principal element p, these formulas read:

$$\left.\begin{aligned} N^X &= \sum_a^A M(A_a)_p^\pi \cdot M^{x_a} \\ N_X &= \sum_a^{A^*} M(A_a)_p^\pi \cdot M_{x_a} \end{aligned}\right\} \quad N = M(A)_p^\pi.$$

In particular, one obtains the equation $N = N^P + P + N_P$ or

$$M(A)_p^\pi = \sum_a^A M(A_a)_p^\pi \cdot M^P + P + \sum_a^{A^*} M(A_a)_p^\pi \cdot M_p.$$

§16
Cantor's Powers and Products

We now suppose that the *argument is the inverse of a well-ordered set*, thus (in the case of powers) that the exponent be well-ordered. Then each well-ordered subset of the argument is finite, and all the product classes and power classes coincide with the first one, which already contains the maximal product and maximal power (§13; it is the case that $T = T^0 = T^1 = \cdots$). The last equation of §15 becomes

$$M((A))_p = \sum_a^A M((A_a))_p \cdot M^P + P + \sum_a^{A^*} M((A_a))_p \cdot M_p$$

or in type form

$$\mu((\alpha))_p = \sum_a^\alpha \mu((\alpha_a))_p \cdot \mu^p + 1 + \sum_a^{\alpha^*} \mu((\alpha_a))_p \cdot \mu_p.$$

We are going to set $\alpha = \beta^*$ in this, so that the exponent β is an ordinal number and the places a are likewise denoted by ordinal numbers

$$\beta = (0\ 1\ 2\ \cdots\ \eta\ \cdots), \quad \eta < \beta,$$

$$\alpha = (\cdots\ \eta\ \cdots\ 2\ 1\ 0);$$

then $\alpha_\eta = \eta^*$ and it follows that

$$\mu((\beta^*))_p = \sum_\eta^{\beta^*} \mu((\eta^*))_p \cdot \mu^p + 1 + \sum_\eta^{\beta} \mu((\eta^*))_p \cdot \mu_p,$$

i.e., a recursion formula that expresses a power with exponent β through powers with lower exponents η.

Here now is the most important special case: the base μ is also well-ordered and its first element is the principal element p. Since in this specialization the power only depends on the base and the exponent, we write

μ^β for $\mu((\beta^*))_p$. Thereby *) $\mu^p = 0$, $\mu_p = -1 + \mu$, and consequently

$$\mu^\beta = 1 + \sum_\eta^\beta \mu^\eta \cdot (-1 + \mu).$$

By induction, it follows from this that μ^β itself is an ordinal number since μ^β is an ordinal number if all μ^η ($\eta < \beta$) are ordinal numbers. If β is a limit number, then for each initial segment of μ^β there is a power μ^η that surpasses it, thus $\mu^\beta = \lim\{\mu^\eta\}$. In addition, since the power rule $\mu^{\beta+\gamma} = \mu^\beta \cdot \mu^\gamma$ holds and $\mu^1 = \mu$, these powers completely agree with those that Herr Cantor defined by induction (Math. Ann. 49, pp. 231–35). The following sums up their place in our general system of powers:

Theorem XIII. *Cantor's powers are those where the argument is the inverse of an ordinal number, thus the exponent is an ordinal number, and the base is likewise an ordinal number whose first element is the principal element. In this case, there are only maximal powers.*

The corresponding holds for products with inverse ordinal numbers as arguments

$$\left(\left(\prod_\eta^{\beta^*} \mu_\eta\right)\right)_P = ((\mu_0 \mu_1 \mu_2 \cdots \mu_\eta \cdots))_P,$$

which are thus maximal products; if the factors are ordinal numbers and the principal elements p_η are their first elements, then they turn into "Cantor's products" $\mu_0 \mu_1 \mu_2 \cdots \mu_\eta \cdots$, which also like the powers can be defined by induction (for example, $\mu_0 \mu_1 \mu_2 \cdots$ is the limit of $\mu_0, \mu_0 \mu_1, \mu_0 \mu_1 \mu_2, \ldots$) and themselves are ordinal numbers.

Cantor's powers where the base is an initial number are called *principal numbers* [[*Hauptzahlen*]]. A principal number is similar to all its end pieces and vice versa.

§17
The Case of Well-Ordered Arguments

Let the ordinal number σ be the type of a well-ordered set whose elements we denote by $0\ 1\ 2 \cdots \varrho \cdots$ ($\varrho < \sigma$); we form the element combinations with argument σ

$$X = \sum_\varrho^\sigma x_\varrho = x_0 + x_1 + x_2 + \cdots + x_\varrho + \cdots,$$

and for each place ϱ, in order to be able to distinguish between the places before and after ϱ in case of need, we think of them decomposed into

$$X = X^\varrho + x_\varrho + X_\varrho \quad (= x_0 + X_0).$$

---

*) $-\lambda + \mu$ denotes the uniquely determined ordinal number that satisfies the equation $\lambda + \xi = \mu$ for $\lambda < \mu$ (λ, μ ordinal numbers). If the equation $\xi + \lambda = \mu$ is solvable, call its smallest solution $\mu - \lambda$.

If each x_ϱ runs through the set M_ϱ, the X ordered by first differences form the complete product $((\prod_\varrho^\sigma M_\varrho))$, and in the case of equal factors, the complete power $M((\sigma))$. However, later (§21) we need a further generalization that gives up the character of a product; namely, we will allow the x_ϱ to run through a set $M(X^\varrho)$ that can depend not only on the place ϱ alone but also on the preceding elements $x_0\, x_1 \cdots$. Such a combination set ordered by first differences may (not quite suitably) be designated as a *product with variable factors*. Since the first differing elements of two combinations always belong to *one* set $M(X^\varrho)$, only the types of the variable factors $M(X^\varrho)$ come into consideration for the type of the combination set, whereas what relationship we assume between two such different sets $M(X^v)$ and $M(Y^\tau)$ plays absolutely no role; these sets may have elements in common or not; they may have interrelated rank orderings or not; they may be separated or any arbitrarily situated subsets of an ordered set. — For $\sigma = 2$, where x_0 runs through a set M_0 and x_1 runs through a set $M(x_0)$ dependent on x_0, the combination set is similar to the type sum $\sum_{x_0}^{M_0} M(x_0)$ (in which, of course, we had to assume that all the sets $M(x_0)$ have no elements in common); thus the product with variable factors is to the product $((\prod_\varrho^\sigma M_\varrho))$, perhaps, as a transfinite type sum is to the product $M_1 M_0$ of two factors, into which it turns with the equality of all summands.

If we represent the variable factors as subsets of one and the same ordered set M (for example, by assuming any two factors are without common elements, M can be formed by the summation of all the $M(X^\varrho)$ with some generator), then the product is a subset of the complete power $M((\sigma))$.

Let us now consider only coverings of σ by M; thus let us assume that all the elements x_ϱ of X belong to a single set M. Let

$$A < B < C < \cdots < L < \cdots$$

be a well-ordered sequence of coverings with type the regular initial number ω_ν; we are going to investigate the possible structures of such a *sequence of coverings*. At first sight, two special cases come to the fore; namely, the place of the first difference between any two consecutive coverings is continually growing and gives an ω_ν-sequence in the *argument* σ, or the place of the first difference between any two coverings is always the same and the first differing elements run through an ω_ν-sequence of the *base* M. In the first case,

(A) $\begin{cases} A = X^\alpha + a_\alpha + A_\alpha & (\alpha < \beta < \gamma < \cdots < \lambda < \cdots) \\ B = X^\beta + b_\beta + B_\beta & (a_\alpha < x_\alpha, \\ C = X^\gamma + c_\gamma + C_\gamma & b_\beta < x_\beta, \\ \cdots\cdots\cdots & c_\gamma < x_\gamma, \ldots \\ L = X^\lambda + l_\lambda + L_\lambda & l_\lambda < x_\lambda, \ldots), \\ \cdots\cdots\cdots & \end{cases}$

and in the second case,

(B)
$$\begin{cases} A = X^\varrho + a_\varrho + A_\varrho \\ B = X^\varrho + b_\varrho + B_\varrho \\ C = X^\varrho + c_\varrho + C_\varrho \\ \cdots\cdots\cdots \\ L = X^\varrho + l_\varrho + L_\varrho \\ \cdots\cdots\cdots \end{cases} \qquad (a_\varrho < b_\varrho < c_\varrho < \cdots < l_\varrho < \cdots).$$

Thus for the first case, let it be assumed that ω_ν is contained in the argument; we say the sequence of coverings is an *argument-like* [[*argumentale*]] sequence. For the second case, let it be assumed that ω_ν is contained in the base; we say the sequence of coverings is a *base-like* [[*basische*]] sequence.

We now prove:

Theorem XIV. *Each ω_ν-sequence of coverings is cofinal with either an argument-like ω_ν-sequence or a base-like ω_ν-sequence.*

Let us mention as preliminary: if there exists a function $\varphi(\alpha)$ on the sequence of ordinal numbers $< \omega_\nu$

$$0\ 1\ 2\ \cdots\ \alpha\ \cdots \qquad (\alpha < \omega_\nu)$$

that assigns to each number α a greater number of the sequence ($\alpha < \varphi(\alpha) < \omega_\nu$), then this function defines certain subsets

$$\alpha_0\ \alpha_1\ \alpha_2\ \cdots\ \alpha_\beta\ \cdots$$

according to the following rules:

$$\alpha_0 = 0, \quad \alpha_{\beta+1} = \varphi(\alpha_\beta),$$
$$\alpha_{\lim\{\beta\}} = \lim\{\alpha_\beta\}.$$

The most comprehensive subset of this kind (the union of all of them) is thus uniquely defined by $\varphi(\alpha)$, and since the total set is cofinal with it, this subset is likewise of type ω_ν if ω_ν is a regular initial number; hence the index β again runs through all ordinal numbers $< \omega_\nu$.

Now let

$$X(0)\ X(1)\ X(2)\ \cdots\ X(\alpha)\ \cdots$$

be an ω_ν-sequence of coverings where $X(\alpha) < X(\beta)$ for $\alpha < \beta$. We denote the place of the first difference between between $X(\alpha)$ and $X(\beta)$ by (α, β); (α, β) is an ordinal number $< \sigma$ (σ is the argument of the coverings). According to an earlier observation (§13), for $\alpha < \beta < \gamma$, the number (α, γ) is equal to the smaller of the two numbers (α, β) and (β, γ), thus

$$(\alpha, \gamma) \leqq (\alpha, \beta), \quad (\alpha, \gamma) \leqq (\beta, \gamma).$$

For fixed α, the ordinal numbers

$$(\alpha, \alpha+1)\ (\alpha, \alpha+2)\ \cdots\ (\alpha, \beta)\ \cdots$$

never increase and hence must have a minimum $\pi(\alpha)$ that first occurs for $\beta = \varphi(\alpha)$; thus

$$(\alpha, \beta) = \pi(\alpha) \quad \text{for } \beta \geqq \varphi(\alpha) > \alpha.$$

As above, with the help of this function $\varphi(\alpha)$, form the accompanying subset $\alpha_0\, \alpha_1\, \cdots\, \alpha_\beta\, \cdots$, so that the original sequence of coverings is cofinal with the following sequence
$$X(\alpha_0)\ X(\alpha_1)\ X(\alpha_2)\ \cdots\ X(\alpha_\beta)\ \cdots,$$
which we now also denote by
$$Y(0)\ Y(1)\ Y(2)\ \cdots\ Y(\beta)\ \cdots.$$
In this sequence, for $\beta < \gamma < \delta$
$$\begin{aligned}\alpha_\gamma &\geqq \alpha_{\beta+1} = \varphi(\alpha_\beta),\\ (\alpha_\beta, \alpha_\gamma) &= \pi(\alpha_\beta) = \varrho(\beta),\\ (\alpha_\beta, \alpha_\delta) &\leqq (\alpha_\gamma, \alpha_\delta), \quad \text{i.e.,} \quad \varrho(\beta) \leqq \varrho(\gamma);\end{aligned}$$
hence each $Y(\beta)$ has the same first differing place $\varrho(\beta)$ with all the sequences that follow it, and the numbers $\varrho(\beta)$ never decrease as the index grows. Here, it is now to be determined whether these numbers have a maximum or not. *If they have no maximum*, for each $\varrho(\beta)$ there exists a $\varrho(\gamma) > \varrho(\beta)$; let $\psi(\beta)$ be the smallest value of γ for which this occurs, thus
$$\varrho(\gamma) > \varrho(\beta) \quad \text{for} \quad \gamma \geqq \psi(\beta) > \beta.$$
With this function $\psi(\beta)$, from the sequence of all indices β, we again form the above defined subsequence $\beta_0\, \beta_1\, \cdots\, \beta_\gamma\, \cdots$; and we obtain a third sequence of coverings
$$Y(\beta_0)\ Y(\beta_1)\ Y(\beta_2)\ \cdots\ Y(\beta_\gamma)\ \cdots,$$
with which the second sequence is cofinal and, consequently, with which the first sequence is cofinal too. In these coverings, for $\gamma < \delta$
$$\beta_\delta \geqq \beta_{\gamma+1} = \psi(\beta_\gamma),$$
hence
$$\varrho(\beta_\delta) > \varrho(\beta_\gamma);$$
so the first differing place of the covering $Y(\beta_\gamma)$ with all the coverings that follow it grows with increasing γ, and we have an argument-like sequence before us. *If the numbers $\varrho(\beta)$ have a maximum ϱ that first occurs for the index β*, then the remainder of the above sequence of coverings
$$Y(\beta)\ Y(\beta+1)\ \cdots\ Y(\gamma)\ \cdots$$
forms a base-like sequence since for any two coverings the first differing place ϱ always continues to be the same and the original sequence is cofinal with this one. Q. E. D.

Sequences of coverings of the inverse type ω_ν^* can be handled in precisely the same way; an argument-like ω_ν^*-sequence arises from an ω_ν-sequence in the argument (not an ω_ν^*-sequence!), a base-like ω_ν^*-sequence arises from an ω_ν^*-sequence in the base, and each ω_ν^*-sequence of coverings is coinitial with either an argument-like or a base-like ω_ν^*-sequence of coverings.

In the complete power $M((\sigma))$, i.e., in the totality of coverings of σ by M, the aggregate of those elements beginning with

$$X^\varrho = x_0 + \cdots + x_\alpha + \cdots + x_\beta + \cdots + x_\gamma + \cdots + x_\lambda + \cdots,$$

where ϱ ($\leqq \sigma$) is the limit of the places $\alpha, \beta, \gamma, \ldots, \lambda, \ldots$, follows immediately after the argument-like sequence (A); for $\varrho = \sigma$ there is only one such element X. If the sequence $a_\varrho\, b_\varrho\, c_\varrho \cdots l_\varrho \cdots$ of base elements has an upper limit x_ϱ in M, the aggregate of those elements beginning with $X^\varrho + x_\varrho$ follows after the base-like sequence (B); for $\varrho = \sigma - 1$ there is a single element; a base sequence with a subsequent gap produces a base-like sequence with a corresponding gap; if the base is cofinal with the base sequence, very complicated circumstances ensue. These propositions give a sufficient grounding for the investigation of the complete power $M((\sigma))$ and its subsets.

Dense Sets

§18

Species and Genera

First of all, the results about the construction of arbitrary sets in §11 refer us to a more careful study of dense sets, for which the fundamentals shall now be given.

To begin with, we consider a dense set free of possible boundary elements; call an unbounded dense set an η-*set* (after the simplest of these sets, the set of rational numbers of type η), and call its type an η-*type*. In an η-set each element is a two-sided limit element (§3); an element produces a decomposition

$$M = A + m + B,$$

where A is cofinal with a particular regular initial number ω_α and B is coinitial with the inverse ω_β^* of such a number; thus m is an $\omega_\alpha \omega_\beta^*$-element or, as we are more briefly going to say, a $c_{\alpha\beta}$-element; call $c_{\alpha\beta}$ the *character* of the element m. Correspondingly, each gap $M = C + D$ is a $\omega_\gamma \omega_\delta^*$-gap or a $c_{\gamma\delta}$-gap, and $c_{\gamma\delta}$ is its character. We then employ element characters and gap characters for a classification of η-sets; namely, we form the sets

$$U = \{c_{\alpha\beta}\}, \quad V = \{c_{\gamma\delta}\}$$

of element characters and gap characters, and we classify two sets with the same U, V to be of the same *species*, which we call the species (U, V). The membership of M or its type μ in the species (U, V) may be expressed by the equation

$$(M) = (\mu) = (U, V).$$

$V = 0$ signifies a species of *continuous* types.

Examples. The linear continuum λ belongs to the species $(c_{00}, 0)$. The types η, ι (the type of the irrational numbers), $\lambda + \lambda$, λ^2, and $\eta\omega_1$ belong to the species (c_{00}, c_{00}). The type $\eta\omega_1 + \eta\omega_1^*$ belongs to $(c_{00}, c_{00}\, c_{11})$, and the type $\eta\omega_1 + 1 + \eta\omega_1^*$ belongs to $(c_{00}\, c_{11}, c_{00})$. The type $\eta\omega_\omega$ belongs to the

species (c_{00}, V), where
$$V = (c_{00}\, c_{10}\, c_{20}\, \cdots) = \{c_{n0}\}.$$

We write
$$(M) < (M'), \quad (U, V) < (U', V')$$
if U is a subset of U' and V is a subset of V' and in at least one of these instances the first is a proper subset of the second (thus not $U = U', V = V'$); we then say that (M) is the *lower* species and (M') is the *higher* species. For example, $(\lambda) < (\eta)$, namely, $(c_{00}, 0) < (c_{00}, c_{00})$. If $(M) \leq (M')$ and at the same time $(M') \leq (M)$, then $(M) = (M')$. If N is a segment or unbounded piece of M, then $(N) \leq (M)$. If M contains a middle segment of a lower species, then M is called *reducible*; if all its middle segments (and with them, all unbounded pieces) belong to the species (M), M is called *irreducible*. For example, $\lambda + \lambda$ is reducible since it belong to the species (c_{00}, c_{00}), but contains a middle segment that belongs to the species $(c_{00}, 0)$; λ and η are irreducible.

Furthermore, we combine species into *genera* [[*Geschlechtern*]] by forming the union of *all* characters
$$W = \mathfrak{M}(U, V) = \{c_{\alpha\beta}, c_{\gamma\delta}\}$$
and by counting two sets with the same W to be of the same genus. Since one and only one species of continuous types $(W, 0)$ belongs to each genus, a special notation for genera can be dispensed with. For example, η, ι, and λ belong to the same genus $(c_{00}, 0)$.

From the considerations in §6, it immediately follows:

If M belongs to the species (U, V), its filling $[M]$ belongs to the species $(W, 0)$; moreover, if M is everywhere discontinuous, then its completion $\overline{M}$ belongs to the species (V, U). A set and its filling belong to the same genus, as perhaps does its completion. If a set is dense in another set, they both belong to the same genus (because they have the same filling).

If one understands by $M_{\alpha\beta} = \{m_{\alpha\beta}\}$ the set of $c_{\alpha\beta}$-elements in M and by $\overline{M}_{\gamma\delta} = \{\overline{m}_{\gamma\delta}\}$ the set of those elements of $\overline{M}$ that correspond to the $c_{\gamma\delta}$-gaps of M (in the case of an everywhere discontinuous M, thus the set of $c_{\gamma\delta}$-elements of $\overline{M}$), then the following easily results

Theorem XV. *For the irreducibility of the discontinuous set M, it is necessary and sufficient that each set $M_{\alpha\beta}$ and each set $M_{\gamma\delta}$ be dense in $[M]$; for the irreducibility of the continuous set M, it is necessary and sufficient that each set $M_{\alpha\beta}$ be dense in M.*

§19
The Derivation of Discontinuous Irreducible Types from Continuous Types

The last theorem in combination with the earlier one (§7, X), in which each dense set can be split into two subsets that are dense in it, allows an important conclusion:

Theorem XVI. *If an irreducible (continuous) set of species $(W, 0)$ exists, then there also exist irreducible sets of any species of the same genus.*

Proof. Let M be an irreducible η-set of species $(W, 0)$, where $W = \{c_{\alpha\beta}\}$, thus each set $M_{\alpha\beta}$ is dense in M. Let (U, V) be a prescribed species of the same genus; thus $W = \mathfrak{M}(U, V)$, whereby the sets U, V can contain characters in common. Let W_0 be the set of these common characters; with this, one obtains the following decompositions

$$U = (W_0, W_1), \quad V = (W_0, W_2), \quad W = (W_0, W_1, W_2),$$

where the sets W_0, W_1, W_2 have no common elements (one of them or both W_1, W_2 can even be empty). Correspondingly, we set

$$W_0 = \{c_{\alpha_0\beta_0}\}, \quad W_1 = \{c_{\alpha_1\beta_1}\}, \quad W_2 = \{c_{\alpha_2\beta_2}\},$$

and we decompose each set $M_{\alpha_0\beta_0}$ into two subsets $P_{\alpha_0\beta_0}$ and $Q_{\alpha_0\beta_0}$ that are dense in it. Then we split M into the following subsets

$$P = \{P_{\alpha_0\beta_0}, M_{\alpha_1\beta_1}\}, \quad Q = \{Q_{\alpha_0\beta_0}, M_{\alpha_2\beta_2}\},$$

and we claim that P is an irreducible set of the desired species (U, V) and that Q is likewise one of the species (V, U).

In fact, from §§5,6 we conclude the following. The set $M_{\alpha_1\beta_1}$ is dense in M, thus all the more so in P; each of its elements $m_{\alpha_1\beta_1}$ is a $c_{\alpha_1\beta_1}$-element in M, thus also in $M_{\alpha_1\beta_1}$, thus also in P. The set $P_{\alpha_0\beta_0}$ is dense in $M_{\alpha_0\beta_0}$, thus also in M, and thus also in P. Each of its elements $p_{\alpha_0\beta_0}$ is a $c_{\alpha_0\beta_0}$-element in M, thus also in P. Therefore P actually contains the prescribed element characters from U and only these; likewise, Q contains the element characters of V and only these. Since P contains subsets that are dense in M, P itself is dense in M, likewise for Q; since M is continuous, $M = [P] = [Q]$ and P and Q are completions of each other and M is their filling. Then of course, the gaps of one set correspond to the elements of the other, so P is of species (U, V) and Q is of species (V, U). Finally, since all sets $P_{\alpha_0\beta_0}, M_{\alpha_1\beta_1}, Q_{\alpha_0\beta_0}, M_{\alpha_2\beta_2}$ are dense in M, P and Q are then irreducible by Theorem XV.

§20
Existence Conditions

The sets U, V, and W cannot be chosen completely at random if there are actually supposed to be sets of the species (U, V). Clearly, we can conclude from §19 that an existence condition will affect only the genus, thus the set W. For example, a set of the species $(c_{11}, 0)$ cannot exist; such a set would have to contain ω_1-sequences and thus ω_0-sequences, and these would have to define ω_0-elements or ω_0-gaps. In general, if ω_κ is the *first regular initial number not contained within M* (§2, i.e, in middle segments of M), so that also no regular initial number $\geqq \omega_\kappa$ is contained within M, then each regular initial number ω_α ($\alpha < \kappa$) is contained within M and must give rise to ω_α-elements or ω_α-gaps, i.e., for each α, W must contain at least one character $c_{\alpha\beta}$. The same holds for the *first regular initial number ω_λ not* *contained within* M^*; for each index $\beta < \lambda$ (ω_β regular), W must contain at least one character $c_{\alpha\beta}$. Finally, Theorem VI (§4) requires the existence of at least one *symmetric* character $c_{\alpha\alpha}$. We are going to call a character set that meets these conditions *complete* [[*vollständig*]]. Hence if we form the tableau

$$\mathfrak{W}: \begin{array}{|ccccc|c|} \hline c_{00} & c_{01} & \cdots & c_{0\beta} & \cdots & c_{0\lambda} \\ c_{10} & c_{11} & \cdots & c_{1\beta} & \cdots & c_{1\lambda} \\ \multicolumn{5}{|c|}{\dotfill} & \cdots \\ c_{\alpha 0} & c_{\alpha 1} & \cdots & c_{\alpha\beta} & \cdots & c_{\alpha\lambda} \\ \multicolumn{5}{|c|}{\dotfill} & \cdots \\ \hline c_{\kappa 0} & c_{\kappa 1} & \cdots & c_{\kappa\beta} & \cdots & c_{\kappa\lambda} \\ \end{array} \quad \begin{array}{c} (\alpha < \kappa,\ \beta < \lambda) \\ (\omega_\alpha, \omega_\beta, \omega_\kappa, \omega_\lambda \text{ regular}) \end{array}$$

of *all* characters $c_{\alpha\beta}$ (the framed border sequences, row κ and column λ that were added for the sake of distinctness, are not supposed to belong to $\mathfrak{W}$), a complete character set with "*boundary numbers*" κ, λ must contain at least one element from each row and each column and from the principal diagonal of $\mathfrak{W}$.

We will now derive the important result that these simple conditions are not only necessary but also sufficient for the existence of irreducible sets of species $(W, 0)$ and, with this, (by §19) for the existence of irreducible sets of any species in the same genus; i.e., we will prove the theorem:

Theorem XVII. *For each complete character set W whose boundary numbers κ, λ are not limit numbers, there exist irreducible continuous sets of species $(W, 0)$.*

As for the added restriction that the regular initial numbers $\omega_\kappa, \omega_\lambda$ should not have limit indices, it is evocative of the reflections in §3, according to which the existence of regular initial numbers with limit index is altogether questionable; the remarks there show that in any case the smallest among them would have to be of an exorbitant cardinality, one that likely exceeds all known sets. The uncertainty as to whether or not Theorem XVII also holds for such numbers has to be seen as no important loss in generality.

§21
The Existence Proof for Genera with Complete Character Sets

In order to prove Theorem XVII, we next bring the prescribed character set W into a particular order. We are going to assume that κ is the larger of the two boundary numbers ($\kappa \geqq \lambda$), else we would consider the inverse set M^* instead of M. We then give W a well-ordering by ordering the existing characters in the rows from left to right and, after that, by ordering the rows from top to bottom; hence

$$W = \sum_\alpha \sum_\beta c_{\alpha\beta}.$$

The type of W is a subset of
$$\lambda\kappa \leq \kappa^2 < \omega_\kappa;$$
since ω_κ as an initial number of non-limit index is greater than its index κ and consequently also of higher cardinality, $\kappa^2 < \omega_\kappa$. Therefore W contains no higher ω_α-sequences than those that ought to occur*) in M.

In addition, we set
$$\varphi_{\alpha\beta} = \omega_\alpha + 1 + \omega_\beta^*,$$

$$\Phi = \sum_{c_{\alpha\beta}}^{W} \varphi_{\alpha\beta} = \sum_\alpha \sum_\beta \varphi_{\alpha\beta},$$

and finally for any values α and β, we set
$$\mu(\beta,\alpha) = 1 + \omega_\beta^* + \Phi + \omega_\alpha + 1.$$

479 Each such type $\mu(\beta,\alpha)$ is *scattered* (§11) but gap free; each of its subsets has either a last (first) element or an upper (lower) limit. There are elements with two neighbors (= consecutive elements), one-sided limit elements with one neighbor, and elements without neighbors; the boundary elements and the two-sided limit elements in the components $\varphi_{\alpha\beta}$ belong to this last category. For $\alpha < \kappa$, $\beta < \lambda$, all the sets $\mu(\beta,\alpha)$ can be viewed as subsets (intervals) of a single
$$\mu = \mu(\lambda - 1, \kappa - 1) = 1 + \omega_{\lambda-1}^* + \Phi + \omega_{\kappa-1} + 1$$
of cardinality $\aleph_{\kappa-1}$.

Now we take notice of the completeness of the character set W, which has not so far been mentioned. As a result, W contains at least one symmetric character; let the smallest of these be $c_{\sigma\sigma}$. For each regular ω_α $(< \omega_\kappa)$, W contains at least one character $c_{\alpha\beta}$; let the smallest of these be $c_{\alpha\beta_\alpha}$. For each regular ω_β $(< \omega_\lambda)$, W contains at least one character $c_{\alpha\beta}$; let the smallest of these be $c_{\alpha_\beta\beta}$. Thereupon, we form a product of argument ω_σ *with variable factors* (§17); i.e., we consider the element combinations
$$X = \sum_\varrho^{\omega_\sigma} x_\varrho = x_0 + x_1 + x_2 + \cdots + x_\varrho + \cdots$$
$$= X^\varrho + x_\varrho + X_\varrho \quad (\varrho < \omega_\sigma),$$
in which the sets $M(X^\varrho)$ that x_ϱ runs through may depend on the preceding elements X^ϱ. In particular, we determine that either $M(X^\varrho)$ should be a set of type $\mu(\beta,\alpha)$, in which case call x_ϱ an *active* [[*bewegliches*]] element, or that $M(X^\varrho)$ should be of type 1, i.e., it should be reduced to a single element, in which case call x_ϱ a *constrained* [[*gebundenes*]] element. By the remarks in §17, since the connection that we introduce between different variable factors is of no consequence, it is allowable to set all the constrained elements equal

---

*) Here is the place that necessitates the restriction mentioned above. If $\omega_\kappa = \kappa$, the construction method performed here appears no more likely than any other to produce a means for stopping the occurrence of ω_κ-sequences.

to a single fixed element p ($x_\varrho = p$). After this, we fix the choice of the set $M(X^\varrho)$ by the following rules:

(1) x_0 runs through some $\mu(\beta, \alpha)$, say the set μ itself.

(2) If ϱ is a limit number and all the elements of X^ϱ are active, then x_ϱ is active and it runs through the set $\mu(\beta_\tau, \alpha_\tau)$, provided that ϱ is cofinal with the regular initial number ω_τ ($< \omega_\sigma$).

(3) Only constrained elements follow after a constrained element, i.e., for $x_\varrho = p$, also $x_{\varrho+1} = x_{\varrho+2} = \cdots = x_\xi = p$ ($\xi \geq \varrho$).

(4) If x_ϱ is active and an element without neighbors (thus either a boundary element or a two-sided limit element in its set), then $x_{\varrho+1} = p$.

(5) If x_ϱ is active and has no left neighbor but has a right neighbor and x_ϱ is an ω_α-element, then $x_{\varrho+1}$ runs through the set $\mu(\beta_\alpha, \sigma)$.

(6) If x_ϱ is active and has no right neighbor but has a left neighbor and x_ϱ is an ω_β^*-element, then $x_{\varrho+1}$ runs through the set $\mu(\sigma, \alpha_\beta)$.

(7) If x_ϱ is active and has two neighbors, a predecessor and a successor, then $x_{\varrho+1}$ runs through the set $\mu(\sigma, \sigma)$.

Call the set of all combinations satisfying these conditions and ordered by first differences N. On the basis of the discussion in §17, we can now investigate the type of this set. N is a subset of the complete power $\mu((\omega_\sigma))$, and thus the sequences contained in N are either cofinal (respectively, coinitial) with an argument-like sequence or with a base-like sequence, so that it suffices to consider these special sequences. First, we remark that if ϱ is the first differing place for two combinations

$$X = X^\varrho + x_\varrho + X_\varrho, \quad Y = X^\varrho + y_\varrho + Y_\varrho, \quad (x_\varrho < y_\varrho),$$

then obviously x_ϱ and all the elements of X^ϱ are active by (3). It follows immediately from this that each argument-like sequence or base-like sequence in N has a limit, and thus generally each sequence in N has a limit. For after an argument-like ω_ν-sequence, there follows, as remarked in §17, the aggregate of all those combinations beginning with X^ϱ where ϱ is a limit number ($\leq \omega_\sigma$) and where, as it can now be added, all the elements of X^ϱ are active. For $\varrho < \omega_\sigma$, x_ϱ has a specified interval to run through; let a_ϱ be its first element, then

$$X = X^\varrho + a_\varrho + p + p + \cdots$$

is the first element beginning with X^ϱ and thus the upper limit of the argument-like sequence. For $\varrho = \omega_\sigma$, $X^\varrho = X$ itself is a whole combination and the only one beginning with X^ϱ; so again it is the limit of the argument-like sequence. After a base-like ω_ν-sequence, however, there follows the totality of elements beginning with $X^\varrho + x_\varrho$ if x_ϱ is the limit of the corresponding base sequence. And this case always occurs since each sequence in our sets $\mu(\beta, \alpha)$ has a limit. Again, the totality mentioned above always has a first element

$$X = X^\varrho + x_\varrho + a_{\varrho+1} + p + p + \cdots$$

or
$$X = X^\varrho + x_\varrho + p + p + p + \cdots,$$
depending on whether $x_{\varrho+1}$ is active (and $a_{\varrho+1}$ is the first element of its interval) or $x_{\varrho+1}$ is constrained and X is the upper limit of the base-like sequence. It also follows just like this that argument-like ω_ν^*-sequences and base-like ω_ν^*-sequences always have lower limits. So the set N certainly does not contain any gaps but rather only jumps and cuts (§1).

We now investigate the elements of N, and we especially intend to show that the two-sided limit elements have only those characters that occur in W. Either a combination X consists of just active elements, or if $x_{\varrho+1}$ is the first constrained element (which by (2) corresponds to a non-limit place and by (1) is also not x_0), it is of the form
$$X^\varrho + x_\varrho + p + p + \cdots;$$
in the first case, all the elements of X are at the same time inner elements of their sets and they have at least one consecutive element; in the second case, all the elements of X^ϱ have these properties too.

If all the elements of X are active, then X is the upper limit of an argument-like ω_σ-sequence, e.g., the sequence consisting of all the combinations
$$A(\varrho) = X^\varrho + a_\varrho + p + p + \cdots,$$
where a_ϱ ($< x_\varrho$) is the first element of the interval allowed for x_ϱ. Likewise, X is the lower limit of an argument-like ω_σ^*-sequence, thus a $c_{\sigma\sigma}$-element.

The second case
$$X = X^\varrho + x_\varrho + p + p + \cdots$$
splits into different subcases, in that x_ϱ can be a two-sided limit element or a boundary element of its interval.

(a) Let x_ϱ be a two-sided limit element of character $c_{\alpha\beta}$. *This character belongs to W* since the $c_{\alpha\beta}$ elements belong to the component $\Phi = \sum \varphi_{\alpha\beta}$, and certainly the element enclosed by ω_α and ω_β^* in
$$\varphi_{\alpha\beta} = \omega_\alpha + 1 + \omega_\beta^*$$
is the only $c_{\alpha\beta}$-element. Thus the assertion is correct since the sum Φ only extends over the characters in W. Now since base-like sequences that enclose the aggregate of all combinations beginning with $X^\varrho + x_\varrho$ correspond to base sequences that enclose the element x_ϱ and since X is the only element in this aggregate, then X is also a $c_{\alpha\beta}$-element of N with the same character as x_ϱ.

(b) Let x_ϱ be a boundary element; it suffices to prove the case where x_ϱ is the *first* element a_ϱ of its interval, thus
$$X = X^\varrho + a_\varrho + p + p + \cdots.$$

Next we have to decide whether ϱ is a limit number or whether it has a predecessor (we can immediately disregard the case $\varrho = 0$ where X is the first element of N).

(ba) Let ϱ be a limit number cofinal with ω_τ ($\omega_\tau < \omega_\sigma$). The element X, as the first combination beginning with X^ϱ, is then the upper limit of an argument-like ω_τ-sequence. Now in this case, by (2) x_ϱ had to run through the interval $\mu(\beta_\tau, \alpha_\tau)$, whose first element a_ϱ is thus the lower limit of a base sequence of type $\omega^*_{\beta_\tau}$. To this base sequence, there corresponds a base-like $\omega^*_{\beta_\tau}$-sequence that immediately precedes the totality of combinations beginning with $X^\varrho + a_\varrho$, i.e., the single combination X of course. Thus X is the lower limit of a base-like $\omega^*_{\beta_\tau}$-sequence and therefore a $c_{\tau\beta_\tau}$-element.

(bb) Let ϱ have a predecessor $\varrho - 1$; by replacing ϱ by $\varrho + 1$, we now write

$$X = X^\varrho + x_\varrho + a_{\varrho+1} + p + p + \cdots,$$

where x_ϱ is an inner element of its interval (as are all the elements of X^ϱ) and has at least one neighbor. This case splits yet again: x_ϱ can be an upper limit element in its set or it can have a left neighbor.

(bba) Let x_ϱ be an ω_α-element; then it has a right neighbor. To the sequence in the base of type ω_α whose limit is x_ϱ, there corresponds a base-like ω_α-sequence whose limit is X (as the first combination beginning with $X^\varrho + x_\varrho$). Now here, by (5), $x_{\varrho+1}$ has to run through the set $\mu(\beta_\alpha, \sigma)$, whose first element $a_{\varrho+1}$ is thus an $\omega^*_{\beta_\alpha}$-limit, from which it follows, as in (ba), that X is the lower limit of a base-like $\omega^*_{\beta_\alpha}$-sequence and thus a $c_{\alpha\beta_\alpha}$-element in N.

(bbb) Let x_ϱ have a left neighbor y_ϱ that is an inner element of its interval, as is x_ϱ. Then the aggregate of combinations $X^\varrho + y_\varrho + y_{\varrho+1} + \cdots$ beginning with $X^\varrho + y_\varrho$ is found immediately before X, and if $b_{\varrho+1}$ is the last element of the interval available for $y_{\varrho+1}$, then the element

$$Y = X^\varrho + y_\varrho + b_{\varrho+1} + p + p + \cdots$$

is the immediate predecessor of X (hence the set N actually includes jumps). Now in this case, by (6) and (7), $x_{\varrho+1}$ has to run through an interval $\mu(\sigma, \alpha_\beta)$ or an interval $\mu(\sigma, \sigma)$; likewise, by (5) and (7), $y_{\varrho+1}$ has to run through an interval $\mu(\beta_\alpha, \sigma)$ or an interval $\mu(\sigma, \sigma)$; in any case, $a_{\varrho+1}$ is an ω^*_σ-limit in its interval and $b_{\varrho+1}$ is an ω_σ-limit in its interval. Therefore X is again the lower limit of a base-like ω^*_σ-sequence, and Y is the upper limit of a base-like ω_σ-sequence; so of the two consecutive elements Y and X, the left is an ω_σ-element and the right is an ω^*_σ-element.

With this, we have finished the case where the last active element in X is the left boundary element of its interval; the corresponding treatment of a right boundary element can be omitted. The result of the somewhat ramified casuistry is this: the set N is gap free and the characters of its two-sided limit elements belong to W; in addition, N still contains one-sided limit elements, namely, two boundary elements and pairs of consecutive elements in which respectively, the left is an ω_σ-element and the right is an ω_σ^*-element.

We still have to show that N contains *all* the $c_{\alpha\beta}$-elements from W, in fact, within each (non-empty) middle segment. Let
$$X = X^\varrho + x_\varrho + X_\varrho, \quad Y = X^\varrho + y_\varrho + Y_\varrho, \quad x_\varrho > y_\varrho$$
be two non-consecutive combinations with first differing place ϱ. *If x_ϱ and y_ϱ are not consecutive*, there lies between them an element u_ϱ that is not a two-sided limit element, and the totality of combinations U beginning with $X^\varrho + u_\varrho$ lies between X and Y. The next element $u_{\varrho+1}$ runs through some set $\mu(\beta, \alpha)$; this set contains the component Φ, and Φ contains all $c_{\alpha\beta}$-elements of W. If however $u_{\varrho+1}$ is a $c_{\alpha\beta}$-element, then by (a) the combination
$$U = X^\varrho + u_\varrho + u_{\varrho+1} + p + p + \cdots$$
is a $c_{\alpha\beta}$-element in N; thus $c_{\alpha\beta}$-elements of all the prescribed characters lie between X and Y. *If x_ϱ and y_ϱ are consecutive*, let
$$A = X^\varrho + x_\varrho + a_{\varrho+1} + p + p = \cdots, \quad B = X^\varrho + y_\varrho + b_{\varrho+1} + p + p + \cdots$$
be the first element beginning with $X^\varrho + x_\varrho$ and the last element beginning with $X^\varrho + y_\varrho$, respectively; thus A and B are consecutive combinations and
$$Y \leqq B < A \leqq X,$$
but it is not the case that simultaneously $X = A$ and $Y = B$. So if $A < X$, say, then an element $u_{\varrho+1}$ that is not a two-sided limit element lies between the non-consecutive $a_{\varrho+1}$ and $x_{\varrho+1}$. And the totality of combinations beginning with $X^\varrho + x_\varrho + u_{\varrho+1}$ lies between A and X or Y and X, with regard to which the above conclusion repeats itself.

Finally, if we now let pairs of consecutive elements of N collapse into single elements (through which a pair becomes a $c_{\sigma\sigma}$-element) and we remove the boundary elements, N turns into an unbounded, dense, gap free, thus continuous, set N' that contains only the element characters of W, and it contains all of them within each middle segment. So N' is an irreducible η-set of species $(W, 0)$. With this, Theorem XVII is proved.

As a subset of the complete power $\mu((\omega_\sigma))$, the set so constructed has cardinality $\leqq \aleph_{\kappa-1}^{\aleph_\sigma}$, and in fact, it has exactly cardinality $\aleph_{\kappa-1}^{\aleph_\sigma}$; for if one considers only the combinations with just active elements, already in these each x_ϱ runs through a set of cardinality $\aleph_{\kappa-1}$. We will see (§24) that a cardinality smaller than $\aleph_{\kappa-1}^{\aleph_\sigma}$ is not possible under our assumptions.

It follows from our last considerations that the set P of those combinations $X^\varrho + x_\varrho + p + p + \cdots$ where x_ϱ is a $c_{\alpha\beta}$-element of its interval is relatively dense in N'. Again, we remind ourselves that the mutual relation of different variable factors plays no role, so nothing stops us from taking an element of the set $\mu(\lambda - 1, \kappa - 1)$ itself for p, and it matters not which. Then P becomes a subset of those covering sets of ω_σ by μ where the elements different from p only occur in a set of places $< \omega_\sigma$, i.e., of the power $\mu(\omega_\sigma)_p^\sigma$ of the $(1 + \sigma)$th class with principal element p (§13). (In order to avoid misunderstanding: this power, however not the type of P, depends of course on which element of μ we identify with p.) For what follows (§25), we make note of the fact that here we have constructed an irreducible set of species $(W, 0)$ in which a subset of the power $\mu(\omega_\sigma)_p^\sigma$ *with scattered base μ is relatively dense.*

Finally, we do not want to fail to point out that in special cases the general method of construction in this section is of course capable of substantial simplification. For many species, instead of products with variable factors, ordinary products with fixed factors suffice and frequently just powers suffice, be they complete powers or powers of lower class. One can also try to form discontinuous types directly, without pursuing the indirect route through continuous types of the same genus that was sketched in §19. Various special investigations along these lines are found in my "Investigations into Order Types," cited at the start of this article.

§22
The Number of Genera and Species

The existence proof for (irreducible) types of each species whose character set W is complete and has non-limit numbers κ, λ as boundary numbers is carried out in Theorems XVI, XVII. The importance of this fundamental result will become clear through an enumeration of the smallest cases.

All the pertinent species (U, V) are obtained from W, as in §19, by splitting W in all possible ways into three subsets W_0, W_1, W_2, with the exception of the case $0, 0, W$. The number of species belonging to a prescribed W that contains n characters is thus $3^n - 1$, where n can be finite or an aleph.

Let the number of complete character sets with boundary numbers κ, λ be
$$\varphi(\kappa, \lambda, n) = \varphi(\lambda, \kappa, n);$$
then the number of genera is

485

$$G(\kappa, \lambda) = \sum_n \varphi(\kappa, \lambda, n),$$

and the number of species is
$$S(\kappa, \lambda) = \sum_n \varphi(\kappa, \lambda, n) \cdot (3^n - 1).$$

As before, we assume $\kappa \geq \lambda$, and we call k, l the cardinalities of κ, λ ($k \geq l$). Then in each case, the tableau $\mathfrak{W}$ has k rows (although for transfinite κ, those numbers $\alpha < \kappa$ that belong to singular ω_α correspond to no row) and l columns.

An always existing complete character set is the total set $\mathfrak{W}$ with kl characters; thus
$$S(\kappa, \lambda) \geq 3^{kl} - 1.$$

Hence if even only one of the boundary numbers is transfinite, then the number of species is $\geq 3^{\aleph_0}$, so at least equal to the cardinality of the continuum.

Moreover, if one thinks of the first row and first column as belonging to W, then certainly W is complete no matter which subset of the remaining tableau (with $k-1$ rows and $l-1$ columns) is added to W. Thus
$$G(\kappa, \lambda) \geq 2^{(k-1)(l-1)}.$$
If one of the boundary numbers κ is transfinite and the other is $\lambda > 1$, then the number of genera is also $\geq 2^{\aleph_0}$, so at least equal to the cardinality of the continuum; for $\lambda = 1$, there is always just one existing genus.

We are going to handle the case of *finite boundary numbers* somewhat more precisely. In order to obtain a complete character set of n elements, one has to place n stones onto a rectangular board with $\kappa \cdot \lambda$ squares in such a way that at least one stone appears in each row and in each column and at least one stone appears in one (hatched) diagonal. If we introduce the generating function

$$\Phi(\kappa, \lambda, x) = \sum_n \varphi(\kappa, \lambda, n) x^n$$

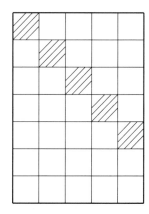

for the number of these arrangements $\varphi(\kappa, \lambda, n)$, an investigation, of which we are going to only give the result, shows that, always assuming that $\kappa \geq \lambda$, the polynomial Φ is defined by the following valid formulas:

$$\Phi(\kappa, \lambda, x)$$
$$= F(\kappa, \lambda, x) - \lambda F(\kappa, \lambda - 1, x) + \frac{\lambda(\lambda-1)}{1 \cdot 2} F(\kappa, \lambda - 2, x) -$$
$$\cdots + (-1)^{\lambda-1} \cdot \lambda F(\kappa, 1, x);$$
$$F(\kappa, \lambda, x) = [(1+x)^\lambda - 1]^\kappa - [(1+x)^{\lambda-1} - 1]^\lambda [(1+x)^\lambda - 1]^{\kappa-\lambda}.$$

486 The auxiliary function
$$F(\kappa, \lambda, x) = \sum_n f(\kappa, \lambda, n) x^n$$

is the generating function for the number $f(\kappa, \lambda, n)$ of arrangements in which each row and the diagonal is occupied by at least one stone.

With this, the functions $\Phi(\kappa, \lambda, x) = \Phi(\lambda, \kappa, x)$ can be calculated for each number pair κ, λ, and they immediately give the number of genera and species, namely,
$$G(\kappa, \lambda) = \Phi(\kappa, \lambda, 1),$$
$$S(\kappa, \lambda) = \Phi(\kappa, \lambda, 3) - \Phi(\kappa, \lambda, 1).$$
For the smallest cases, one obtains

κ	λ	$G(\kappa,\lambda)$	$S(\kappa,\lambda)$
1	1	1	2
2	1	1	8
2	2	6	192
3	1	1	26
3	2	22	3164
3	3	247	236786

For dense sets with countable sequences, there is only one genus $(c_{00}, 0)$ with two species $(c_{00}, 0)$ and (c_{00}, c_{00}) represented by the continuum and the set of rational numbers. For dense types within which sequences of up to the second infinite cardinality occur, there are $G(1,1) + G(1,2) + G(2,1) + G(2,2) = 9$ genera with 210 species; for those with sequences up to the third infinite cardinality, there are 302 genera with 243376 species.

In the smallest cases, there are the following complete character sets:

κ	λ	W			
1	1	c_{00}			
1	2	c_{00}	c_{01}		
2	1	c_{00}	c_{10}		
2	2	c_{00}	c_{11}		
		c_{00}	c_{01}	c_{10}	
		c_{00}	c_{01}	c_{11}	
		c_{00}	c_{10}	c_{11}	
		c_{01}	c_{10}	c_{11}	
		c_{00}	c_{01}	c_{10}	c_{11}

from which one can immediately write down all 210 species of types with sequences up to the third infinite cardinality; for example, the eight species of the genus $W = c_{00}\, c_{11}$:

$$(c_{00}, c_{11}) \quad (c_{00}, c_{00}\, c_{11}) \quad (c_{11}, c_{00}) \quad (c_{11}, c_{00}\, c_{11})$$

$$(c_{00}\, c_{11}, 0) \quad (c_{00}\, c_{11}, c_{00}) \quad (c_{00}\, c_{11}, c_{11}) \quad (c_{00}\, c_{11}, c_{00}\, c_{11}).$$

§23
e_π-Sets

For the properties of an η-set M, especially for its cardinality, beside the boundary numbers κ, λ (the indices of the first regular initial numbers $\omega_\kappa, \omega_\lambda$ not contained within M, M^*), there are yet certain other significant numbers $\mu, \nu, \pi, \varrho, \sigma$, all indices of *regular* initial numbers, that are defined as follows:

M is cofinal with ω_μ and coinitial with ω_ν^*.

$c_{\sigma\sigma}$ is the lowest *symmetric* character of the set M, which therefore contains neither elements nor gaps of characters $c_{00}\, c_{11} \cdots c_{\alpha\alpha} \cdots (\alpha < \sigma)$.

If, as before,

$$U = \{c_{\alpha\beta}\}, \quad V = \{c_{\gamma\delta}\}$$

are the sets of the element characters, respectively, gap characters and ε is the maximum of both numbers γ, δ, then ϱ is to be the minimum of all the numbers $\alpha, \beta, \varepsilon$;

$$\alpha \geqq \varrho, \quad \beta \geqq \varrho, \quad \varepsilon \geqq \varrho,$$

where the equality sign holds in at least one of these formulas.

Finally, π is the minimum of the three numbers μ, ν, ϱ.

To save space, we going to call M a $c_{\sigma\sigma}$-set, a d_ϱ-set, and an e_π-set. The numbers κ, λ, σ are characteristic of the genus, the number ϱ is characteristic of the species, and the numbers μ, ν, π are characteristic of the individual set. Again, if we assume $\kappa \geqq \lambda$, the following inequalities exist

$$\pi \leqq \varrho \leqq \sigma < \lambda \leqq \kappa,$$
$$\pi \leqq \mu \leqq \kappa, \quad \pi \leqq \nu \leqq \lambda.$$

Obviously, $\varrho = \pi = 0$ for a continuous set; such a set is thus a d_0-set and a e_0-set.

If M' is a segment or unbounded piece of M, then M' is of a lower or equal species than M (§18), and thus

$$\kappa' \leqq \kappa, \quad \lambda' \leqq \lambda, \quad \varrho' \geqq \varrho, \quad \sigma' \geqq \sigma.$$

In particular, if M' is a *middle segment* of M, then the inequalities

$$\mu' \geqq \varrho, \quad \nu' \geqq \varrho, \quad \pi' \geqq \varrho$$

hold for it.

488 Two adjacent subsets (§1) of a d_ϱ-set are never simultaneously of cardinality $< \aleph_\varrho$; the larger of the two is always $\geqq \aleph_\varrho$, and at least in one case exactly $= \aleph_\varrho$. For an e_π-set, the larger of two adjacent sets, as well as each set with which M is cofinal or coinitial, is $\geqq \aleph_\pi$ and at least in one case exactly $= \aleph_\pi$.

We put forward the following definition:

An η-set is called an η_τ-set if it is not cofinal or coinitial with any set of cardinality $< \aleph_\tau$ and if it contains no pair of adjacent subsets each of cardinality $< \aleph_\tau$,

where τ need not necessarily be the index of a regular initial number, so all these conditions are fulfilled for $\tau \leqq \pi$. Each e_π-set is thus simultaneously an $\eta_0, \eta_1, \cdots, \eta_\pi$-set, and for each η_τ-set, its π-index π is $\geqq \tau$. The η_0-sets are identical to the η-sets, the dense, unbounded sets themselves. For singular ω_τ, an η_τ-set is simultaneously an $\eta_{\tau+1}$-set. Each middle segment of a d_ϱ-set is an η_ϱ-set.

Example. Among the η-sets with sequences up to the second infinite cardinality, there is only one genus with c_{11}-sets, namely, $W = c_{01} c_{10} c_{11}$. The d_1-sets of this genus belong to the species where $U = c_{11}$, thus to the two species

$$(c_{11}, c_{01} c_{10}) \quad \text{and} \quad (c_{11}, c_{01} c_{10} c_{11}).$$

The sets of these species can still be e_0-sets or e_1-sets; their middle segments are certainly e_1-sets or η_1-sets.

Next, we prove the following theorems:

Theorem XVIII. *An e_π-set contains each arbitrary ordered set of cardinality $\leq \aleph_\pi$ as a subset.*

Theorem XIX. *There is at most one e_π-type of cardinality $\aleph_\pi$.*

Let A be an ordered set of cardinality $\leq \aleph_\pi$. We can thus give its elements a well-ordering

$$a_0\, a_1\, a_2 \cdots a_\alpha \cdots \quad (\alpha < \omega_\pi).$$

Let M be an e_π-set. All elements a_α are to be assigned elements $\varphi(a_\alpha)$ of M in such a way that $\varphi(a_\alpha)$ has the same rank ordering with respect to $\varphi(a_\beta)$ in M as a_α has with respect to a_β in A. We prove that this is possible by induction: if all the elements of $A^\alpha = \{a_\xi\}$, $\xi < \alpha$ are mapped, a mapping for a_α is also possible. Indeed, let A^α be decomposed by a_α into the two summands $A_1^\alpha + A_2^\alpha$, i.e.,

$$A_1^\alpha < a_\alpha < A_2^\alpha \quad \text{in } A,$$

one summand of which can be empty. The already found image set $\varphi(A^\alpha)$ decomposes correspondingly into $\varphi(A_1^\alpha) + \varphi(A_2^\alpha)$; the cardinality of these sets is $< \aleph_\pi$, and since M contains no adjacent subsets $< \aleph_\pi$ nor is it cofinal or coinitial with such a set, there always exists an element m for which

$$\varphi(A_1^\alpha) < m < \varphi(A_2^\alpha)$$

(respectively, $m \gtrless \varphi(A^\alpha)$ in case one of the summands vanishes), and therefore it can be chosen as $\varphi(a_\alpha)$. With this, Theorem XVIII is proved. The remaining ambiguity in the choice of the image can be removed in the well-known way through a well-ordering of the set M,

$$m_0\, m_1\, m_2 \cdots m_\beta \cdots,$$

by taking for $\varphi(a_\alpha)$ the element m_β of lowest index from among the available elements. This can be expressed as follows: for each index α there are subsets

$$M^\alpha = (m_{\beta_0}\, m_{\beta_1} \cdots m_{\beta_\xi} \cdots) \quad (\xi < \alpha)$$

that are similar to A^α if m_{β_ξ} is assigned to a_ξ; imagine the M^α (for fixed α) as *ordered by first differences* of the numbering system $\beta_0\, \beta_1 \cdots \beta_\xi \cdots$ and take the smallest among them; then the union of all these smallest M^α is the sought after image set $\varphi(A)$.

Now let

$$\begin{array}{llllll} A: & a_0 & a_1 & a_2 \cdots & a_\alpha \cdots & (\alpha < \omega_\pi), \\ B: & b_0 & b_1 & b_2 \cdots & b_\beta \cdots & (\beta < \omega_\pi) \end{array}$$

be two e_π-sets of cardinality $\aleph_\pi$, written in well-ordered form. In the way specified above, if we map A onto a subset $\varphi(A)$ of B, then we claim that $\varphi(A) = B$, thus that A is mapped similarly onto the whole set B. Again, we prove this by induction: if all the elements of $B^\beta = \{b_\eta\}$, $\eta < \beta$, are contained in $\varphi(A)$, then b_β is also in $\varphi(A)$. Let α be the smallest index for which B^β is contained in $\varphi(A^\alpha)$ (it surely exists, based on the hypothesis that ω_π is regular). Then either b_β is also in $\varphi(A^\alpha)$, or b_β is the lowest element of B that is not in $\varphi(A^\alpha)$. In the latter case, which alone remains to be discussed, b_β produces a decomposition $\varphi(A^\alpha) = \varphi(A_1^\alpha) + \varphi(A_2^\alpha)$, i.e.,

$$\varphi(A_1^\alpha) < b_\beta < \varphi(A_2^\alpha) \quad \text{in } B.$$

Then also $A^\alpha = A_1^\alpha + A_2^\alpha$ in A. Now since A is an e_π-set, there certainly are elements a between A_1^α, A_2^α, and let a_γ ($\gamma \geqq \alpha$) be the lowest element for which

$$A_1^\alpha < a_\gamma < A_2^\alpha \quad \text{in } A.$$

For the set A^γ of all elements with smaller index ($< \gamma$), a_γ produces the decomposition

$$A_1^\gamma < a_\gamma < A_2^\gamma \quad \text{in } A,$$

and then A_1^γ is obviously cofinal with its subset A_1^α and A_2^γ is coinitial with A_2^α, for otherwise a_γ would not be the earliest element between A_1^α and A_2^α. Then again, $\varphi(A_1^\gamma)$ is cofinal with $\varphi(A_1^\alpha)$ and $\varphi(A_2^\gamma)$ is coinitial with $\varphi(A_2^\alpha)$, thus

$$\varphi(A_1^\gamma) < b_\beta < \varphi(A_2^\gamma) \quad \text{in } B.$$

From this, it certainly follows that b_β is the element b of lowest index that has the same position with respect to $\varphi(A^\gamma)$ that a_γ has with respect to A^γ, i.e., it is b_β that is chosen as the image of a_γ, $b_\beta = \varphi(a_\gamma)$, and therefore b_β is contained in $\varphi(A^{\gamma+1})$. Again, it need not be remarked that one of the summands of A^α can vanish, by which this analysis is correspondingly simplified.

Therefore two e_π-sets of cardinality $\aleph_\pi$ are similar, and with this, Theorem XIX is proved. However, in general, whether there is an e_π-set of cardinality $\aleph_\pi$ or only ones $> \aleph_\pi$ remains undecided here (cf. §24).

In Theorems XVIII and XIX, η_π-set and η_π-type can be substituted for e_π-set and e_π-type.

§24
$c_{\sigma\sigma}$-Sets

Among the ordinals, we are going to distinguish the *even* 2ξ and the *odd* $2\xi + 1$, so that each limit number is even. We consider a set N of type $\omega_\beta^* + \omega_\alpha$ (these initial numbers need not necessarily be regular), and we write its elements in the form

$$\cdots q_\eta \cdots q_2 \, q_1 \, p_0 \, p_1 \, p_2 \cdots p_\xi \cdots \quad (\xi < \omega_\alpha, \ \eta < \omega_\beta).$$

We call the elements with an even index even elements, and we call the elements with an odd index odd elements; thus there are

even elements: $\cdots q_{\omega+2} \, q_\omega \cdots q_4 \, q_2 \, p_0 \, p_2 \, p_4 \cdots p_\omega \, p_{\omega+2} \cdots$,

odd elements: $\cdots q_{\omega+3} \, q_{\omega+1} \cdots q_3 \, q_1 \, p_1 \, p_3 \cdots p_{\omega+1} \, p_{\omega+3} \cdots$.

Each odd element has two immediate neighbors, a predecessor and a successor.

Now we consider coverings of ω_σ with base N,

$$X = x_0 + x_1 + \cdots + x_\gamma + \cdots = X^\gamma + x_\gamma + X_\gamma \quad (\gamma < \omega_\sigma),$$

under the restriction that after an even element $x_\gamma = p_{2\xi}$ or $q_{2\eta}$ there shall always follow $x_{\gamma+1} = p_0 = p$. A covering either has a first even element and is of the form
$$X(\gamma) = X^\gamma + x_\gamma + p + p + \cdots \quad (\gamma < \omega_\sigma),$$
where all the elements of X^γ are odd (and it can already be that $x_\gamma = p$), or the covering contains just odd elements and may be denoted by X. Then let $R(\gamma)$ be the set of all coverings $X(\gamma)$ that have a first even element in place γ, R the set of all coverings without even elements, and $S(\delta)$ the set of all coverings $X(\gamma)$ for $\gamma < \delta$; thus
$$S(\delta) = \{R(0)\, R(1)\, R(2)\, \cdots\, R(\gamma)\, \cdots\} \quad (\gamma < \delta),$$
and in particular, $S(\omega_\sigma)$ is the set of all coverings that contain even elements and $S = (S(\omega_\sigma), R)$ is the set of all admissible coverings. All these sets are subsets of the *complete power* $N((\omega_\sigma))$ and are to be ordered by first differences. We now claim:

Theorem XX. *A $c_{\sigma\sigma}$-set in which each middle segment contains an ω_α-sequence and a ω_β^*-sequence contains a subset similar to $S(\omega_\sigma)$, and in the case that the $c_{\sigma\sigma}$-set is free of $c_{\sigma\sigma}$-gaps, it contains a subset similar to S.*

In order to show that a subset $\varphi S(\omega_\sigma)$ of M similar to $S(\omega_\sigma)$ exists, we have to find a suitable image $\varphi X(\gamma)$ for all the coverings $X(\gamma)$; we prove that this is possible for all the elements of $R(\delta)$ if it has already been done for all the preceding elements, thus those of $S(\delta)$. If we understand by $R(\delta, X^\delta)$ the set of those
$$X(\delta) = X^\delta + x_\delta + p + p + \cdots$$
that start with a fixed complex X^δ, thus which differ only in the value x_δ, then no covering from $S(\delta)$ lies between any two such coverings; hence each element of $S(\delta)$ is $\lessgtr R(\delta, X^\delta)$, and $S(\delta)$ thereby decomposes into
$$S(\delta) = S_1(\delta, X^\delta) + S_2(\delta, X^\delta)$$
in such a way that $R(\delta, X^\delta)$ lies between the two summands. On the other hand: two coverings $X(\delta)$ and $Y(\delta)$ that do not belong to the same $R(\delta, X^\delta)$, but rather have an earlier differing place γ, are surely separated by an element of $S(\delta)$; for if
$$X(\delta) = X^\gamma + x_\gamma + X_\gamma, \quad Y(\delta) = X^\gamma + y_\gamma + Y_\gamma \quad \begin{pmatrix} x_\gamma < y_\gamma \\ \gamma < \delta \end{pmatrix},$$
then x_γ and y_γ are both odd and surely separated by an even element, say $\overset{+}{x}_\gamma$, that is next after x_γ, so that the covering
$$X(\gamma) = X^\gamma + \overset{+}{x}_\gamma + p + p + \cdots$$
lies between $X(\delta)$ and $Y(\delta)$. Thus to each X^δ there corresponds a decomposition of $S(\delta)$, and to two different X^δ there correspond two different decompositions.

Now of what sort are these decompositions? We claim that they represent either a jump or a $c_{\tau\tau}$-gap ($\tau < \sigma$) of $S(\delta)$. Namely, if δ is not a limit number and it is > 0, thus provided with a predecessor $\delta - 1$, then $X^\delta = X^{\delta-1} + x_{\delta-1}$, where the element $x_{\delta-1}$ is odd and has two even neighbors $\overset{+}{x}_{\delta-1}$ (successor) and $\bar{x}_{\delta-1}$ (predecessor). The two coverings

$$\bar{X}(\delta - 1) = X^{\delta-1} + \bar{x}_{\delta-1} + p + p + \cdots,$$

$$\overset{+}{X}(\delta - 1) = X^{\delta-1} + \overset{+}{x}_{\delta-1} + p + p + \cdots$$

are neighbors of all the coverings beginning with X^δ, i.e., they are the last element of $S_1(\delta, X^\delta)$ and the first element of $S_2(\delta, X^\delta)$.

However, if δ is a limit number and cofinal with ω_τ ($\tau < \sigma$), the totality of coverings beginning with X^δ is bounded immediately on the left and right by two *argument-like* sequences (§17)

$$\bar{X}(\gamma) = X^\gamma + \bar{x}_\gamma + p + p + \cdots \quad (\gamma < \delta),$$

$$\overset{+}{X}(\gamma) = X^\gamma + \overset{+}{x}_\gamma + p + p + \cdots \quad (\gamma < \delta),$$

where γ runs through an ω_τ-sequence of places with limit δ; hence $S_1(\delta, X^\delta)$ is cofinal with ω_τ, and $S_2(\delta, X^\delta)$ is coinitial with ω_τ^*.

Let us summarize: it follows that the set $R(\delta)$ decomposes into nothing but separated components $R(\delta, X^\delta)$ of type $\omega_\beta^* + \omega_\alpha$, each of which, and certainly always only one, either lies between two consecutive elements of $S(\delta)$ or in a $c_{\tau\tau}$-gap ($\tau < \sigma$) of $S(\delta)$. Since M as an η-set contains no consecutive elements and as a $c_{\sigma\sigma}$-set contains neither $c_{\tau\tau}$-gaps nor $c_{\tau\tau}$-elements, each of the decompositions under discussion of the already obtained image set $\varphi S(\delta)$ has between its halves an as yet unencumbered piece of M in which lie infinitely many elements of M and in which, on the basis of the full hypothesis of Theorem XX, one can give a sequence of type $\omega_\beta^* + \omega_\alpha$ as the image of $R(\delta, X^\delta)$. The start of the process, corresponding to the as yet untreated case $\delta = 0$, constructs the specification of an arbitrary sequence $\omega_\beta^* + \omega_\alpha$ in M itself as an image for $R(0)$; again, the arbitrariness in the choice of images could be removed through a preliminary well-ordering of M.

With this, the existence of a subset $\varphi S(\omega_\sigma)$ in M is proved. Now each covering X with only odd elements is enclosed by two argument-like sequences of types ω_σ and ω_σ^*, belonging to $S(\omega_\sigma)$; thus X falls into a $c_{\sigma\sigma}$-gap of $S(\omega_\sigma)$, and in fact, two different X's fall into two different gaps. So if M also contains no $c_{\sigma\sigma}$-gaps, then each $c_{\sigma\sigma}$-gap of $\varphi S(\omega_\sigma)$ has at least one element of M in it that can be assigned to X as an image, and in this case, M contains a subset similar to $(S(\omega_\sigma), R)$. With this, Theorem XX is proved.

From this, there follow remarkable results with regard to lower bounds on the cardinality of dense sets. Each element in the complex X^δ runs through the set of all odd elements of N. This set, as well as N itself and the set of all even elements, has cardinality $\aleph_\alpha + \aleph_\beta$; so X^δ runs through a set of cardinality $(\aleph_\alpha + \aleph_\beta)^d$ where d $(< \aleph_\sigma)$ is the cardinality of δ $(< \omega_\sigma)$, and $X(\delta)$ runs through the set $R(\delta)$ of cardinality $(\aleph_\alpha + \aleph_\beta)^{d+1}$, while a covering X with only odd elements runs through the set R of cardinality $(\aleph_\alpha + \aleph_\beta)^{\aleph_\sigma}$. For any cardinality a, let us understand by $(a)_\sigma$ the sum of powers $\sum_\delta^{\omega_\sigma} a^d$ taken over the cardinalities of all the ordinal numbers $< \omega_\sigma$, obviously for which $a > 1$ is assumed; one can also set

$$(a)_\sigma = \sum_d a^d \quad (d < \aleph_\sigma),$$

where the sum is taken over all the different cardinal numbers $< \aleph_\sigma$. Thus

$$(a)_\sigma = a + a^2 + a^3 + \cdots + a^{\aleph_0} + a^{\aleph_1} + \cdots a^{\aleph_\varrho} + \cdots \quad (\varrho < \sigma).$$

Then $S(\omega_\sigma)$ is of cardinality $(\aleph_\alpha + \aleph_\beta)_\sigma$, and S is of cardinality

$$(\aleph_\alpha + \aleph_\beta)_{\sigma+1} = (\aleph_\alpha + \aleph_\beta)^{\aleph_\sigma}.$$

And in general, if M denotes a $c_{\sigma\sigma}$-set and M_1 specifically denotes a $c_{\sigma\sigma}$-set without $c_{\sigma\sigma}$-gaps, then

$$M \geqq (\aleph_\alpha + \aleph_\beta)_\sigma, \quad M_1 \geqq (\aleph_\alpha + \aleph_\beta)^{\aleph_\sigma},$$

provided that each middle segment of M contains sequences of types ω_α and ω_β^*.

But surely M contains ω_σ-sequences and ω_σ^*-sequences in each middle segment; therefore

$$M \geqq (\aleph_\sigma)_\sigma, \quad M_1 \geqq \aleph_\sigma^{\aleph_\sigma}$$

hold under all conditions, and thus the theorem:

Theorem XXI. *The cardinality of a $c_{\sigma\sigma}$-set is at least equal to the sum of all the powers of $\aleph_\sigma$ with exponents smaller than $\aleph_\sigma$ and that of a $c_{\sigma\sigma}$-set without $c_{\sigma\sigma}$-gaps is at least equal to $\aleph_\sigma^{\aleph_\sigma}$.*

For example, all sets of the genus $W = c_{01} c_{10} c_{11}$ have cardinality at least $\aleph_1^{\aleph_0}$ (of the continuum), and in case they contain no c_{11}-gaps, they have cardinality at least $\aleph_1^{\aleph_1}$.

For *irreducible* sets the bounds can perhaps be raised; again, if κ is the greater boundary number and $\alpha < \kappa$, then each middle segment certainly contains (even for singular ω_α) an ω_α-sequence, and for $\kappa = (\kappa - 1) + 1$ one can write

$$M \geqq (\aleph_{\kappa-1})_\sigma, \quad M_1 \geqq \aleph_{\kappa-1}^{\aleph_\sigma}.$$

It is evident from this that the construction given in §21 for an irreducible, continuous $c_{\sigma\sigma}$-set makes do with the smallest possible cardinality $\aleph_{\kappa-1}^{\aleph_\sigma}$.

Should there be a $c_{\sigma\sigma}$-set of cardinality $\aleph_\sigma$, then it must be that $(\aleph_\sigma)_\sigma = \aleph_\sigma$, i.e., $\aleph_\sigma$ is equal to the sum of all its powers with exponents below $\aleph_\sigma$ and is thus itself equal to each of these powers. We are going to call this *Cantor's Aleph Hypothesis* since in the first problematic case $\sigma = 1$ it turns into the Continuum Hypothesis $\aleph_1 = \aleph_1^{\aleph_0}$; for $\sigma = 0$ it is a not question of the hypothetical, but rather a question of a well-known equation

$$\aleph_0 = (\aleph_0)_0 = \aleph_0 + \aleph_0^2 + \aleph_0^3 + \cdots,$$

and in fact, there indeed is a c_{00}-type of cardinality $\aleph_0$, namely, η. For $\sigma = (\sigma - 1) + 1$, the Aleph Hypothesis reads

$$\aleph_\sigma = \aleph_\sigma^{\aleph_{\sigma-1}} = 2^{\aleph_{\sigma-1}}.$$

It must in no way be forgotten that we are speaking here exclusively about *regular* $\aleph_\sigma$.*)

We will see at once that the Aleph Hypothesis is also sufficient for the existence of $c_{\sigma\sigma}$-types of cardinality $\aleph_\sigma$, in particular for an (and by XIX a unique) e_σ-type of this cardinality.

We draw yet a further conclusion from Theorem XX. Since in the theorem one can always take the base N to be of type $\omega^* + \omega$, each $c_{\sigma\sigma}$-set contains a subset similar to $S(\omega_\sigma)$, where $S(\omega_\sigma)$ is the set of all coverings of ω_σ by the set N,

$$N: \cdots q_3 \, q_2 \, q_1 \, p \, p_1 \, p_2 \, p_3 \cdots,$$

in which a first even element appears and after that just the element p. If we restrict the base even further to only three elements

$$3: q_1 \, p \, p_1$$

and call $T(\omega_\sigma)$ the set of all coverings with this base in which the middle element p occurs uninterruptedly from a first place on, thus all coverings

$$X(\gamma) = X^\gamma + p + p + \cdots \quad (\gamma < \omega_\sigma),$$

where X^γ is made up of only p_1 and q_1, then each $c_{\sigma\sigma}$-set contains a subset similar to $T(\omega_\sigma)$ (of course, the covering sets are always to be thought of as ordered by first differences). But this set $T(\omega_\sigma)$ is itself a $c_{\sigma\sigma}$-set and at the same time an e_σ-set. In particular, it is cofinal with an argument-like ω_σ-sequence $P(0), P(1), \cdots, P(\gamma), \cdots$, in which

$$P(\gamma) = P^\gamma + p + p + \cdots$$

and all elements of P^γ are equal to p_1. Correspondingly, it is coinitial with an argument-like ω_σ^*-sequence; the same holds for the totality of coverings that begin with a specific X^γ. Therefore each ω_τ-sequence (here of course, always an argument-like sequence) determines a $c_{\tau\sigma}$-gap, each ω_τ^*-sequence determines a $c_{\sigma\tau}$-gap, and each element $X(\gamma)$ is a $c_{\sigma\sigma}$-element since it is enclosed by the two sets of coverings that begin with $X^\gamma + q_1$ and $X^\gamma + p_1$. Moreover, $c_{\sigma\sigma}$-gaps exist since a covering X (not belonging to $T(\omega_\sigma)$) with just odd elements, in the case in which both p_1 and q_1 occur $\aleph_\sigma$ times, is

---

*) For a singular $\aleph_\beta$ whose initial number ω_β is cofinal with ω_α ($< \omega_\beta$), it is already the case that $\aleph_\beta^{\aleph_\alpha} > \aleph_\beta$, as can be concluded in an obvious generalization of a formula of Herr J. König (Math. Ann. 60, pp. 177–80); the Aleph Hypothesis can then at best be expressed in the form $(\aleph_\beta)_\alpha = \aleph_\beta$.

enclosed by two argument-like $\omega_\sigma, \omega_\sigma^*$-sequences. Thus $T(\omega_\sigma)$ is an η-set of species $(c_{\sigma\sigma}, \mathfrak{W}_\sigma)$, where

$$\mathfrak{W}_\sigma = \left\{ \begin{array}{ccccc} c_{\sigma 0} & c_{\sigma 1} & \cdots & c_{\sigma\tau} & \cdots \\ c_{0\sigma} & c_{1\sigma} & \cdots & c_{\tau\sigma} & \cdots \end{array} \right\} c_{\sigma\sigma}, \quad \tau < \sigma$$

includes the elements of the σth row and column of the tableau $\mathfrak{W}$ (§20) up to their intersection; in this case $\pi = \varrho = \sigma$, i.e., this set is at the same time an e_σ-set, and because it is contained in every $c_{\sigma\sigma}$-set, it is certainly both a $c_{\sigma\sigma}$-set and an e_σ-set of smallest possible cardinality. This cardinality is $(2)_\sigma = (\aleph_\sigma)_\sigma$, and it reduces to $\aleph_\sigma$ under Cantor's Aleph Hypothesis, so that there is then actually one and only one e_σ-type of cardinality $\aleph_\sigma$.

The set $T(\omega_\sigma)$ is obviously a subset of the power of the $(1+\sigma)$th class

$$\eta_\sigma = 3(\omega_\sigma)_p^\sigma,$$

in which the principal element p is the middle element of the base and only a place set $< \omega_\sigma$ is covered by the two other elements p_1, q_1; certainly, those covers in which odd elements still follow after the even element p also belong to η_σ. The above remarks about $T(\omega_\sigma)$, of course, hold mutatis mutandis for η_σ and show that this set is also a $c_{\sigma\sigma}$-set and an e_σ-set of species $(c_{\sigma\sigma}, \mathfrak{W}_\sigma)$. Furthermore, it is easy to see that η_σ is also contained in every $c_{\sigma\sigma}$-set. Namely, if we write the elements of η_σ as coverings of ω_σ by the set $(p_3\, p\, q_3)$, and after that, if we replace each element p that is followed by an odd element by either p_1 or q_1, depending on whether p_3 or q_3 is the next following odd element, then for the rank ordering

$$q_3 < q_1 < p < p_1 < p_3,$$

the rank ordering of those coverings remains unchanged as one easily proves, even though by this alteration the first differing place of two coverings can be shifted to the left. In this way, the covering set η_σ mentioned above is similar to the set of those coverings of ω_σ with base

$$5: \quad q_3\, q_1\, p\, p_1\, p_3,$$

where no odd element may follow an even element; thus it is similar to a subset of $S(\omega_\sigma)$ and therefore to each $c_{\sigma\sigma}$-set.

With this, we have obtained the result:

Theorem XXII. *Each $c_{\sigma\sigma}$-set contains a subset of type*

$$\eta_\sigma = 3(\omega_\sigma)_p^\sigma$$

that is both an e_σ-set and a $c_{\sigma\sigma}$-set of the smallest cardinality of species $(c_{\sigma\sigma}, \mathfrak{W}_\sigma)$. If $\aleph_\sigma$ is equal to each of its powers with smaller exponent, then there exists one and only one e_σ-type of cardinality $\aleph_\sigma$, namely, η_σ.

The fact that each $c_{\sigma\sigma}$-set already contains any set of cardinality $\leqq \aleph_\sigma$ follows from XVIII and XXII and represents an important strengthening of Theorem XIX. For example, a set of species $(c_{01}, c_{10}\, c_{11})$ is itself only an e_0-set, however, it contains an e_1-set as a subset and thereby each type of cardinality $\leqq \aleph_1$. By filling the $c_{\sigma\sigma}$-gaps of η_σ, one can immediately give a

$c_{\sigma\sigma}$-set without $c_{\sigma\sigma}$-gaps that is contained in each such set as a subset and that therefore has the lowest cardinality; thereby, there exists a $c_{\sigma\sigma}$-set and a e_σ-set of species $(c_{\sigma\sigma}, \mathfrak{W}_\sigma - c_{\sigma\sigma})$ and cardinality $2^{\aleph_\sigma} = \aleph_\sigma^{\aleph_\sigma}$. This set, call its type λ_σ (λ_0 is the linear continuum, as η_0 is η), is thus contained in each $c_{\sigma\sigma}$-set without $c_{\sigma\sigma}$-gaps, however, not generally in each $c_{\sigma\sigma}$-set, e.g., not in η_σ. In this regard, we are still going to prove the theorem:

Theorem XXIII. *The power $M(\omega_\sigma)_p^\sigma$ of the $(1+\sigma)$th class with any principal element p contains no $c_{\sigma\sigma}$-set without $c_{\sigma\sigma}$-gaps as a subset, except when the base M already contains such a subset.*

Instead of this we can say: if λ_σ is not contained in M, it is also not contained in the power. First, we note that λ_σ is irreducible because one cannot omit any of its element or gap characters without destroying the *completeness* of the character set $\mathfrak{W}_\sigma$; so each middle segment of λ_σ is of the same species (λ_σ), incidentally, even exactly of type λ_σ.

Now we have to consider the coverings with argument ω_σ

$$X = x_0 + x_1 + \cdots + x_\alpha + \cdots = X^\alpha + x_\alpha + X_\alpha \quad (\alpha < \omega_\sigma),$$

in which the x_α run through the set M and the secondary elements ($\neq p$) occupy a place set $< \omega_\sigma$, thus coverings of the form

$$X = X^\alpha + p + p + \cdots,$$

where α is the first place or any later place that follows the set of all places covered by secondary elements. All these coverings form the power $M(\omega_\sigma)_p^\sigma$; let a part of them form a dense subset N that we assume is of type λ_σ in order to show the absurdity of this assumption. We distinguish whether or not *base-like* sequences occur in each middle segment of N.

If base-like sequences occur in each middle segment of N and if one such sequence with differing place α is indicated by

$$U = Q^\alpha + u_\alpha + U_\alpha, \quad V = Q^\alpha + v_\alpha + V_\alpha, \quad \ldots,$$

then consider the totality in N of all coverings $X = Q^\alpha + x_\alpha + X_\alpha$ beginning with Q^α. Here in any case, x_α runs through a transfinite subset of M because it is supposed to contain base sequences. Now it cannot be that for each x_α there is just *one* covering of N beginning with $Q^\alpha + x_\alpha$ since otherwise N would contain a piece of the type of a subset of M, which by hypothesis is certainly not a $c_{\sigma\sigma}$-set without $c_{\sigma\sigma}$-gaps. So if N is supposed to be of type λ_σ, then for at least one x_α, and what is more, which arises in the same way, for a transfinite set of values x_α the piece of coverings beginning with $Q^\alpha + x_\alpha$ must be a *forbidden* [[ächtes]] piece (> 1) of N. That is enough for us to draw the following conclusion: there are three elements $a_\alpha < q_\alpha < b_\alpha$, of which q_α is a secondary element ($\neq p$), such that there are coverings in N beginning with

$$Q^\alpha + a_\alpha, \quad Q^\alpha + q_\alpha, \quad Q^\alpha + b_\alpha,$$

and certainly for $Q^\alpha + q_\alpha$, there are at least two such coverings, thus infinitely many. The piece of coverings of N that begin with $Q^\alpha + q_\alpha$ should again contain base-like sequences that have the form

$$U = Q^\beta + u_\beta + U_\beta, \quad V = Q^\beta + v_\beta + V_\beta, \quad \ldots$$

$$\beta > \alpha, \quad Q^\beta = Q^\alpha + q_\alpha + Q^\beta_\alpha;$$

and the exact repetition of the above reasoning is permitted. Accordingly, by induction, there exists an element sequence

$$Q = \cdots + q_{\alpha_0} + \cdots + q_{\alpha_1} + \cdots + q_{\alpha_\xi} + \cdots = Q^\alpha + q_\alpha + Q_\alpha$$

with $\aleph_\sigma$ secondary elements $q_{\alpha_0} q_{\alpha_1} \cdots q_{\alpha_\xi} \cdots$ such that for each α there is an infinite set of coverings beginning with Q^α, and each covering beginning with $Q^{\alpha_\xi} + q_{\alpha_\xi}$ is enclosed by two coverings that begin with

$$Q^{\alpha_\xi} + a_{\alpha_\xi}, \quad Q^{\alpha_\xi} + b_{\alpha_\xi} \quad (a_{\alpha_\xi} < q_{\alpha_\xi} < b_{\alpha_\xi})$$

and belong to N. The inductive proof splits into the inference from ξ to $\xi + 1$, just dealt with above, and the inference for the limit of a number sequence; this latter inference, however, depends on the assumption that N should not contain any $c_{\tau\tau}$-gaps nor any $c_{\tau\tau}$-elements ($\tau < \sigma$), according to which, for $\xi = \lim\{\eta\}$ and $\alpha = \lim\{\alpha_\eta\}$, there surely exists in N a forbidden piece (> 1) of coverings beginning with Q^α. Now of course, the assumption has led to an absurdity, for the pair of enclosing coverings mentioned above form two argument-like sequences of types ω_σ and ω_σ^* that enclose only the covering Q; this covering, because of its $\aleph_\sigma$ secondary elements, does not belong to N, and N certainly contains a $c_{\sigma\sigma}$-gap.

Second, if a middle segment N_1 of N is free from base-like sequences, then each of its elements, as $c_{\sigma\sigma}$-elements, must be the limit element of two *argument-like* sequences of types ω_σ and ω_σ^*; for each element $X = X^\alpha + p + p + \cdots$ belonging to N_1, there surely exists another

$$Y = X^\alpha + p + p + \cdots + y_\beta + Y^\beta \quad (\beta \geqq \alpha, \ y_\beta > p)$$

that likewise belongs to N_1, is $> X$, and contains at least one secondary element y_β after the secondary elements of X; in the same way, there is always an element of the same kind $< X$. By an induction that, as in the inference for limit numbers, is again based on the absence of $c_{\tau\tau}$-gaps and $c_{\tau\tau}$-elements in N_1, one then immediately proves the existence of argument-like sequence pairs

$$Q_0 \, Q_2 \, Q_4 \cdots Q_{2\xi} \cdots \mid \cdots Q_{2\xi+1} \cdots Q_5 \, Q_3 \, Q_1$$

of type $\omega_\sigma + \omega_\sigma^*$, in which, for $\xi < \eta$, Q_η always contains all the secondary elements of Q_ξ and at least one more secondary element following them. One of the enclosed coverings Q would have to contain $\aleph_\sigma$ secondary elements and so of course does not belong to N_1; with this, we again arrive at a $c_{\sigma\sigma}$-gap in N. Therefore Theorem XXIII is proved in all cases.

§25
A Strengthening of the Existence Proof

Our previous existence proof (§21, §19) confined itself to the construction of an irreducible type for each admissible species; our current goal is to prove the existence of *infinitely many distinct* irreducible types for each such species. In general, it is trivial that there are distinct types in each species inasmuch as these types can be cofinal with and coinitial with the inverses of different regular initial numbers; e.g., besides η, the types $\eta\omega_1$, $\eta\omega_1^*$, and $\eta\omega_1 + \eta\omega_1^*$ belong to the species (c_{00}, c_{00}). If we do not want to attach any importance to this obvious result, we will have to formulate a more precise concept of distinctness: *we call two dense sets essentially distinct if it is never the case that a middle segment of one is similar to a middle segment of the other.* In this sense, two irreducible sets of different species are essentially distinct; η and ι, the set of rational numbers and the set of irrational numbers, furnish a prime example of essentially distinct types of the same species.

Next we show how one can derive additional irreducible sets of the same species from an irreducible $c_{\sigma\sigma}$-set M of species $(W, 0)$ by product formation with variable factors that are either intervals of M or single elements. We denote the intervals of M by

$$M(u, v) = u + M_u^v + v,$$

and by $u_{\alpha\beta}$ and $v_{\alpha\beta}$, we understand any $c_{\alpha\beta}$-elements of M; furthermore, again let such elements that run through a proper set (> 1) be called active, those that run through a set of type 1 be called constrained, and in this last mentioned case, let them be set equal to a fixed element p. If κ, λ are then the boundary numbers of W and again $\kappa \geq \lambda$, thus λ is the smaller boundary number, and if ϱ is an ordinal number $< \omega_\lambda$ and cofinal with ω_σ, thus

$$\omega_\sigma \leq \varrho < \omega_\lambda,$$

then we consider the element combinations

$$X = x_0 + x_1 + \cdots + x_\xi + \cdots = X^\xi + x_\xi + X_\xi \quad (\xi < \varrho)$$

with argument ϱ, and we determine the set that x_ξ runs through by the following rules:

(1) x_0 runs through the whole set M.

(2) If ξ is a limit number and all the elements of X^ξ are active, then x_ξ is active, and if ξ is cofinal with ω_τ ($< \omega_\lambda$), x_ξ runs through an interval $M(u_{\tau\beta}, v_{\alpha\tau})$.

(3) After a constrained element, there follow only constrained elements, i.e., for $x_\xi = p$, it is also the case that $x_{\xi+1} = x_{\xi+2} = \cdots = p$.

(4) If x_ξ is active and a $c_{\alpha\beta}$-element *within* its set, $x_{\xi+1}$ runs through an interval $M(u_{\alpha\beta}, v_{\alpha\beta})$.

(5) If x_ξ is active and a boundary element of its interval, then $x_{\xi+1} = p$.

The existence of such intervals as required here is obvious; one has only to make a specific choice. Call the set of all the combinations so defined and ordered by first differences $N(\varrho)$; the discussion of this set is similar to, but much simpler than, the corresponding investigation in §21. The combinations without boundary elements are $c_{\sigma\sigma}$-limits enclosed by two argument-like sequences. For a limit number ξ, a combination of the form $X^\xi + a_\xi + p + p + \cdots$, where a_ξ is the left boundary element of its interval, is a $c_{\tau\beta}$-element that is bounded on the left by an argument-like sequence and on the right by a base-like sequence; in the case that $\xi = (\xi - 1) + 1$, it is a $c_{\alpha\beta}$-element bounded on both sides by base-like sequences. The corresponding situation holds for the combination $X^\xi + b_\xi + p + p + \cdots$, whose last active element b_ξ is the right boundary element of its interval. Gaps do not exist since each argument-like or base-like sequence has a limit. All characters of $N(\varrho)$ occur in M, and, conversely, because of the irreducibility of M, all characters of M occur in each middle segment of $N(\varrho)$. In fact, $N(\varrho)$ is an irreducible set of the same species $(W, 0)$ as M itself. The difficulty first begins with the proof that one arrives at distinct and, with a suitable choice of ϱ, even at essentially distinct types of the same species.

If we choose as M the set constructed in §21, by the observation at the conclusion of this section, there is a subset P dense in M that on the other hand is a subset of a certain power of the $(1 + \sigma)$th class *with a scattered base*. By Theorem XXIII, any such set does not contain any $c_{\sigma\sigma}$-set without $c_{\sigma\sigma}$-gaps as a subset since the base contains no such set (in this case, no dense subset whatsoever). Therefore P also contains no such subsets.

Moreover, if we recall (§4) that we call two intervals that have at most one element in common *exclusive* and if we assign to an *exclusive interval set* (as we are going to say more concisely for a set of pairwise exclusive intervals) the order type of the set of their left endpoints, then it is clear that if the dense set M contains an exclusive interval set of type γ, then each subset P dense in M must contain a subset of type γ. Thus if even just one set P dense in M exists that contains no subset of type γ, then M contains no exclusive interval set γ.

Combining this with the above remarks yields the existence of an irreducible $c_{\sigma\sigma}$-set M of species $(W, 0)$ that *has no exclusive interval set of the type of a $c_{\sigma\sigma}$-set without $c_{\sigma\sigma}$-gaps*. This property of M is the basis of the following proof that the above method furnishes only distinct and possibly essentially distinct sets $N(\varrho)$.

The line of reasoning is thereby somewhat intricate because the variable factors of our product can be single elements (which, on the other hand, cannot be avoided if one wants to realize the goal of a construction of *generally* prescribed species $(W, 0)$); it simplifies noticeably if all the factors are proper sets, and it is very much to be recommended that the course of the study be clarified, say, with the special case of the complete power of a single M that contains no exclusive interval set similar to itself. We turn at once to the general case that we formulate even more broadly than would be absolutely necessary in order to throw into bold relief that which is essential.

We consider a variable product with argument α, formed from the combinations
$$X = x_0 + x_1 + \cdots + x_\xi + \cdots = X^\xi + x_\xi + X_\xi \quad (\xi < \alpha)$$
that are ordered by first differences and in which the set $M(X^\xi)$ through which x_ξ runs can depend on the place ξ and the preceding elements. We distinguish whether $M(X^\xi)$ is a proper set (> 1) or a single element p, and in the first case, we again call the element x_ξ active, and in the second case, we call the element x_ξ constrained. Thereupon, we give the following special rules:

(a) After a constrained element, there follow only constrained elements, i.e., for $x_\xi = p$, it is the case that $x_{\xi+1} = x_{\xi+2} = \cdots = p$.

(b) If constrained elements occur, the first appears at a place $\xi+1$ (thus not at place 0 or at a place with a limit number).

Hence, either our combinations contain just active elements, or they are of the form $X^\xi + x_\xi + p + p + \cdots$, where x_ξ is the last active element and $x_{\xi+1} = p$ is the first constrained element.

If x_ξ is an active element that runs through the set $M(X^\xi) > 1$, then we distinguish in $M(X^\xi)$ those elements u_ξ for which $M(X^\xi + u_\xi) = p$, thus for which $x_{\xi+1}$ is constrained, and those elements v_ξ for which $M(X^\xi + v_\xi) > 1$, thus for which $x_{\xi+1}$ is active. The sets of these elements are called $U(X^\xi)$ and $V(X^\xi)$; their union is $M(X^\xi)$. With respect to these sets, we now make the basic assumption:

(c) *Each set $V(X^\xi)$ is a proper set, and it is not similar to any exclusive interval set of any $M(Y^\eta)$.*

Note that here it is only meant that $V(X^\xi)$ is a subset of the *proper* set $M(X^\xi)$; if we wished to also extend the notation to the case where $M(X^\xi) = p$, we would have to set $U(X^\xi) = p$ and $V(X^\xi) = 0$.

Call a variable product with argument α of the sort described $P(\alpha)$. If X and X' are two of its elements with first differing place ξ, we are going to say that the interval (X, X') is *of order* ξ. If $P_1(\alpha)$ is a *piece* (§1) of $P(\alpha)$ and ξ is the lowest order of the intervals contained in it, then call $P_1(\alpha)$ a piece of order ξ; $P(\alpha)$ itself is of order 0. The order of a piece contained in $P_1(\alpha)$ is then $\geqq \xi$.

Call the aggregate of elements of $P(\alpha)$ that begin with X^ξ a *principal piece* [[*Hauptstück*]], the principal piece $[X^\xi]$. If $M(X^\xi) > 1$, this piece is a proper piece of order ξ, or if $M(X^\xi) = p$, it reduces to a single element $X^\xi + p + p + \cdots$. In the first case, it decomposes into the principal pieces $[X^\xi + x_\xi]$ of which again the $[X^\xi + u_\xi]$ are single elements and the $[X^\xi + v_\xi]$ are proper pieces of order $\xi + 1$; the latter form a set of separated pieces that are similar to the set $V(X^\xi)$.

Now we claim:

Theorem XXIV. *If a product $P(\alpha)$ that satisfies conditions* (a), (b), (c) *is similar to another product $Q(\beta)$ of the same kind or to one of its pieces, then it must be that $\alpha \leqq \beta$; thus if $P(\alpha)$ is similar to $Q(\beta)$, it must be that $\alpha = \beta$.*

The proof depends on the lemma:

If $P(\alpha)$ can be mapped onto $Q(\beta)$ or a piece $Q_1(\beta)$, there exists an element

$$A = a_0 + a_1 + \cdots + a_\xi + \cdots = A^\xi + a_\xi + A_\xi \quad (\xi < \alpha)$$

of $P(\alpha)$ with just active elements a_ξ such that each principal piece $[A^\xi]$ of order ξ has a piece of order $\geqq \xi$ as an image.

This lemma is proved by induction:

The inference for a limit number. If ξ is a limit number and all principal pieces $[A^\lambda]$ for $\lambda < \xi$ have images of order $\mu \geqq \lambda$, then according to rule (b), $[A^\xi]$ is a proper piece of order ξ and it is contained in all pieces $[A^\lambda]$. Its image, a proper piece of $Q(\beta)$, is contained in the images of the $[A^\lambda]$, and its order η is thus $\geqq \mu \geqq \lambda$; η is therefore greater than all the numbers λ and at least equal to their limit ξ, thus $\eta \geqq \xi$.

The inference from ξ to $\xi + 1$. Let $[A^\xi]$ of order ξ have an image of order $\eta \geqq \xi$; the elements of this image are thus of the form

$$Y = B^\eta + y_\eta + Y_\eta$$

with fixed beginning B^η, while y_η runs through a subset of $M(B^\eta)$. Among the pieces into which $[A^\xi]$ decomposes, we consider only the proper pieces $[A^\xi + v_\xi]$ of order $\xi + 1$, which form a set of pieces of the type $V(A^\xi)$. The images of these pieces are of order $\geqq \eta$; however, they cannot all be of order η. For if the image of the piece $[A^\xi + v_\xi]$ is of order η, it contains two combinations

$$Y = B^\eta + y_\eta + Y_\eta, \quad Z = B^\eta + z_\eta + Z_\eta \quad (y_\eta < z_\eta)$$

with differing place η, where the differing elements bound an interval (y_η, z_η) of $M(B^\eta)$; if for a second piece $[A^\xi + v'_\xi]$ the image is again of order η, it contains two elements

$$Y' = B^\eta + y'_\eta + Y'_\eta, \quad Z' = B^\eta + z'_\eta + Z'_\eta \quad (y'_\eta < z'_\eta),$$

and for $v_\xi < v'_\xi$, it is the case that $Y < Z < Y' < Z'$; thus

$$y_\eta < z_\eta \leqq y'_\eta < z'_\eta,$$

i.e., both intervals $(y_\eta < z_\eta)$ and $(y'_\eta < z'_\eta)$ are exclusive. Hence were the image for each v_ξ of order η, $M(B^\eta)$ would have to contain an exclusive interval set of type $V(A^\xi)$. Since this is contrary to assumption (c), at least one of the v_ξ, let us call it a_ξ, is such that the piece $[A^\xi + a_\xi]$ has an image of order $\geqq \eta + 1 \geqq \xi + 1$.

With this, the induction has been carried out; its start holds of its own accord since by hypothesis $P(\alpha)$, a principal piece of order 0, is supposed to be mapped onto $Q(\beta)$ or $Q_1(\beta)$, a piece of order $\geqq 0$.

From the lemma, it follows at once that for each number ξ ($< \alpha$) there exists a number η ($\xi \leq \eta < \beta$); thus β must be greater than all the η and therefore it must be that $\beta \geq \alpha$. With this, Theorem XXIV is also proved: two products of the described sort with distinct arguments are certainly distinct.

From the observation that each proper principal piece $[X^\xi]$ of $P(\alpha)$ is itself a product $P(-\xi+\alpha)$ of the same kind with argument $(-\xi+\alpha)$ (an end piece of α), it can be immediately concluded which choice of arguments α will lead to the sought after *essential* difference; however, for this we require the additional hypothesis:

(d) Each set $V(X^\xi)$ is dense in $M(X^\xi)$.

Then let α be a *principal number* (§16), i.e., equal to each of its end pieces, so that each proper principal piece of $P(\alpha)$ itself has the same form $\Pi(\alpha)$ (just without having to be similar to $P(\alpha)$). On the basis of hypothesis (d), between any two elements of $P(\alpha)$

$$X = X^\xi + x_\xi + X_\xi, \quad X' = X^\xi + x'_\xi + X'_\xi \quad (x_\xi < x'_\xi),$$

there certainly lies a principal piece $[X^\xi + v_\xi]$, where $x_\xi < v_\xi < x'_\xi$, i.e., each piece $P_1(\alpha)$ now contains a piece $\Pi(\alpha)$. Hence by Theorem XXIV, if $\alpha > \beta$, it cannot be that $P(\alpha) = Q(\beta)$ or $P(\alpha) = Q_1(\beta)$; in addition, it also cannot be that $P_1(\alpha) = Q(\beta)$ or $P_1(\alpha) = Q_1(\beta)$ since otherwise $\Pi(\alpha)$ would be similar to a piece of $Q(\beta)$. Thus we have the strengthening of XXIV:

Theorem XXV. *A product that satisfies conditions* (a)—(d) *and has a principal number as argument is essentially distinct from all products of the same kind with smaller argument; two products of this kind with different principal numbers as arguments are essentially distinct.*

Among the special cases of this theorem, we note the following: conditions (a), (b), and (c) are fulfilled if generally no constrained elements occur, thus all factors are proper, and none of the factors is similar to an exclusive interval set of any other factor (not even of itself); then one must set $U(X^\xi) = 0$ and $V(X^\xi) = M(X^\xi)$. Condition (d) still demands the density of all factors. Hence:

Theorem XXVI. *A product whose factors (variable or fixed) are always proper sets, none of which are similar to an exclusive interval set of any other of the sets, is distinct from all the products of the same sort with distinct arguments. And in case all the factors are dense, they are essentially distinct if the arguments are principal numbers.*

In particular, this holds for complete powers of a base M that contains no exclusive interval set of its own type; the complete powers of such a set are all distinct; the complete powers of a dense set of this kind with principal number as argument are all essentially distinct. For example, this holds for the linear continuum, the unbounded λ as well as for the bounded $\vartheta = 1 + \lambda + 1$ since each exclusive interval set of these types is countable.

Thus the powers
$$\vartheta = \vartheta((1)), \vartheta((\omega)), \vartheta((\omega^2)), \ldots, \vartheta((\omega^\alpha)), \ldots$$
with principal number arguments from the second number class are $\aleph_1$ essentially distinct*) types of species $(c_{00}, 0)$ (after the removal of boundary elements); moreover, if one goes to principal numbers of the third number class, one obtains $\aleph_2$ essentially distinct types of species $(c_{00}\ c_{01}\ c_{10}, 0)$ and $(c_{01}\ c_{10}\ c_{11}, 0)$; in the general case, if the argument is any principal number of cardinality $\aleph_\gamma$ that is cofinal with ω_σ, one obtains $\aleph_{\gamma+1}$ essentially distinct representatives of species $(W, 0)$ whose character set comprises the σth row and σth column of the tableau with boundary numbers $\gamma + 1, \gamma + 1$:
$$W = \{c_{\alpha\sigma}, c_{\sigma\alpha}\}, \quad (\alpha \leqq \gamma).$$
This example illustrates the remark at the conclusion of §21.

If we now return to the general construction at the beginning of the present section, we recognize the applicability of Theorem XXV. The rules there about constrained elements correspond to conditions (a) and (b). The proper sets $M(X^\xi)$ are intervals of M and the only intervals whose boundary elements make the next element into a constrained element; if we denote these boundary elements by a_ξ and b_ξ, then
$$U(X^\xi) = (a_\xi, b_\xi), \quad M(X^\xi) = a_\xi + V(X^\xi) + b_\xi.$$
As a middle segment of M, $V(X^\xi)$ is an irreducible set of species $(W, 0)$, thus a $c_{\sigma\sigma}$-set without $c_{\sigma\sigma}$-gaps. Since neither M nor any of its intervals $M(Y^\eta)$ contain an exclusive interval set of this sort, condition (c) is satisfied. Finally, (d) is also fulfilled. Hence if we restrict ourselves to principal number arguments ϱ that are cofinal with ω_σ and $< \omega_\lambda$, our construction furnishes $\aleph_\lambda$ essentially distinct, irreducible representatives $N(\varrho)$ of species $(W, 0)$, namely, besides $M = N(1)$, the products
$$N(\omega_\sigma), N(\omega_\sigma^2), \ldots, N(\omega_\sigma^{\omega+1}), \ldots, N(\omega_\sigma^\gamma), \ldots \quad (\gamma < \omega_\lambda),$$
where the exponent γ of Cantor's power is either a non-limit number or cofinal with ω_σ.

By §19, the hereby strengthened existence proof for the species $(W, 0)$ can be extended to discontinuous species of the same genus with a stroke of the pen. If N is continuous and P is dense in N, then $N = [P]$ is the filling of P and is uniquely determined by P. Thus if N is distinct from N', then surely P is distinct from P'. And the application to middle segments shows that here instead of "distinct" we may even say "essentially distinct." With this, we have obtained the theorem:

Theorem XXVII. *Each species with a complete character set whose boundary numbers κ, λ are not limit numbers is, for $\kappa \geqq \lambda$, represented by at least $\aleph_\lambda$ essentially distinct, irreducible types.*

---

*) Since essentially distinct types are independent (§10) of one another, the basis of the ring of all types of cardinality $\leqq 2^{\aleph_0}$ must include at least $\aleph_1$ types; cf. the remark at the conclusion of §11.

If we call $\aleph_\tau$ the cardinality of ϱ ($\aleph_\sigma \leqq \aleph_\tau < \aleph_\lambda$), then the cardinality of $N(\varrho)$ is $\aleph_{\kappa-1}^{\aleph_\sigma \aleph_\tau} = \aleph_{\kappa-1}^{\aleph_\tau}$, and it lies between $\aleph_{\kappa-1}^{\aleph_\sigma}$ and $\aleph_{\kappa-1}^{\aleph_{\lambda-1}}$; in particular, there are surely $\aleph_{\sigma+1}$ essentially distinct sets $(W, 0)$ of the smallest cardinality $\aleph_{\kappa-1}^{\aleph_\sigma}$. Since $N(\varrho)$ contains an exclusive interval set of type $N(1) = M$ and of cardinality $\aleph_{\kappa-1}^{\aleph_\sigma}$, each set dense in $N(\varrho)$ is of cardinality $\geqq \aleph_{\kappa-1}^{\aleph_\sigma}$. The $\aleph_\lambda$ representatives of species (U, V) obtained here that contain $c_{\sigma\sigma}$-gaps no longer need to have the attainable, in this case, minimal cardinality $(\aleph_{\kappa-1})_\sigma$; one can only force this upon the relatively dense subsets formed from the lowest set $N(1) = M$ itself by the rule of §19. If, for example, $\aleph_\sigma$ satisfies Cantor's Aleph Hypothesis, our method certainly furnishes $\aleph_{\sigma+1}$ essentially distinct representatives of species $(c_{\sigma\sigma}, \mathfrak{W}_\sigma)$; however, only one is $= \aleph_\sigma$; the rest are $\geqq \aleph_\sigma^{\aleph_\sigma} > \aleph_\sigma$. In particular, for the species (c_{00}, c_{00}) of the rational number set, it furnishes $\aleph_1$ essentially distinct exemplars of which, however, only one is countable, namely, η, while the rest are of the cardinality of the continuum.

Introduction to "Graduation by Final Behavior"

Hausdorff's last major work on ordered sets, *Graduation by Final Behavior* [[*Die Graduierung nach dem Endverlauf*]], is clearly a continuation of *On Pantachie Types*, the article that ended [H 1907a].[1] In a July 15, 1907, letter[2] to Hilbert about the writing of [H 1908], Hausdorff parenthetically remarks that he has "just now finished some addenda to the work on pantachie types." This may actually refer to the material in §2 and §5 of [H 1909a] where some results from *On Pantachie Types* are repeated and shown to hold when limited to the case of monotonic sequences. But in §3 and §4, Hausdorff, going beyond the supplementation of his earlier article, presents new phenomena: the existence of a pantachie closed under rational operations and the existence (without the use of CH) of a pantachie with an $\Omega\Omega^*$-gap.[3] The former is the first application of a maximal principle in algebra; the latter is the primordial *Hausdorff gap*.

In *On Pantachie Types*, Hausdorff began with a critical examination of Paul Du Bois-Reymond's infinitary (partial) ordering of the monontonic real functions that go to $+\infty$ with increasing x. The "infinitary relations" $<, \sim, >$ among such functions f and g are defined in terms of the behavior of the quotient f/g as x goes to $+\infty$. If this quotient's limit is 0, then $f < g$; if its limit is finite and $\neq 0$, then $f \sim g$; if its limit is $+\infty$, then $f > g$. (A similar definition provides an infinitary ordering for monotonic real sequences that converge to $+\infty$.) Of course, incomparability is the default fourth possibility.

Because of incomparability, Hausdorff dismissed Du Bois-Reymond's "infinitary pantachie," a purported universal scale for the infinities associated with monotonic real functions that go to $+\infty$ with increasing x, and he initiated his own program of investigating the maximal linearly ordered subsets of this infinitary ordering; he adopted Du Bois-Reymond's term *pantachies* for such subsets. Almost immediately, he replaced the study of function pantachies with that of "numerical sequence" pantachies by associating the sequence $(f(n))$, with the function f. He then dropped the monotonicity requirement and changed the objects to be ordered to arbitrary sequences of positive reals. His final step was to consider a different partial ordering of such sequences (and functions), the "final rank ordering": for sequences A and B, he defined $A < B, A \sim B$, and $A > B$ if *finally* (for all but finitely many n) $a_n < b_n, a_n = b_n$, and $a_n > b_n$, respectively.

The final ordering of sequences (and functions) possessed what Hausdorff saw as the "essence of Du Bois-Reymond's idea," a dependence upon *final behavior*. Also in the final ordering, Hausdorff showed his uncanny sense for the telling simplification; in practice, the order properties of its pantachies are much easier to establish, and since each infinitary relation can be represented by a countably infinite conjunction of final relations ([H 1907a, 116n]), these order properties are shared by the panatachies of the infinitary order. Understandably, the pantachies of numerical sequences under the final partial ordering are the main focus of *On Pantachie Types*.

Graduation by Final Behavior, too, is mainly concerned with the final ordering of numerical sequences.[4] *On Pantachie Types* and *Graduation by Final Behavior* have remarkably similar beginnings:

> The idea of graduating the convergence of functions to a limit, e.g., their becoming zero or infinite, has been carried out in a systematic way by P. Du Bois-Reymond in particular ... [H 1907a, 105]

> The idea of graduating the behavior of functions $f(x)$ for a particular passage to a limit of the independent variable (say for $\lim x = +\infty$), especially placing the going to zero or infinity of the functions in a metric or at least ordinal series is certainly one of the most obvious problems of analysis ... [H 1909a, 297].

But as we can see, even at its start [H 1909a] signals Hausdorff's movement toward a more generic view of graduation problems and away from a focus on the works of one individual. In the remainder of the introduction, the mention of the controversial aspects of Du Bois-Reymond's work is kept to a minimum, and the tone is decidedly non-polemical.[5] The phenomenon of incomparable elements in any sufficiently comprehensive ordering of real functions or of real sequences that claims to be predicated on final behavior is portrayed as a fact of nature—not a natural catastrophe— that leads as a matter of course to the investigation of the maximal linearly ordered subsets of such orderings. A major question is the existence of such subsets.

In *On Pantachie Types*, Hausdorff established the existence of pantachies for sequences of positive reals under the final ordering with two proofs in which he made his first uses of Zermelo's Well-Ordering Theorem [H 1907a, 117–118]. His first proof began with a well-ordering of the set of all sequences of positive reals and then proceeded with an inductive whittling down of the set so ordered to a pantachie. His second proof was an argument by contradiction: the non-existence of a pantachie allowed the construction of a strictly increasing, well-ordered sequence $\{B_\alpha\}$ of arbitrary length, consisting of sets of sequences B_α that are linearly ordered by the final rank ordering. (Hausdorff called any linearly ordered set of sequences a *Bereich*, which we translate as *ordered domain*.) The existence of a strictly increasing sequence of ordered domains of arbitrary length led to the conclusion that $2^{\aleph_0}$, the cardinality of the set of all positive sequences, is not an aleph, contradicting the Well-Ordering Theorem.

This second proof is revisited in the introduction to [H 1909a] and given a purely set theoretic form which is that of a recognizable *maximal principle*. Hausdorff is the first to have published such a statement. Some care must be taken here; Hausdorff is not offering a new axiom, and he certainly makes no attempt to derive the Well-Ordering Theorem from his purely set theoretic formulation. What is now referred to as *Hausdorff's maximal principle* is intended, in this instance, to be the distillation of a previous proof for broader application. This being said, it is still of historical significance.[6]

The relevant objects in this new existence proof are the subsets of an infinite set M that satisfy some property ε of an unspecified nature.[7] A well-ordered sequence of length ξ of strictly increasing ε-subsets is called a *chain* of type ξ. Hausdorff makes two assumptions: (1) M has at least one ε-subset; (2) any chain of limit type can be continued. He explicitly notes that if the union of a chain of limit type is an ε-set then the chain is continuable. (Requiring that the ε-subsets in a chain only be linearly ordered by proper set inclusion and that the union of a chain be an ε-set, one has Zorn's lemma [Zo 1935].)

Hausdorff observes that since the cardinality of M is an aleph (by the Well-Ordering Theorem), the number of entries in any chain is bounded by the cardinality of M. Thus there must be non-continuable chains; if every chain were continuable, there would be chains of *every* ordinal length. Under assumption (2), any non-continuable chain is indexed by a non-limit ordinal; that is, it must have a last entry, which of course is an ε-subset that is not properly contained in any other ε-subset, i.e., a *maximal ε-set*. Hausdorff does not use the adjective *maximal*; he calls such a non-extendible entry a *closed ε-set*.[8]

Hausdorff applies his *maximal principle* to show that pantachies exist. In particular, for $\varepsilon = $ *ordered domain*, assumption (1) trivially holds since any single element is an ordered domain, and assumption (2) holds since, as Hausdorff shows, the union of a chain of limit type is again an ordered domain. Thus pantachies exist. Although this argument is general enough to prove that there are maximal linearly ordered sets in any partially ordered set, it is clear from Hausdorff's presentation that he is only thinking about sets of real functions and sets of real sequences partially ordered by the infinitary and final orderings or their variants. One possible explanation for this particularity, aside from the context within which he is working, is that he did not yet recognize the category of arbitrary partially ordered sets. Hausdorff only considered partially ordered sets in a general way in Chapter VI, §1 of [H 1914a], and there he did prove the theorem that an arbitrary partially ordered set has a maximal linearly ordered subset. However, he did not invoke his maximal principle. Instead he used transfinite induction and a *choice function* to well-order the given partially ordered set in such a way that the initial segment determined by the first element in the well-ordering that is not comparable (in the original partial ordering) with all its predecessors is a "greatest linearly ordered subset" [H 1914a, 140–141].

Going beyond existence, Hausdorff quickly concentrates on the order properties that these "most comprehensive" linearly ordered sets of sequences of cardinality $2^{\aleph_0}$ have in common. These properties were first enumerated in [H 1907a, 118–120], where Hausdorff derived them from *interpolation* theorems for countable sets of sequences. In [H 1909a, 305], Hausdorff gathers up these earlier theorems and states them as the *Fundamental Theorem*:

> *If $\mathfrak{A}$ is an at most countable set of numerical sequences, there always exists a sequence $X > \mathfrak{A}$ and a sequence $Y < \mathfrak{A}$. If $\mathfrak{A}$ and $\mathfrak{B}$ are two at most countable sets of numerical sequences and $\mathfrak{A} < \mathfrak{B}$, there always exists a sequence X such that $\mathfrak{A} < X < \mathfrak{B}$.*[9]

Hausdorff recapitulates the straightforward proof of this result for numerical sequences under the final ordering and takes up the question of how broadly the fundamental theorem holds. He easily extends its proof to finally ordered real functions, and then, displaying the power of good notation, he makes transparent the relationship between the infinitary ordering and final ordering and establishes the fundamental theorem for real sequences under the infinitary ordering. In considering special subclasses under the final ordering, he observes that the theorem does not hold for positive sequences that decrease monotonically, but he gives a brief, clever proof that the theorem holds for positive sequences that decrease monotonically to 0. He already noted that the fundamental theorem held for the class of positive real sequences converging to 0. He then gets the desired result by showing that any sequence converging to 0 can be uniquely permuted to a monotonic sequence converging to 0 and that such a permutation preserves final relationships.

Hausdorff derives the following common order properties of pantachies directly from the fundamental theorem:

> *Each pantachie is an unbounded, dense set that is neither cofinal with ω nor coinitial with ω^*, and it contains neither ω-elements nor ω^*-elements, nor does it contain $\omega\omega^*$-gaps.*
> [H 1907a, 119–120; H 1909a, 305]

In §3 of *On Pantachie Types*, he used these attributes to define the class of H-types. He proved that H-types are *universal* for ordered sets of cardinality $\leq \aleph_1$ and that any two H-types of cardinality $\aleph_1$ are similar. The latter proof is the occasion for his introduction of the *back-and-forth* construction.[10] Hausdorff then generalized H-types by introducing the η_α-sets; an H-type is the type of an η_1-set. (The η_α-sets have turned out to be of lasting importance. See [Fe 2002] for their applications in various areas of mathematics.)

The pantachie types of concern to Hausdorff all turn out to be H-types of cardinality $2^{\aleph_0}$. Under CH, the question of their structural diversity has a complete and simple answer: all pantachies types are similar, homogeneous (any two segments, be they initial, middle, or final, are similar), and equal to the second class power $3'_0(\Omega)$ [H 1907a, 128–132]. Furthermore, since he had earlier proved that an everywhere dense type of cardinality $\aleph_1$ without $\omega\omega^*$-limits and without $\omega\omega^*$-gaps has a $\Omega\Omega^*$-gap ([H 1907a, 85–86]), CH implies that all pantaches have $\Omega\Omega^*$-gaps.[11] Because of the unsettled status of CH and his perception that progress on this conjecture seems unlikely "in the present state of set theory," Hausdorff undertakes to establish the existence, without the use of CH, of pantachies that have just one of the salient properties that all must have under CH. In [H 1909a], he proves the existence of a homogeneous pantachie (or more accurately reproves it); he also produces a pantachie that has an $\Omega\Omega^*$-gap. Both are done only with the aid of AC, through the maximal principle.

It was in [H 1907a, 140–146] that Hausdorff first proved the existence of a homogeneous pantachie without using CH. The key to establishing homogeneity is having enough transformations to provide the needed similarity mappings. He noted that if he could construct a pantachie that was simultaneously a field, i.e., closed under rational operations (namely, $+, -, \times, \div$), then certain simple rational transformations would yield homogeneity. At the time, however, obtaining a pantachie closed under rational operations seemed blocked by algebraic obstructions. Nevertheless, Hausdorff ingeniously crafted a construction that used sets of sequences that were closed under a limited array of rational operations, the so-called *semifield domains*.[12] After showing how to extend semifield domains that are not yet maximal, he applied his second existence proof for pantachies to obtain a semifield pantachie. With a little more work, he was able to prove that such pantachies have all their middle segments similar; any middle segment is thus homogeneous and since such a middle segment is similar to a pantachie (which need no longer be a semifield), there are homogeneous pantachies. More importantly, CH had been eliminated from the proof of their existence.[13] (See the *Introduction to "Investigations into Order Types IV, V,"* pp. 105–107.)

In §3 of [H 1909a], Hausdorff establishes the existence of a pantachie of real sequences that is closed under $+, -, \times, \div$, thereby incidentally reproving the existence of a homogeneous pantachie. He defines the addition, multiplication, and division of real sequences (up to final equality) by the corresponding term by term operations.[14] In the resulting partially ordered ring, the sequences with infinitely many 0s are divisors of zero. Division by such sequences is not allowed. The ordinary reals are embedded in this algebra of sequences by thinking of an arbitrary real r as the constant sequence (r). Hausdorff calls an ordered domain a *rational ordered domain* if it is closed under $+, -, \times, \div$. In a rational ordered domain, 0 is the only zero divisor; so in such a domain only division by 0 is forbidden. The embedded rationals are a rational ordered domain, and any non-trivial rational ordered domain always contains them.

In order to make use of his maximal principle to produce a pantachie closed under rational operations, Hausdorff must find a way to enlarge a rational ordered domain that is not yet maximal to a larger rational ordered domain. For a rational ordered domain $\Re$ and an element Ξ not in it, the set of rational functions in Ξ with coefficients in $\Re$ is a larger domain that is closed under rational operations, but as he points out, it may no longer be an ordered domain.

To remove this apparent obstacle, Hausdorff is led to the study of the polynomial ring $\Re[X]$, where $\Re$ is a rational ordered domain. He quickly establishes that $\Re[X]$ is a unique factorization domain (in modern terms). In his presentation, he makes an important conceptual move: he considers the indeterminate X to be a sequence of indeterminates (x_n). Then a polynomial $F(X)$ of degree r over $\Re$ can be thought of as a sequence of single-variable real polynomials $(f_n(x_n))$, each eventually of degree r, whose coefficients are the appropriate terms from the sequences that make up the coefficients of $F(X)$. Actually, we recognize in hindsight that Hausdorff is making intuitive use of the proper isomorphic embedding of the polynomial ring with coefficients in the reduced power of the reals $\mathbb{R}$ over the index set of the natural numbers $\mathbb{N}$, modulo the filter $Cof(\mathbb{N})$ of cofinite subsets of $\mathbb{N}$ into the reduced product of the polynomial rings $\mathbb{R}[x_n]$ over the index set $\mathbb{N}$, again modulo $Cof(\mathbb{N})$. The embedding is obtained by sending the monomial $[(a_n)]X^r$ to the element $[(a_n x_n^r)]$ and then extending linearly.

Without proof, but with unerring insight, Hausdorff uses the following consequences of his identification of X with (x_n): for $F(X) = (f_n(x_n))$ and $G(X) = (g_n(x_n))$, the resultant of $F(X)$ and $G(X)$ is eventually the sequence of resultants of $f_n(x_n)$ and $g_n(x_n)$; if $Q(X) = (q_n(x_n))$ and $R(X) = (r_n(x_n))$ are the quotient and remainder, respectively, of $F(X)$ divided by $G(X)$, then eventually $q_n(x_n)$ and $r_n(x_n)$ are the quotient and remainder when $f_n(x_n)$ is divided by $g_n(x_n)$, and so on for other entities (discriminants) and other procedures such as the Euclidean algorithm.

Continuing, Hausdorff defines the sequence of reals $\Xi = (\xi_n)$ to be a *real root* of $F(X) = (f_n(x_n))$ if eventually ξ_n is a real root of $f_n(x_n)$. Using the example of $X^2 - 1 = (x_n^2 - 1)$, he points out a striking difference between the polynomials of $\Re[X]$ and ordinary real polynomials. The polynomial $X^2 - 1$ has uncountably many roots: any sequence of ± 1s is a root. But sanity returns by observing that only two of these roots, the ones determined by the constant sequences (-1) and (1), are comparable with 0. These are what Hausdorff goes on to call the *regular roots* of $X^2 - 1$.

In his formal definition of regular root, Hausdorff arguably makes the cleverest use of Sturm's Theorem for counting the number of distinct real roots of a real polynomial until Tarski employs his own variants of Sturm's Theorem in his decision procedure for the first order theory of the real field.[15] The definition of regular root allows Hausdorff to create an appropriate theory for extending rational ordered domains to larger ordered domains by the adjunction of such roots.

Hausdorff begins by considering an $F(X)$ with non-zero discriminant, i.e., an $F(X)$ with only simple roots. Thus eventually the real polynomials $f_n(x_n)$ in the sequence of real polynomials representing $F(X)$ have non-zero discriminants, and eventually they, too, have only simple roots. A standard Sturm sequence of polynomials in $\mathfrak{R}[X]$ for $F(X)$ is generated from $F(X)$ and its derivative $F'(X)$ by the Euclidean algorithm. The nth entries in the sequence representations of the Sturm polynomials for $F(X)$ are eventually the standard Sturm sequences for the real polynomials $f_n(x_n)$, and the polynomials that make up the Sturm sequences for the $f_n(x_n)$ eventually have the same degrees as and their lead coefficients have the same signs as the corresponding polynomials in the Sturm sequence for $F(X)$. So eventually the number of real roots of an $f_n(x_n)$ is constantly some value ϱ, where $\varrho \leq r =$ the degree of $F(X)$. The sequences $\Xi^{(1)} < \Xi^{(2)} < \cdots < \Xi^{(\varrho)}$ formed from the 1st, 2nd, ..., and ϱth roots of the $f_n(x_n)$, taken in order, are said to be the regular roots of $F(X)$. Sturm's Theorem also yields that the regular roots of $F(X)$ are comparable with all the elements of $\mathfrak{R}$. The crucial theorem is that for $F(X)$ a polynomial irreducible in $\mathfrak{R}[X]$ and Ξ a regular root of $F(X)$, the set of rational functions in Ξ with coefficients in $\mathfrak{R}$ is a rational ordered domain.

After extending the notion of regular root to polynomials that have non-simple roots, Hausdorff introduces the concept of an *algebraic ordered domain*: it is a rational ordered domain $\mathfrak{R}$ that contains all the regular roots of the polynomials $F(X)$ in $\mathfrak{R}[X]$. For $\mathfrak{R}$ a rational ordered domain, the ordered domain $\mathfrak{A}$ consisting of all the regular roots of the polynomials in $\mathfrak{R}[X]$ turns out to be an algebraic ordered domain. Further, the union of any chain of limit type of algebraic ordered domains is again an algebraic ordered domain. So taking $\varepsilon = algebraic\ ordered\ domain$, Hausdorff applies his maximal principle to obtain an algebraic ordered domain domain that cannot be extended, i.e., a pantachie that is closed under rational operations. This is the first application of a maximal principle in algebra.[16]

In §4, entitled *Relations with the Second Number Class*, Hausdorff poses four questions about the structure of pantachies in the absence of CH. The first three are: *Does there exist a pantachie that* (1) *is cofinal with Ω or that* (2) *contains Ω-elements or that* (3) *contains $\Omega\omega^*$-gaps?* Hausdorff succeeds in showing that these questions are equivalent, in that a "yes" to any one of them means a "yes" to the other two.[17] Before establishing this equivalence, Hausdorff proves that his fourth question, *Does there exist a pantachie with a $\Omega\Omega^*$-gap?*, has an outright "yes" answer.

Hausdorff begins his pursuit of the positive answer to his "gap" question by investigating a rather simple entity, the *critical number* (AB) that is defined for sequences $A < B$ as the smallest n_0 such that $a_n < b_n$ for all $n \geq n_0$.[18] (Critical numbers are used strategically in proving the equivalence of questions (1) and (3) above.) For A belonging to the infinite set of sequences $\mathfrak{A}$ and B a sequence such that $\mathfrak{A} < B$, Hausdorff scrutinizes the possible behavior of the critical numbers (AB). Out of the thicket of terms that he introduces to sort out the possiblilities, the following special case is the most important: for each infinite subset $\mathfrak{A}'$ of $\mathfrak{A}$, the set $\{(AB) \mid A \in \mathfrak{A}'\}$ is an unbounded set of integers. In this event, he says that "B *non-uniformly* surpasses $\mathfrak{A}$." He then proves the key Theorem II: for a countable well-ordered set of sequences $\mathfrak{A}$ and a sequence C, if C non-uniformly surpasses each initial segment of $\mathfrak{A}$, then there is a $B \leq C$ such that B non-uniformly surpasses $\mathfrak{A}$ [H 1909a, 322]. This result, in conjunction with the Fundamental Theorem, allows him to inductively construct an Ω-sequence $\mathfrak{A} = \{A_\alpha\}$ and an Ω^*-sequence $\mathfrak{B} = \{B_\alpha\}$ of real sequences such that $\mathfrak{A} < \mathfrak{B}$ and such that each B_α non-uniformly surpasses $\mathfrak{A}_\alpha$, the initial segment of $\mathfrak{A}$ determined by A_α. These properties are enough to guarantee that no real sequence lies between $\mathfrak{A}$ and $\mathfrak{B}$, and when extended to a pantachie, they provide the desired example of a pantachie with an $\Omega\Omega^*$-gap [H 1909a, 323].

Hausdorff is particularly proud of this result, which he deems "important" and "a conclusion that can be characterized as the first fairly close relationship between the continuum and the second number class" [H 1909a, 303]. Earlier he had hoped that a proof, without CH, of the existence of an Ω-scale would provide such a close connection [H 1907a, 155]. That was not to be (nor could it be in ZFC), but surprisingly he did achieve such a relationship, to his satisfaction, with the gap theorem. In Ulrich Felgner's opinion, "it is rightly counted among Hausdorff's best results in set theory" [Fe 2002, 651].

There is now a large literature and much activity in the subject of Hausdorff gaps in partial orderings. Scheeper's 1993 survey article *Gaps In ω^ω* is an excellent source for the history and the modern theory up through the early 90s. (See also [Fe 2002, 649–655].) In reality, Hausdorff's 1909 gap theorem was overlooked by his peers. A quarter of a century later, he published a similar result for sequences of 0s and 1s under the final ordering in *Summen von $\aleph_1$ Mengen* [[*Sums of $\aleph_1$ Sets*]][H 1936b], whose translation appears in the Appendix.[19] Explaining in [H 1936b, 244n] why he is giving a full reworking of his [H 1909a] gap theorem in this new context, Hausdorff candidly writes of [1909a], "Since this work is undoubtedly little known ..." Adding to the panache of [H 1936b] is the first application of the existence of an $\Omega\Omega^*$-gap; using such a gap, Hausdorff shows that Cantor space (the countable direct product of the discrete spaces $\{0, 1\}$ with the product topology) can be represented as an ascending union of G_δ sets.

Finally in §5 of [H 1909a], Hausdorff returns to the topic of convergent and divergent series of positive reals, which he considered in [H 1907a]. The sequence $A = (a_n)$ is called *convergent* or *divergent* depending on whether the associated series, $\sum a_n$, is convergent or divergent. In [H 1907a, 122–125], Hausdorff presented and proved several *interpolation* theorems that he now revisits, streamlines, and reproves as a "strengthened Fundamental Theorem" :

> *If $\mathfrak{A}$ is an at most countable set of convergent sequences, there always exists a convergent sequence $X > \mathfrak{A}$; If $\mathfrak{B}$ is an at most countable ordered domain of divergent sequences, there always exists a divergent sequence $Y < \mathfrak{B}$; if in addition $\mathfrak{A} < \mathfrak{B}$, there exist infinitely many convergent X and divergent Y such that $\mathfrak{A} < X, Y < \mathfrak{B}$.*[20]

For the latter two parts, Hausdorff repeats his comment of [H 1907a, 123n1] on the need for some condition on $\mathfrak{B}$ beyond countability. The same comment to the effect that two divergent sequences P and Q whose odd and even termed subsequences, respectively, are convergent have the property that any $X < P, Q$ must be convergent appears in [Ha 1894, 325]. Hausdorff gives a detailed example of two divergent monotonic sequences P and Q for which this is the case.

Hausdorff's proof of the strengthened theorem is appealing for its simplicity, but the first two parts have essentially the same content as [Ha 1894, 326, 328]. (See the *Introduction to "Investigations into Order Types IV, V,"* p. 104.) The last part of the strengthened theorem follows easily from the first two parts and the Fundamental Theorem, and it leads Hausdorff to observe that "scales" intended to determine convergence or divergence of series by comparison with fixed countable sets of known convergent and divergent series are bound to fail. Previously this had only been pointed out for Bonnet's logarithmic scale ([Du 1873, 88–91], [Pr 1890, 351–356]).

As Hausdorff notes, the strengthened theorem leaves open the possibility that no further sequence, convergent or divergent, may lie between some uncountable set $\mathfrak{C}$ of convergent sequences and some divergent D with $\mathfrak{C} < D$. Furthermore, were $\mathfrak{C}$ an ordered domain, then in a pantachie extending $\mathfrak{C}$ and D, the sequence D would be the first divergent sequence. Hausdorff uses this observation as an incentive to explore the divergent and convergent elements in a pantachie. (He had earlier considered such elements in [H 1907a, 149–151].) For $\mathfrak{P}$ a pantachie, we always have the decomposition $\mathfrak{P} = \mathfrak{P}_c + \mathfrak{P}_d$ where those elements that are convergent are in $\mathfrak{P}_c$, while those that are divergent are in $\mathfrak{P}_d$. In [H 1907a, 150], Hausdorff points out that $\mathfrak{P}_c = 0$ is possible but $\mathfrak{P}_d = 0$ is impossible.

Hausdorff touched upon pantachie decompositions, $\mathfrak{P} = \mathfrak{P}_c + \mathfrak{P}_d$, briefly in [H 1907a, 149–151]. There he viewed his results as providing a context where Du Bois-Reymond's (problematic) boundary between convergence and divergence could be realized.[21] In particular, he sketched arguments that there are pantachies where there is a last convergent element and pantachies where there are first divergent elements, and in fact he argued that any "cut-form" that occurs in a pantachie segment can be realized by the cut made up of the convergent and divergent sequences of some other pantachie segment. In [H 1909a, 331–332], Hausdorff gives a precise statement and proof of this result. He then goes on to provide a much more involved proof that a given cut in a pantachie segment can be be realized as $\mathfrak{M}_c + \mathfrak{M}_d$ where the pantachie $\mathfrak{M}$ consists of strictly monotonic sequences decreasing to 0. (Scheepers gives a detailed exposition of this proof in [Sc 1993, 550–556].) Hausdorff remarks that $\mathfrak{M}_d = 0$ is still impossible, but he leaves undecided whether one can have $\mathfrak{M}_c = 0$.

Notes

1. The phrase "Graduierung nach dem Endverlauf" appears in *On Pantachie Types* [H 1907a, 116]. Chapter V of [Du 1882] is entitled *Ueber den Endverlauf der Functionen*.

2. See the *Introduction to "The Fundamentals of a Theory of Ordered Sets,"* Note 2, p. 193.

3. A study [[Studie]], dated May 19, 1908, dealing with the existence of a pantachie closed under rational operations and that has algebraic content similar to that of §3 of [H 1909a] appears in Hausdorff's *Nachlaß* [Kapsel 31: Fasz. 111]. It is entitled Algebra and Final Rank Ordering [[Algebra und finale Rangordnung]] [Pu 1995, 112].

   There is also a study from the beginning of January, 1908 (Pantachie Problem [[Pantachieproblem]]), that has a penciled note from January 23/25 stating that "the existence of an $\Omega\Omega^*$-gap can be proved" [Kapsel 31: Fasz. 116]. No material dated January 23/25 with such a proof is found in the *Nachlaß* [Pu 1995, 114].

4. Some of the notation and terms used in [H 1907a] are changed in [H 1909a]. Final equality was denoted by $\sim$ and now is denoted by $=$. That something holds *for all but finitely many n* was expressed by *finally* [[*final*]]; now it is expressed by *eventually* [[*schließlich*]].

   In [H 1909a, 306], apparently for the first time, Hausdorff refers to ordering sequences by the principle of first differences as ordering them "lexicographically" [["lexikographisch"]] (his quotation marks). In [H 1914a] *lexicographic ordering* is the preferred term.

5. A reader of either *On Pantachie Types* or *Graduation by Final Behavior* would be well-advised to peruse the other article's introduction.

6. A good source for the tangled history of *maximal principles* is [Mo 1982]. See also [Ca 1978].

7. This may show the influence of Zermelo's formulation of his Axiom of Separation (Axiom III in [Ze 1908b]). Zermelo speaks of a set $M_{\mathfrak{E}}$ whose members belong to a given set M and in addition satisfy an unspecified propositional function $\mathfrak{E}(x)$ which is "definite" for all the elements of M.

8. In [H 1927a, 173–176], Hausdorff uses the terms *maximal sets* [[Maximalmengen]] and *minimal sets* [[Minimalmengen]] for the first time in his writings and indicates, under suitable assumptions, how to derive the existence of maximal and minimal sets. His derivation employs transfinite induction and implicitly uses AC.

9. See the *Introduction to "Investigations into Order Types IV, V,"* Note 17, p. 111.

10. See the *Introduction to "Investigations into Order Types IV, V,"* Note 19, p. 111.

11. In [H 1907a, 128–132], Hausdorff proves that an H-type without $\Omega\Omega^*$-gaps has cardinality $\geq 2^{\aleph_1}$. Thus not only CH but also the weaker (as we now know) hypothesis $2^{\aleph_0} < 2^{\aleph_1}$ imply that all pantachies have $\Omega\Omega^*$-gaps.

12. The definition of semifield domain and the proof that there are semifield pantachies that appears in [H 1907a, 143–144] is amended by [H 1909a, 310n1]. This correction of a previous article is a unique event in these papers.

13. See the *Introduction to "Investigations into Order Types IV, V,"* Note 20, p. 111.

14. Instead of taking equivalence classes of real sequences under the relation of final equality as his basic objects, Hausdorff opts to work with a set $\mathfrak{Z}$ of representative sequences chosen from each such class. This means that equations between sequences have to be interpreted as statements about final equality. In any case, the resulting structure would be recognized today by model theorists as the reduced power of the real field $\mathbb{R}$ over the index set $\mathbb{N}$ modulo the filter $Cof(\mathbb{N})$ of cofinite subsets of $\mathbb{N}$.

15. By the late nineteenth century, Sturm's 1829 theorem was textbook material, e.g., it appears in [We 1895]. Hausdorff had lectured on the theorem prior to 1909. Notes for such lectures are found in his *Nachlaß*: from winter semester 1903/1904 at Leipzig [Kapsel 2: Fasz. 9]; from summer semester 1907 at Leipzig [Kapsel 5: Fasz. 22]—as cataloged in [Pu 1995].

    For Sturm's Theorem, see [vdW 1963, 218–221]. The last paragraph on p. 221 explains the role of the leading coefficients and the degrees of the Sturm polynomials.

    In Note 12 of [Ta 1951, 51–52], Tarski lays out in detail the relationship of his work to Sturm's Theorem. Tarski's bibliography lists both the texts of Weber and van der Waerden.

16. The first application of the Well-Ordering Theorem in algebra is Hamel's 1904 proof of the existence of a basis for the reals (as a vector space) over the rationals (*Eine Basis aller Zahlen und die unstetigen Lösungen der Funktiongleichung*: $f(x+y) = f(x) + f(y)$, Mathematische Annalen **60** (1904), 459–462). As for the use of maximal principles in algebra, Hausdorff's proof of the existence of a pantachie closed under rational operations is the first.

    Campbell ([Ca 1978, 81]) claims there were no uses of maximal principles in algebra before [Zo 1935].

17. Question (1) is a particular case of what Hausdorff called "the Scale Problem" in [H 1907a, 152–155]. There he proved that CH implies a positive answer to Question (1).

    The equivalence of these three questions was rediscovered by F. Rothenberger (*Sur les familles indénomerables de suites de nombres naturels et les problèmes concernant la propriété* **C**, Proceedings of the Cambridge Philosophical Society **37** (1941), 109–126.) See [Sc 1993, §1.4].

18. In [H 1907a, 89–90], Hausdorff used the index (β, γ) to denote the place of first difference for elements x^β and x^γ of the power $\mu(\alpha)$, where α is an ordinal. He employed indices to prove that an ω_ν-sequence in $\mu(\alpha)$ is cofinal with either an ω_ν-sequence associated with the argument α or with an ω_ν-sequence associated with the base μ. Such indices appeared earlier in [H 1906b, 133, 160].*

19. Cf. [H 1909a, 321–322] and [H 1936b, 245–246]. Theorem II of [H 1909a, 321] becomes the *Second Interpolation Theorem* of [H 1936b, 245]. In passing from [H 1909a] to [H 1936b], the concept of "B non-uniformly surpasses $\mathfrak{A}$" for $\mathfrak{A} < B$, which is defined in terms of the *critical numbers* (AB) for $A \in \mathfrak{A}$, is replaced by the analogous concept of "b lies near to A" for $A < b$, which is defined in terms of the numbers (ab) for $a \in A$. However, the definition of (ab) for $a < b$ in final ordering of dyadic sequences is not the same as the definition of (AB) for $A < B$ in the final ordering of positive real sequences.

20. See the *Introduction to "Investigations into Order Types IV, V,"* pp. 103–104 and Note 18 on p. 111.

21. See the *Introduction to "Investigations into Order Types IV, V,"* p. 100.

Graduation by Final Behavior

By

F. HAUSDORFF

§1
Graduation Problems

The idea of *graduating* [[zu *graduieren*]] the behavior of functions $f(x)$ for a particular passage to a limit of the independent variable (say for $\lim x = +\infty$), especially placing the going to zero or infinity of the functions in a metric or at least ordinal series, is certainly one of the most obvious problems of analysis, but significant difficulties, which at present have still not been overcome, stand in the way of its realization. The first steps in this direction are indeed almost obvious: one ascribes to the function x^2 a more intensive way of becoming infinite, a "stronger infinity," than to the function x, one ascribes to the function $x^{\frac{1}{2}}$ a weaker infinity than to the function x, and one characterizes the going to infinity or zero of x^α for $\lim x = +\infty$ by the positive or negative "ordinal number" α. One immediately recognizes as well that the scale hereby obtained not only requires expansion at the ends, but it also requires the insertion of intermediate stages at each position; the function e^x becomes more strongly infinite and the function lx more weakly infinite than each power with positive exponent, and xlx certainly becomes more strongly infinite than x but more weakly infinite than each $x^{1+\alpha}$, with α however small and positive. A continuation of these elementary considerations shows that any possible scale that is supposed to include all degrees of going to infinity would have to be constructed in an essentially different way than the series of real numbers; its elements would violate the Archimedean Axiom and can turn out to be infinitely large or infinitely small relative to one another; countable sequences of its elements would not define limit elements, etc. However, all this is still not a difficulty, rather it is an incentive to become better acquainted with the remarkable properties of this scale, and at best, yet a reason to prefer leaving out the number concept and, abandoning the *metric* symbols that are supposed to represent the degree of infinity in the manner of those exponents α, to directly consider the *ordering* [[*Anordnung*]] of functions by equally infinite, more strongly infinite, and more weakly infinite. A real difficulty, however, on which the entire theory threatens to founder, lies in the fact that in each fairly extensive collection of functions (e.g., in the domain of all functions for which $f(+\infty) = +\infty$ or in the domains of all continuous functions, of all continuous monotone functions, or of all entire transcendental functions of this type), there always exists two functions each of which has an infinity that is neither equal to, nor stronger than, nor weaker than the other, thus two *incomparable* functions that cannot belong to one and the same scale. Stated in this form, the assertion seems a bit broad and in need of restriction to the scales conceived of so far; for by assigning all functions the same infinity, after all nothing prevents this, one could take the assertion to be disproved. Meanwhile, we will impose on each graduation, i.e., on each definition of the three cases of comparability

$$f < g, \quad f = g, \quad f > g,$$

the requirement that the three signs $<=>$ satisfy the well-known formal conditions and that $f < g$ occurs only if eventually [[schließlich]] $f(x) < g(x)$ (i.e., for each $x \geq x_0$). If we then choose our function domain to be so extensive that there exist two functions $f < g$ for which a third function h can be found so that for two numerical sequences a_n, b_n converging to $+\infty$

$$h(a_n) < f(a_n), \quad h(b_n) > g(b_n),$$

then each of the assumptions $h \gtreqless f, h \gtreqless g$ leads to a contradiction, and h has a fourth relation to f and g that we call incomparability and that we are going to express by the symbols $h \| f, h \| g$. This general consideration finds its full confirmation in the graduations proposed so far. If one simply chooses the eventual sign of the difference $f(x) - g(x)$ to be the basis, so that trichotomy determines

$$f(x) < g(x), \quad f(x) = g(x), \quad f(x) > g(x) \quad (\text{for } x \geq x_0),$$

then that this difference has no fixed eventual sign remains as a fourth case. If, as is customary, one lets $\lim \dfrac{f(x)}{g(x)}$ be the deciding factor, inasmuch as this limit can be

$$0, \quad \text{positive}, \quad +\infty,$$

the nonexistence of the limit remains as the fourth case. If, instead of this, one relies upon the numbers

$$\lambda = \liminf \frac{f(x)}{g(x)}, \quad \mu = \limsup \frac{f(x)}{g(x)}$$

for the decision and defines the three cases by

$$0 = \lambda \leq \mu < +\infty, \quad 0 < \lambda \leq \mu < +\infty, \quad 0 < \lambda \leq \mu = +\infty,$$

then $\lambda = 0$, $\mu = +\infty$ remains as the fourth case. One can conceive of countless other graduations by replacing the difference or quotient of the functions by another critical expression or by generalizing the limit concept, and the like. However, if the above stated conditions are fulfilled, the fourth case will always occur, and in fact, as one can add, as a *rule* it actually does occur, as opposed to the exceptional case of comparability that is associated with an unusual behavior of two functions.

Essentially the same phenomenon appears in the comparison of numerical sequences a_n, b_n where, say, the eventual behavior of the difference $a_n - b_n$ or of the quotient $a_n : b_n$ provides the criterion for comparability. In the main, be they related to functions of x or of n, to their final behavior [[Endverlauf]] per se or to their convergence to a limit, or to convergence and divergence of series or definite integrals, all these graduation problems point at the same type with secondary differences that are based on the definition of the standard of comparison and the domain of the compared elements. Thus monotone numerical sequences or continuous functions behave somewhat differently than arbitrary numerical sequences or functions. The common nature of these problems can be designated as *graduation by final behavior* [[*Graduierung nach dem Endverlauf*]] or as *final rank ordering* [[*finale Rangordnung*]], inasmuch as it is always the eventual behavior (for $x \geqq x_0$ or $n \geqq n_0$) that decides the issue. Again, the fourth case of incomparability appears each time. This proves that the hope of including all elements of an extensive domain in a universal, coherent scale is illusory. The "infinitary pantachie" of P. DU BOIS-REYMOND in which each functional infinity takes its determined place like a point on a line or like a number in the series of real numbers does not exist, neither does an all-inclusive scale of convergent or divergent series nor anything similar.

Thus the situation is formally as follows: between any two elements f and g in a set M there exists one and only one of the four relations

$$f < g, \quad f = g, \quad f > g, \quad f \| g.$$

Let us call a subset N in which between every two elements one of the two relations $f \lesseqgtr g$ always exists an *ordered domain* [[*Bereich*]]; any such ordered domain is a *simply ordered set* in the sense of G. CANTOR. A set of pairwise comparable elements, where of course the relation $f = g$ is also allowed, becomes an ordered domain only if a single element is kept from each class of equal elements. Such ordered domains or sets of comparable elements can easily be given: the most well-known examples are the set of rational functions of x with rational or real coefficients, or the set of functions

$$x^\alpha (lx)^{\alpha_1} (l^2 x)^{\alpha_2} \ldots (l^k x)^{\alpha_k}$$

with real exponents, or the set of all functions which arise from x and arbitrary real coefficients through the four fundamental operations of arithmetic with the addition of the operations e^x and lx (this last for $x > 0$). But all these sets or ordered domains are extendible, and therefore they give us no complete picture of the laws of the final rank ordering, and this would not change if we wished to leave the realm of the exponential-logarithmic and to drag in functions of still more rapid growth such as $e^x, e^{x^x}, \ldots$. In fact, although no *all-embracing* domain of comparable elements exists, the original intent of the graduation problem presses us to at least consider the *most comprehensive* [[*meistumfassenden*]] domains of comparable elements, those that are in no way extendible. We are going to call an ordered domain N that is contained as a subset in no other ordered domain a *closed ordered domain* or, for the sake of brevity, a *pantachie*, retaining DU BOIS-REYMOND's term, but abandoning the illusion of a unique pantachie since, in any case, two incomparable elements belong to different pantachies.

The proof that such pantachies exist falls to abstract *set theory* and can be carried out in a rather general form as follows. Each subset N of an infinite overall set M may or may not have a certain property ε (i.e., we decompose the set of all subsets of M into two complementary parts); those that have the property shall be called ε-sets. Let a well-ordered sequence of ε-sets of type ξ,

$$N_0 \, N_1 \, \ldots \, N_\alpha \, \ldots \qquad (\alpha < \xi),$$

where N_α is a proper subset of N_β for $\alpha < \beta$, be called a *chain* [[*Kette*]] of type ξ, and call this chain *continuable* [[*fortsetzbar*]] if one can add a further ε-set N_ξ in such a way that $N_0 \ldots N_\alpha \ldots N_\xi$ is a chain of type $\xi + 1$. We assume that there is at least one ε-set, thus that there is a chain having type 1. Now if each chain were continuable, there would be chains with each arbitrary ordinal number as type. But if $\aleph_\mu$ is the cardinality[1]) of M, there can only be chains of such types whose cardinalities are $\leqq \aleph_\mu$. Thus there are certainly chains that are not continuable. We now explicitly make the assumption *that a chain of limit number type is definitely continuable*; e.g., this is the case if the *union* [[*Vereinigungsmenge*]] $N = \mathfrak{M}(N_0 \, N_1 \, \ldots \, N_\alpha \, \ldots)$ of each endless chain is again an ε-set. The smallest ordinal number that appears as the type of a non-continuable chain is then $\geqq 1$ and not a limit number. Thus it is of the form $\xi + 1$, and there is a non-continuable chain of the form $N_0 \ldots N_\xi$; i.e., there is a ε-set N_ξ that is not contained as a proper subset in any other ε-set, so a *closed* ε-set.

If we identify ε-sets with ordered domains, then it is the case that the union $N = \mathfrak{M}(N_0 \, N_1 \, \ldots \, N_\alpha \, \ldots)$ of an endless chain of ordered domains is again an ordered domain. For if f and g are two elements of N and N_α and N_β are the ordered domains of least index in which f and g are contained, then for $\alpha \leqq \beta$, f as well as g is in the ordered domain N_β, thus $f \lessgtr g$. Consequently, the above conclusion comes into effect, and there exists an ordered domain N_ξ that is not contained in any greater ordered domain, i.e., a pantachie. Since the initial ordered domain N_0 can be chosen arbitrarily (even as a single element of M), there are infinitely many different pantachies.

With this, the graduation problem has acquired a definitive form: it has turned into the question of the structure of pantachies. While in this way the troublesome case of incomparability could be eliminated by the pantachie concept, now however difficulties arise the overcoming of which can hardly

---

[1]) E. ZERMELO, Beweis, daß jede Menge wohlgeordnet werden kann, Math. Ann. 59 (1904). Neuer Beweis für die Möglichkeit einer Wohlordnung, Math. Ann. 65 (1908).

be hoped for in the present state of set theory. In particular, pantachies (whose elements are functions $f(x)$ or number sequences $f(n)$) turn out to be ordered sets of the cardinality of the continuum in which *uncountable* well-ordered sets and their inverses are contained as subsets. The question of what type these well-ordered sets are or can be is very closely connected with the question of the cardinality of the continuum, and presently it is as far from an answer as is the latter question. Should the continuum have the second infinite cardinality, the pantachie problem would be completely decided: all pantachies then have a common, fairly simple type H that, analogously, is to the type η of the set of rational numbers as the type Ω of the second number class is to the type ω of the first number class. That is at least a hypothetical explanation. The reverse attempt to find, independently of CANTOR's Continuum Hypothesis, a single pantachie with even one of the properties that all pantachies must share under the validity of that hypothesis already requires complicated lemmas. Its success in each individual case may claim a certain importance as an advance or rather a first step into an unknown area. In this spirit, I take the liberty to return to the pantachie problem that I have already treated in the reports of this society[1]) and to communicate some new investigations into this problem. The rank ordering of real sequences based on the eventual sign of the difference is the simplest model of all graduation problems. In order not to take anything for granted, I recapitulate in §2 the proof of the "Fundamental Theorem," and I also go into the case of monotone numerical sequences, which was not handled earlier. I develop an algebra of numerical sequences in §3 and with it prove the existence of rational pantachies, i.e., those in which the rational operations can be carried out. In §4, in the sense of CANTOR's Continuum Hypothesis, I seek to limit the scope of the uncountable that the Fundamental Theorem leaves open. And in particular, I find the important result that there are pantachies with $\Omega\Omega^*$-gaps, a conclusion that may be characterized as the first fairly close relationship between the continuum and the second number class. In §5, which again repeats some of my earlier article, graduation is applied to convergent and divergent positive sequences, and in connection with DU BOIS REYMOND's idea of a "boundary between convergence and divergence," the distribution of both series types within a pantachie is studied, whereby each cut-form is proved to be possible, even in the rather complicated case of monotone sequences.

We mention beforehand the following explication of some terms that occur from time to time. The ordered set M is said to be *cofinal* [[*konfinal*]] (resp. *coinitial*) with its subset A if it contains no element $> A$ (resp. $< A$); two subsets $A < B$ between which no element lies are called *adjacent* [[*benachbart*]]. A set without a first and last element is called *unbounded* [[*unbegrenzt*]]; a set without adjacent elements is called *dense* [[*dicht*]]. The

---

[1]) Über Pantachietypen, Leipz. Ber. 59 (1907), 105–159. I believe that a repetition of the references given there may be omitted here.

aggregate of elements of M that come after an element or before an element or that lie between two elements is called a *segment* [[*Strecke*]]; if the boundary elements are added, we speak of an *interval*. A set that is similar to all its segments is called *homogeneous*. The smallest ordinal number of cardinality $\aleph_\alpha$ is called the *initial number* [[*Anfangszahl*]] ω_α; an initial number that is not cofinal with a smaller ordinal number is called *regular*. Each set without a last element is cofinal with one and only one regular initial number. The smallest (regular) initial numbers are $\omega_0 = \omega$ and $\omega_1 = \Omega$. If m immediately follows a subset A that has no last element, m is called the *upper limit* of A; if A is cofinal with the regular initial number ω_α, then m is called an ω_α-element. Lower limit, ω_β^*-element, and $\omega_\alpha \omega_\beta^*$-element are defined correspondingly. If A and B are adjacent ($A < B$) and A has no last element and B has no first element, we say that A and B determine a *gap*, which is called an $\omega_\alpha \omega_\beta^*$-gap if A is cofinal with ω_α and B is coinitial with ω_β^*.

§2
The Rank Ordering of Numerical Sequences

To begin with, let the stuff to be graduated consist of all *numerical sequences*[1])

$$A \equiv (a_1\, a_2\, a_3\, \ldots\, a_n\, \ldots)$$

in which the a_n are arbitrary real numbers, and let the final ordering result from the eventual sign of the difference; i.e., we define

$$\begin{aligned} A < B &\quad \text{if eventually} \quad a_n < b_n, \\ A = B &\quad \text{if eventually} \quad a_n = b_n, \\ A > B &\quad \text{if eventually} \quad a_n > b_n, \\ A \,\|\, B &\quad \text{in all other cases.} \end{aligned}$$

"Eventually" means for all values of n with the exception of a finite number, thus for $n \geqq n_0$.

For the sake of completeness, we group the formal properties of these symbols together, where ϱ can denote any of them:

(1) $\qquad\qquad\qquad A = A;$

(2) $\qquad\qquad B > A, \quad B = A, \quad B < A, \quad B\,\|\,A$
$\quad\text{follows from}\quad A < B, \quad A = B, \quad A > B, \quad A\,\|\,B;$

(3) $\qquad\qquad A \varrho C \quad \text{follows from} \quad A = B, \quad B \varrho C,$
$\qquad\qquad A \varrho C \quad \text{follows from} \quad A \varrho B, \quad B = C,$
$\qquad\qquad A < C \quad \text{follows from} \quad A < B, \quad B < C,$
$\qquad\qquad A > C \quad \text{follows from} \quad A > B, \quad B > C.$

If $\mathfrak{A}$ and $\mathfrak{B}$ are arbitrary sets of numerical sequences, then the relation $\mathfrak{A}\varrho\mathfrak{B}$ shall denote the fact that each element A of $\mathfrak{A}$ is in the relation $A\rho B$

---

[1]) In contrast to $=$, the symbol $\equiv$ denotes not only the eventual agreement but the agreement everywhere of two numerical sequences.

with each element B of $\mathfrak{B}$ (no matter what relations the A may have among themselves and the B may have among themselves).

A set of numerical sequences that are pairwise in the relations $\lessgtr$ is called an ordered domain; an ordered domain that is not contained in any larger ordered domain is called a pantachie.

The common properties of all pantachies follow from the *Fundamental Theorem*:

I. *If $\mathfrak{A}$ is an at most countable set of numerical sequences, there always exists a numerical sequence $> \mathfrak{A}$ and a numerical sequence $< \mathfrak{A}$. If $\mathfrak{A}$ and $\mathfrak{B}$ are two at most countable sets of numerical sequences and $\mathfrak{A} < \mathfrak{B}$, there always exists a numerical sequence between both sets.*

The proof is extremely simple. Let $\mathfrak{A}$ consist of the numerical sequences $A, B, C, \ldots$ (among which one may imagine that if $\mathfrak{A}$ is only a finite set, a sequence is repeated infinitely often). If X is now chosen so that

$$x_1 > a_1; \quad x_2 > a_2, b_2; \quad x_3 > a_3, b_3, c_3; \quad \ldots,$$

then $X > A$, $X > B$, $X > C$, $\ldots$, thus $X > \mathfrak{A}$. One can analogously construct an $X < \mathfrak{A}$.

Furthermore, let $\mathfrak{B}$ consist of the numerical sequences $P, Q, R, \ldots$; by assumption, each element of $\mathfrak{A}$ is finally smaller than each element from $\mathfrak{B}$. Consequently, one can determine a sequence of increasing indices $\lambda < \mu < \nu < \cdots$ so that

$$\begin{aligned}
a_n &< p_n & \text{for} \quad & n \geqq \lambda, \\
a_n, b_n &< p_n, q_n & \text{for} \quad & n \geqq \mu, \\
a_n, b_n, c_n &< p_n, q_n r_n & \text{for} \quad & n \geqq \nu,
\end{aligned}$$

and so on. If one then chooses X so that

$$\begin{aligned}
a_n &< x_n < p_n & \text{for} \quad & \lambda \leqq n < \mu, \\
a_n, b_n &< x_n < p_n, q_n & \text{for} \quad & \mu \leqq n < \nu, \\
a_n, b_n, c_n &< x_n < p_n, q_n r_n & \text{for} \quad & \nu \leqq n < \pi,
\end{aligned}$$

and so on, then $A < X < P$, $B < X < Q$, $C < X < R$, $\ldots$; hence $\mathfrak{A} < X < \mathfrak{B}$.

Thus in a pantachie, before and after any countable (or finite) subset as well as between two such sets, there are further elements; i.e., a pantachie is not cofinal[1]) or coinitial with any countable set, and in a pantachie two countable sets are never adjacent to each other. Therefore if we enumerate the cases in which we are interested, a pantachie contains no last element and no first element; it is neither cofinal with ω nor coinitial with ω^*; it contains no adjacent elements and no limit elements for ω-sequences or ω^*-sequences and no $\omega\omega^*$-gaps. We formulate this result as follows:

II. *Each pantachie is an unbounded, dense set that is neither cofinal with ω nor coinitial with ω^*; it contains neither ω-elements nor ω^*-elements, nor does it contain $\omega\omega^*$-gaps.*

As a dense set without $\omega\omega^*$-gaps, each pantachie is at least of the car-

---
[1]) Cf. the explanation at the end of §1.

dinality of the continuum; as a subset of the set of all numerical sequences, it is at most of the cardinality of the continuum; thus each pantachie has exactly the cardinality of the continuum. According to II, it certainly contains subsets of type Ω and Ω^*, but the question is whether it contains any sequences of higher cardinality.

I add the following hypothetical assertions, whose proof, insofar as it does not follow from the general properties of pantachies, can be found in my earlier works.[1]) If ordered domains that are well-ordered are at most of the second infinite cardinality, then each pantachie is cofinal with Ω and coinitial with Ω^*, and each of its elements is an $\Omega\Omega^*$-element; moreover, it certainly contains $\omega\Omega^*$-gaps and $\Omega\omega^*$-gaps. Then, if it does not contain any $\Omega\Omega^*$-gaps, it is of cardinality $2^{\aleph_1} > \aleph_1$. In particular, if the continuum is of the second infinite cardinality, then it must contain $\Omega\Omega^*$-gaps. However, in this case one can say more. Namely, on the one hand, there exists a fairly simple type H of the cardinality of the continuum that has all the properties listed in Theorem II; one obtains it from an ordered set of three elements $l < m < n$ by forming all element sequences of type Ω

$$x = (x_0\, x_1\, x_2\, \ldots\, x_\omega\, \ldots\, x_\alpha\, \ldots) \qquad (x_\alpha = l, m, n),$$

in which the outer elements l and n occur with at most countable frequency, and by ordering these sequences "lexicographically," i.e., in such a way that x and y have the same rank order as their first pair of distinct elements x_α and y_α. On the other hand, there exists a unique type *of the second infinite cardinality* with the properties of II, if there exists any at all. So if the continuum is of the second infinite cardinality, then all pantachies (and pantachie segments) have the same homogeneous type H, with which the pantachie problem would be completely solved.

Up to now, we have based the graduation problem on a specifically simple form (which we also later keep as a model) by comparing real number sequences according to the eventual sign of their difference; however, we would like to indicate in a few words how to proceed in other cases and how the validity of a Fundamental Theorem analogous to I can be established.

If we are dealing not with numerical sequences but with real functions of a positive real variable t and with their graduation that likewise is the result of the eventual sign of the difference as $\lim t = +\infty$, then one proves the Fundamental Theorem by division into the intervals $i_1\, i_2\, \ldots\, i_n\, \ldots$,

$$i_n: \quad n-1 < t \leqq n.$$

For example, in order to finally surpass the sequence of functions $a(t), b(t), c(t), \ldots$ with $x(t)$, for instance, one has to choose:

$$\begin{aligned}
x(t) &> a(t) & &\text{in} & &i_1, \\
x(t) &> a(t), b(t) & &\text{in} & &i_2, \\
x(t) &> a(t), b(t), c(t) & &\text{in} & &i_3,
\end{aligned}$$

---

[1]) Über Pantachietypen (loc. cit.), §3. Further: Grundzüge einer Theorie der geordneten Mengen, Math. Ann. 65 (1908), §§23, 24.

and so on. We further note that a function pantachie also has only the cardinality of the continuum ($2^{\aleph_0}$), even though the set of all functions is of higher cardinality ($2^{2^{\aleph_0}}$). For if one assigns to each function $a(t)$ the numerical sequence

$$A = (a(1)\, a(2)\, a(3) \ldots a(n) \ldots),$$

it then follows from $a \lesseqgtr b$ that also $A \lesseqgtr B$; a function pantachie is thus similar to an ordered domain or a pantachie of numerical sequences and is at most of the cardinality of the continuum, but as a consequence of the Fundamental Theorem, it is also of at least the cardinality of the continuum. — Restrictions of function domains in part require special modifications (e.g., if only continuous or analytic functions are to be considered), which do not interest us here, in part they can be treated according to examples of analogous restrictions on the domains of numerical sequences (e.g., for functions with $\lim f(t) = +\infty$ or for monotone functions).

We return to numerical sequences, but under the restriction to sequences of positive numbers (for which the Fundamental Theorem holds without alteration) we change the graduation criterion by defining

$$A \prec B, \quad A \sim B, \quad A \succ B$$

according to whether $\lim(a_n : b_n) = 0$, positive, or infinity. This "infinitary" graduation can easily be reduced to our final one; namely, if one assigns to each numerical sequence A the countable set $\mathfrak{A}$ of numerical sequences

$$\ldots, \tfrac{1}{3}A, \tfrac{1}{2}A, A, 2A, 3A, \ldots,$$

where of course by ϱA the numerical sequence $(\varrho a_1, \varrho a_2, \ldots, \varrho a_n, \ldots)$ is understood, then the infinitary relation $A \prec B$, for whose existence $\mathfrak{A} < B$ or $A < \mathfrak{B}$ already suffices, is equivalent to the final relation $\mathfrak{A} < \mathfrak{B}$. In order to infinitarily surpass a countable set $A, B, C, \ldots$, one has only to finally surpass the likewise countable set $\mathfrak{A}, \mathfrak{B}, \mathfrak{C}, \ldots$, by which the proof itself of the Fundamental Theorem transfers directly to the infinitary rank ordering.

If we ultimately stay with numerical sequences and their final graduation according to $sg(a_n - b_n)$, then it would still have to be proved whether the Fundamental Theorem also retains its validity in restricted domains of numerical sequences. A quite simple remark suffices in this regard: namely, if $\mathfrak{A}$ and $\mathfrak{B}$ are finite or countable sets of numerical sequences for which $\mathfrak{A} < \mathfrak{B}$ and if one denotes by

- $\mathfrak{Z}$ the domain of all numerical sequences,
- $\mathfrak{Z}_\mathfrak{A}$ the domain of all numerical sequences $> \mathfrak{A}$,
- $\mathfrak{Z}^\mathfrak{B}$ the domain of all numerical sequences $< \mathfrak{B}$,
- $\mathfrak{Z}^\mathfrak{B}_\mathfrak{A}$ the domain of all numerical sequences between $\mathfrak{A}$ and $\mathfrak{B}$,

then the Fundamental Theorem holds not only within $\mathfrak{Z}$, but also within the other three domains. For example, if $\mathfrak{A}$ belongs to the domain $\mathfrak{Z}^\mathfrak{B}$, i.e., if $\mathfrak{A} < \mathfrak{B}$, then according to I, there exists a numerical sequence X such that $\mathfrak{A} < X < \mathfrak{B}$; i.e., there exists in $\mathfrak{Z}^\mathfrak{B}$ a numerical sequence $X > \mathfrak{A}$, and similarly in the remaining cases. If one understands by a the numerical sequence $(aaa\ldots)$, by $\mathfrak{E}$ the countable set of numerical sequences

$$\ldots, \tfrac{1}{3}, \tfrac{1}{2}, 1, 2, 3, \ldots,$$

and by $\overline{\mathfrak{E}}$ the set of numerical sequences

$$\ldots, -3, -2, -1, -\tfrac{1}{2}, -\tfrac{1}{3}, \ldots,$$

then, for example, the validity of the Fundamental Theorem can be asserted for the following domains:

for the domain $\mathfrak{Z}_0$ of all (eventually) positive numerical sequences $a_n > 0$;

for the domain $\mathfrak{Z}_\mathfrak{E}$ of all numerical sequences with $\lim a_n = +\infty$;

for the domain $\mathfrak{Z}_0^\mathfrak{E}$ of all positive numerical sequences with $\lim a_n = 0$;

for the domain $\mathfrak{Z}_\mathfrak{E}^\mathfrak{E}$ of all numerical sequences with $\lim a_n = 0$;

and moreover, its validity follows, for example, from this for the domain of all numerical sequences that converge to a specified limit a or for the domain of all convergent and divergent (to $\pm\infty$) numerical sequences. By comparison, it does not hold for the domain of all convergent sequences since there does not exist a convergent sequence $> \mathfrak{E}$. Nor does it hold for the domain of all monotonically decreasing numerical sequences; for if $\mathfrak{A}$ is a countable set of such sequences $A, B, C, \ldots$ with $\lim a_n = 1$, $\lim b_n = 2$, $\lim c_n = 3$, $\ldots$, then there does not exist any monotonically decreasing sequence $> \mathfrak{A}$.

However, the Fundamental Theorem again holds for the domain of all sequences decreasing monotonically to zero. To prove this, we note that it held for the domain of all *positive sequences converging to zero*. Such a sequence X ($x_n > 0$, $\lim x_n = 0$), which we are going to briefly call a *null sequence*, can be permuted to a *monotonic* sequence

$$\Xi \equiv (\xi_1\, \xi_2\, \ldots\, \xi_n\, \ldots) \qquad (\xi_n \geqq \xi_{n+1},\ \lim \xi_n = 0)$$

that is uniquely determined by X; while conversely, infinitely many non-monotonic permutations X belong to a monotonic sequence Ξ. With that, the following theorem holds:

III. *If X and Y are two null sequences, Ξ and H the associated monotonic sequences, and if $X \lesseqgtr Y$, then also $\Xi \lesseqgtr H$.*

Since each final relation $X \rho Y$ is preserved by one and the same permutation that is simultaneously applied to X and Y, for the proof of Theorem III, we can assume that one of the numerical sequences is already monotonic. First of all, if $X = Y$, thus

$$Y \equiv (y_1\, y_2\, \ldots\, y_l\, x_{l+1}\, x_{l+2}\, \ldots),$$

and if X is already a monotonic null sequence, then let ε be the smallest of the numbers $y_1\, y_2\, \ldots\, y_l$, and after that, let x_{m+1} ($m \geq l$) be the first among the numbers of Y that is $< \varepsilon$. Then it is the case that

$$y_1, y_2, \ldots, y_l, x_{l+1}, \ldots, x_m \geqq \varepsilon > x_{m+1}, x_{m+2}, \ldots,$$

and upon permuting Y to H, only the first m terms can undergo rearrangement; hence $H = Y = X$.

Second, let $X > Y$ and let X already be monotonic. Then for $n \geq n_0$, it is the case that $x_n > y_n$, and consequently, $x_n > y_n, y_{n+1}, y_{n+2}, \ldots$; so at most $n - 1$ components of Y (namely, $y_1\, y_2\, \ldots\, y_{n-1}$) can be $\geq x_n$. Thus also $\eta_n < x_n$ for $n \geq n_0$, since for $\eta_n \geq x_n$ the n components $\eta_1, \eta_2, \ldots, \eta_n$ would all have to be $\geq x_n$; i.e., it is the case that $X > H$.

With this, Theorem III is proved. If now $\mathfrak{A}$, say, is a countable set of monotonic null sequences, then to begin with there exists an ordinary null sequence $X > \mathfrak{A}$; the monotonic null sequence Ξ associated with X is then similarly $> \mathfrak{A}$; so with this, the correctness of the Fundamental Theorem for the domain of positive, monotonic sequences that converge to zero has been verified.

If one wants to take the monotonicity requirement in the stronger sense that $\xi_n > \xi_{n+1}$ is supposed to hold rather than just $\xi_n \geqq \xi_{n+1}$, one has to prevent the occurrence of equal terms in the null sequence X, and accordingly, one has to require

$$x_n \neq x_1, x_2, \ldots, x_{n-1}.$$

In the proof of the Fundamental Theorem, this requirement is always satisfiable since one freely chooses each x_n inside an interval of real numbers; thus one can avoid a finite (or even countable) set of values.

Of course, the Fundamental Theorem also holds in the domain of the monotonic or strictly monotonic sequences growing to $+\infty$, as one can realize by switching to reciprocal numerical sequences $1 : x_n$.

§3
The Algebra of Numerical Sequences; Rational Pantachies

If the continuum is of the second infinite cardinality, then all pantachies and pantachie segments are similar (of type H). This remark leads to the problem of showing the existence of a *homogeneous* pantachie, i.e., all its segments are similar, independently of the Continuum Hypothesis. And we immediately set ourselves the more specific task of proving the existence of a pantachie in which *rational operations* are possible. Indeed, the required auxiliary investigation of the algebra of numerical sequences offers a certain interest in itself, whereas the mere existence proof for a homogeneous pantachie can also be carried out with simpler methods[1]) (in which only a part of the rational operations are needed).

We again take as basic the domain of real number sequences with graduation according to $sq(a_n - b_n)$, but we keep only a single element from each class of finally equal sequences; the ordered domains and pantachies of this

---

[1]) Über Pantachietypen (loc. cit.), p. 141 ff. The proof there needs a small correction: understand by $\mathfrak{H}_0$ the ordered domain of all rational functions of the A_α given in no. 1, then strengthen the requirement (β) of the "Semifield" $\mathfrak{H}$, so that multiplication by arbitrary elements of $\mathfrak{H}_0$ shall be feasible, and for the elements $AX + B$ in no. 2, let A run through only the elements of $\mathfrak{H}_0$ and let B run through all elements of $\mathfrak{H}$.

section shall be taken from the domain $\mathfrak{z}$ so reduced, in which now only the relations $<> \parallel$ hold. If $(a_1 \, a_2 \, \ldots \, a_n \, \ldots)$ is an arbitrary numerical sequence, belonging to $\mathfrak{z}$ or not belonging to $\mathfrak{z}$, then there exists in $\mathfrak{z}$ one and only one numerical sequence A that satisfies the final equation

$$A = (a_1 \, a_2 \, \ldots \, a_n \, \ldots),$$

so that the n-th component of A *eventually* becomes a_n; this holds also if we allow among the a_n a finite number of objects that are not real numbers (for example, complex numbers or symbols like $\frac{1}{0}, \frac{0}{0}$). In this sense we define, in case a is any real number, the real numerical sequence a by the equation

$$a = (a \, a \, a \, \ldots \, a \, \ldots)$$

and further the sum, difference, product and quotient of two numerical sequences by the equations

$$\begin{aligned}
A + B &= (a_1 + b_1, a_2 + b_2, \ldots, a_n + b_n, \ldots), \\
A - B &= (a_1 - b_1, a_2 - b_2, \ldots, a_n - b_n, \ldots), \\
AB &= (a_1 b_1, a_2 b_2, \ldots, a_n b_n, \ldots), \\
\frac{A}{B} &= (\frac{a_1}{b_1}, \frac{a_2}{b_2}, \ldots, \frac{a_n}{b_n}, \ldots),
\end{aligned}$$

whereby the numerical sequence on the left is always uniquely determined by its membership in $\mathfrak{z}$. The quotient $\dfrac{A}{B}$ is defined only if eventually $b_n \neq 0$, thus only if the numerical sequence B does not contain infinitely many zeroes; so division by a "divisor of zero" is forbidden.

With this, each rational function $f(A, B, \ldots, L)$ of finitely many numerical sequences with rational or real coefficients is also defined (in case no division by a divisor of zero occurs) and indeed as the numerical sequence belonging to $\mathfrak{z}$ whose components are eventually $f(a_n, b_n, \ldots, l_n)$.

A domain [[Gebiet]] (ordered domain, pantachie) contained in $\mathfrak{z}$ is called a *rational domain (rational ordered domain, rational pantachie)* if, in addition to A and B, the numerical sequences $A+B$, $A-B$, AB and $\dfrac{A}{B}$ also always belong to it, forbidden divisions excluded. A rational domain in which there is at least one numerical sequence that is not a divisor of zero contains all rational sequences $r = (r \, r \, \ldots \, r \, \ldots)$. A rational ordered domain (that does not consist of the single numerical sequence 0) contains only numerical sequences $\lesseqgtr 0$ and no divisor of zero besides zero itself; only division by zero is forbidden in it.

Our objective is to produce a rational pantachie; for this we must be able to extend a rational ordered domain that is not yet a pantachie to a larger rational ordered domain. However, this is not possible in just anyway as it is for number fields since here the case of incomparability comes into consideration; by arbitrary adjunction, one would obtain an ordered domain or a rational domain but not necessarily a rational ordered domain. For example, one can adjoin to the rational ordered domain of numerical sequences r a numerical sequence A that is comparable with all its elements, where $\lim a_n = \sqrt{2}$; however, if the a_n converge to $\sqrt{2}$ from both sides, then it is the case that $A^2 \| 2$, and so A does not belong to any rational ordered domain. As one sees, this case would be avoided if *before* additional adjunctions one first enlarged the ordered domain of sequences r by the $\sqrt{2}$ itself and generally by all the algebraic real numbers $a = (a\,a\,\ldots\,a\,\ldots)$ and if one tried the corresponding thing for each rational ordered domain. For this task, the rudiments of the algebra of numerical sequences must be developed.

Let $\mathfrak{R}$ be a rational ordered domain, and let

$$F(X) = A^{(0)} X^r + A^{(1)} X^{r-1} + \cdots + A^{(r)} \qquad (A^{(0)} \lessgtr 0)$$

be a "polynomial in $\mathfrak{R}$," i.e., one whose coefficients $A^{(0)} \ldots A^{(r)}$ belong to $\mathfrak{R}$. If one substitutes an indeterminate numerical sequence $(x_1\, x_2\, \ldots\, x_n\, \ldots)$ for X, then $F(X)$ is a numerical sequence with the eventual components

$$f_n(x_n) = a_n^{(0)} x_n^r + a_n^{(1)} x_n^{r-1} + \cdots + a_n^{(r)}.$$

In quite the usual way, formal algebra can now be built up in the rational ordered domain $\mathfrak{R}$. If

$$G(X) = B^{(0)} X^s + B^{(1)} X^{s-1} + \cdots + B^{(s)}$$

is a second polynomial in $\mathfrak{R}$ and correspondingly

$$g_n(x_n) = b_n^{(0)} x_n^s + b_n^{(1)} x_n^{s-1} + \cdots + b_n^{(s)},$$

then first of all, the quotient and remainder for the division $F(X) : G(X)$ are definable as polynomials in $\mathfrak{R}$. From this comes the notion of divisibility, the Euclidean Algorithm, the greatest common divisor, and the existence of two polynomials $\Phi(X)$ and $\Psi(X)$ of degrees below s and r whose coefficients do not all vanish in $\mathfrak{R}$ and with the property that

$$F(X)\Phi(X) + G(X)\Psi(X) = 1 \text{ or } 0,$$

according to whether the polynomials are relatively prime or not. If $R(A, B)$ is the resultant of the polynomials $F(X)$ and $G(X)$, which as an element of $\mathfrak{R}$ can only be $\lessgtr 0$, then eventually $R(a_n, b_n)$ is the resultant of $f_n(x_n)$, $g_n(x_n)$; since eventually

$$f_n(x_n)\phi_n(x_n) + g_n(x_n)\psi_n(x_n) = 1 \text{ or } 0,$$

$R(a_n, b_n)$ is eventually $\neq 0$ or $= 0$. The non-vanishing of $R(A, B)$ follows as a necessary and sufficient condition for $F(X)$ and $G(X)$ to be relatively prime. The corresponding holds for the discriminant $D(A)$ of the polynomial $F(X)$. Finally, if one define the notions "reducible or irreducible in $\mathfrak{R}$," then it follows from the above given representation of the numerical sequence 1 by relatively prime polynomials that a product is divisible by an irreducible polynomial only when one of its factors is so divisible. The unique factorization of a polynomial into irreducible factors follows from this.

A real numerical sequence $\Xi = (\xi_1\, \xi_2 \ldots \xi_n \ldots)$ (belonging to $\mathfrak{Z}$) that is a member of $\mathfrak{R}$ and for which $F(\Xi) = 0$, thus eventually $f_n(\xi_n) = 0$, is called a *real root* of the equation $F(X) = 0$. One sees immediately that in general the set of real roots of an equation is infinite (of the cardinality of the continuum); e.g., for each combination of signs the numerical sequence

$$(\pm 1, \pm 1, \pm 1, \ldots, \pm 1, \ldots)$$

is a real root of the equation $X^2 = 1$. In the following way, we distinguish among these a finite number of real roots that we are going to call *regular roots*.

To begin with, let $F(X)$ be relatively prime to its derivative $F'(X)$; thus the discriminant $D(A) \lessgtr 0$ and eventually $D(a_n) \lessgtr 0$. Hence the equation $f_n(x_n) = 0$ eventually has just simple roots, and among these, there is eventually a *constant* number ϱ of real roots that is independent of n. In order to see this, by means of the Euclidean Algorithm, we form the STURM polynomials $F(X), F'(X), G(X), H(X), \ldots$ from $F(X)$ in the usual way:

$$F(X) = F'(X) \cdot Q_1(X) - G(X),$$
$$F'(X) = G(X) \cdot Q_2(X) - H(X), \quad \text{etc.}$$

For instance, let

$$F(X) = A^{(0)} X^r + \cdots,$$
$$F'(X) = r A^{(0)} X^{r-1} + \cdots,$$
$$G(X) = B^{(0)} X^s + \cdots,$$
$$H(X) = C^{(0)} X^t + \cdots, \quad \text{etc.,}$$
$$r - 1 > s > t > \cdots.$$

Since the above equations also eventually hold for the n-th components, which eventually are polynomials of degrees $r, r-1, s, t, \cdots$, these components eventually form the STURM chain for $f_n(x_n)$, and since the highest coefficients of these component polynomials eventually have fixed sign (namely, those of $A^{(0)}, A^{(0)}, B^{(0)}, C^{(0)}, \ldots$) and the number ϱ of real roots of the equations $f_n(x_n) = 0$ depends in a known way on these signs, ϱ is eventually independent of n. As soon as ϱ has attained its definitive value, we denote the ϱ real roots of $f_n(x_n) = 0$, according to size, by

$$\xi_n^{(1)} < \xi_n^{(2)} < \cdots < \xi_n^{(\varrho)},$$

and we define thereby numerical sequences

$$\Xi^{(i)} = (\xi_1^{(i)}\, \xi_2^{(i)} \cdots \xi_n^{(i)} \cdots) \qquad (i = 1, 2, \ldots, \varrho)$$

belonging to $\mathfrak{Z}$, so that also $\Xi^{(1)} < \Xi^{(2)} < \cdots < \Xi^{(\varrho)}$; we call these the *regular roots* of the equation $F(X) = 0$.

These regular roots are comparable to all the numerical sequences U of the rational ordered domain $\mathfrak{R}$. For if one substitutes U into the STURM polynomials, then $F(U)$, $F'(U)$, $G(U)$, ... are elements of $\mathfrak{R}$ with definite signs $\lessgtr 0$, which they eventually impart to their components $f_n(u_n)$, $f'_n(u_n)$, $g_n(u_n)$, ... The number of real roots of the equations $f_n(x_n) = 0$ that are $\lessgtr u_n$ depend on these components and the signs of the highest coefficients; these numbers are eventually constant. If $i - 1$ is the specific number of roots $< u_n$ and $F(U) \lessgtr 0$, then eventually $\xi_n^{(i-1)} < u_n < \xi_n^{(i)}$, thus $\Xi^{(i-1)} < U < \Xi^{(i)}$; if $F(U) = 0$, then $U = \Xi^{(i)}$; so in each case, U is comparable with all Ξ. We record the results so far in the theorem

I. *If $F(X)$ is a polynomial in the rational ordered domain $\mathfrak{R}$ with non-vanishing discriminant, then the regular roots of the equation $F(X) = 0$ are comparable to each other and to all numerical sequences in $\mathfrak{R}$.*

We further note that the definition of regular roots is independent of the rational ordered domain to which the coefficients of the polynomial belong. Thus if $\mathfrak{R}'$ is a rational ordered domain that contains $\mathfrak{R}$ or any rational ordered domain that contains the coefficients of $F(X)$, then the regular roots of the equation $F(X) = 0$ are also comparable to all the numerical sequences in $\mathfrak{R}'$.

In addition, we now have to show that the regular roots of two equations are comparable. First we take note of the evident theorem

II. *If $F(X) \cdot G(X)$ is a polynomial with non-vanishing discriminant and if the regular roots of $F(X) = 0$ and $G(X) = 0$ are comparable with each other, then each regular root of the equation $F(X) \cdot G(X) = 0$ is either a regular root of $F(X) = 0$ or a regular root of $G(X) = 0$.*

Indeed, the equation $f_n(x_n) \cdot g_n(x_n) = 0$ eventually has $\varrho + \sigma$ simple real roots $\zeta_n^{(1)} \cdots \zeta_n^{(\varrho+\sigma)}$, and among these are the ϱ roots $\xi_n^{(1)} \cdots \xi_n^{(\varrho)}$ of the equation $f_n(x_n) = 0$ and the σ roots $\eta_n^{(1)} \cdots \eta_n^{(\sigma)}$ of the equation $g_n(x_n) = 0$; let all three systems be ordered by magnitude. By assumption, the $\xi_n^{(i)}$ and $\eta_n^{(k)}$ eventually have a positional relationship that is independent of n, from which it follows that $\zeta_n^{(l)}$ is either a $\xi_n^{(i)}$ or an $\eta_n^{(k)}$; thus each regular root $Z^{(l)}$ of the equation $F(X) \cdot G(X) = 0$ is either a $\Xi^{(i)}$ or an $H^{(k)}$.

Moreover, the following theorem holds

III. *If U is a numerical sequence that is comparable to all the regular roots of the equation $F(X) = 0$, then $F(U)$ has a definite sign $\lessgtr 0$.*

By assumption $U \lessgtr \Xi^{(i)}$ for $i = 1, 2, \ldots, \varrho$. If we exclude the case $U = \Xi^{(i)}$ and $F(U) = 0$, then let, say, $\Xi^{(i-1)} < U < \Xi^{(i)}$; thus eventually $\xi_n^{(i-1)} < u_n < \xi_n^{(i)}$. Since the equation $f_n(x_n) = 0$ eventually has the simple roots $\xi_n^{(1)} \cdots \xi_n^{(\varrho)}$ (the definition of regular roots was certainly up to now restricted to this case), then in these positions $f_n(x_n) = a_n^{(0)} x^r + \cdots$ alternately changes from positive to negative and from negative to positive; therefore $f_n(u_n)$ eventually has the sign $(-1)^{r+i+1} sg\, a_n^{(0)} = (-1)^{r+i+1} sg\, A^{(0)}$, which is independent of n. Thus $F(U) \lessgtr 0$. — In particular, if $F(X) = 0$ does not even have regular roots, then for each real number sequence U, $F(U)$ has one and the same sign, namely, that of $A^{(0)} \lessgtr 0$; $F(X)$ is "definite" and of course its degree r is even.

Now we prove the crucial theorem

IV. *If $F(X)$ is irreducible in $\mathfrak{R}$ and $G(X)$ is an arbitrary polynomial in $\mathfrak{R}$ and Ξ is a regular root of the equation $F(X) = 0$, then $G(\Xi) \lessgtr 0$. That is, by adjoining Ξ, there arises from $\mathfrak{R}$ a rational ordered domain $(\mathfrak{R}, \Xi)$ consisting of all the rational functions of Ξ with coefficients from $\mathfrak{R}$.*

The theorem is true when $F(X)$ is of the first degree since then Ξ itself belongs to $\mathfrak{R}$. We prove it through recursion by assuming that it is already proved for all irreducible polynomials of degree $< r$. To begin with, if $G(X)$ is an irreducible polynomial of degree $< r$ and if H is one of its regular roots, then according to the assumption that was made, $\mathfrak{R}' = (\mathfrak{R}, H)$ is a rational ordered domain and $F(X)$ is a polynomial in $\mathfrak{R}'$ with a non-vanishing discriminant. According to I, each regular root Ξ of $F(X) = 0$ is comparable with all elements of $\mathfrak{R}'$, thus in particular with H; since $F(X)$ and $G(X)$ are relatively prime, it is the case that $\Xi \lessgtr H$. So both the equations $F(X) = 0$ and $G(X) = 0$ have distinct but comparable regular roots, and according to III, $G(\Xi) \lessgtr 0$. Should $G(X) = 0$ have no regular roots, then $G(X)$ is definite and likewise $G(\Xi) \lessgtr 0$. Moreover, if $G(X)$ is reducible but still of degree $< r$, then each of its irreducible factors, thus also $G(X)$ itself, takes on a definite sign $\lessgtr 0$ for $X = \Xi$. Finally, if $G(X)$ is of degree $\geq r$ and $G_0(X)$ is the remainder of the division $G(X) : F(X)$, then $G(\Xi) = G_0(\Xi)$ and $G(X)$ has a fixed sign $\lessgtr 0$, except when the remainder is zero and $G(X)$ is divisible by $F(X)$, in which case $G(\Xi) = 0$.

With this, Theorem IV is proved. From it immediately follows:

V. *The regular roots of two equations irreducible in $\mathfrak{R}$ are always comparable with each other.*

For if Ξ and H are two such roots, then $(\mathfrak{R}, H)$ is a rational ordered domain and Ξ is comparable to each of its elements, hence comparable with H.

If one decomposes a reducible polynomial with non-vanishing discriminant into its irreducible factors $F(X) = F_1(X) \cdot F_2(X) \cdot \ldots$, then the regular roots of $F_1(X) = 0$, $F_2(X) = 0$, $\ldots$ are pairwise comparable, and according to II, they form in their totality the regular roots of $F(X) = 0$. This state of affairs finally permits one to also define the regular roots of equations with vanishing discriminants; if decomposed into irreducible factors $F(X) = F_1(X)^{p_1} \cdot F_2(X)^{p_2} \cdot \ldots$, then the regular roots of $F_k(X) = 0$ shall be declared the p_k-fold regular roots of $F(X) = 0$.

Through this, full correspondence with ordinary algebra is achieved for regular roots, and the following statements need no further proof. Each regular root Ξ of any equation in $\mathfrak{R}$ is a regular root of an irreducible equation in $\mathfrak{R}$ (the equation of least degree in $\mathfrak{R}$ that Ξ satisfies), and in any rational ordered domain it is a regular root of each equation that it satisfies at all. A regular root of $F(X) \cdot G(X) = 0$ is a regular root of $F(X) = 0$ or a regular root of $G(X) = 0$. Each integral or rational function of Ξ with coefficients in $\mathfrak{R}$ is $\lessgtr 0$; a rational ordered domain $(\mathfrak{R}, \Xi)$ arises by the adjunction of Ξ. Two regular roots are always comparable ($\Xi \lessgtr H$).

Moreover, since the equation in $\mathfrak{R}$ of which H is a regular root is also an equation in $(\mathfrak{R}, \Xi)$, from the adjunction of H to $(\mathfrak{R}, \Xi)$ there arises yet again a rational ordered domain $(\mathfrak{R}, \Xi, H)$, consisting of all rational functions of Ξ and H with coefficients from $\mathfrak{R}$. This can be continued, i.e., if Ξ, H, Z, ... are regular roots of equations in $\mathfrak{R}$ (finite in number), then from their adjunction to $\mathfrak{R}$, there arises the rational ordered domain $\mathfrak{R}' = (\mathfrak{R}, \Xi, H, Z, \ldots)$ of all rational functions of Ξ, H, Z, ... with coefficients in $\mathfrak{R}$.

Then if $G(X) = B^{(0)}X^s + \cdots + B^{(s)}$ is a polynomial in $\mathfrak{R}'$ whose coefficients are thus rational or, as we can assume, integral functions of Ξ, H, Z, ..., then from the equation $G(X) = 0$ an equation $F(X) = 0$ with coefficients in $\mathfrak{R}$ can be produced by resultant formation (elimination of Ξ, H, Z, ...). Then each regular root of $G(X) = 0$ is also a root of $F(X) = 0$ and certainly a regular root as we have seen; i.e., the regular roots of equations in $\mathfrak{R}'$, in particular the elements of $\mathfrak{R}'$ themselves, are at the same time regular roots of equations in $\mathfrak{R}$.

After this, we finally consider the domain $\mathfrak{A}$ that consists of *all* regular roots of equations in $\mathfrak{R}$ (to which of course the elements of $\mathfrak{R}$ itself belong; furthermore, it suffices to restrict oneself to equations irreducible in $\mathfrak{R}$). Since two numerical sequences Ξ and H in $\mathfrak{A}$ are always comparable, $\mathfrak{A}$ is a *ordered domain*. Furthermore, $\Xi + H$, $\Xi - H$, ΞH and (for $H \lessgtr 0$) $\Xi : H$ belong to the rational ordered domain $(\mathfrak{R}, \Xi, H)$ and are thus regular roots of equations in $\mathfrak{R}$ and consequently elements of $\mathfrak{A}$, i.e., $\mathfrak{A}$ is a *rational ordered domain*. Certainly, this ordered domain also has the further property that all regular roots of equations in $\mathfrak{A}$ themselves belong to $\mathfrak{A}$; for an equation in $\mathfrak{A}$ whose left side is $A^{(0)}X^r + \cdots + A^{(r)}$ is also an equation in the rational ordered domain $(\mathfrak{R}, A^{(0)}, A^{(1)}, \ldots, A^{(r)})$, and its regular roots are simultaneously regular roots of equations in $\mathfrak{R}$, i.e., elements of $\mathfrak{A}$. So if we define a rational ordered domain to be an *algebraic ordered domain* when all regular roots of equations in the ordered domain are themselves members of the ordered domain, we then have the theorem:

VI. *To each rational ordered domain $\mathfrak{R}$ is associated an algebraic ordered domain $\mathfrak{A}$ that consists of the regular roots of all the (irreducible) equations in $\mathfrak{R}$.*

The simplest rational ordered domain is that of rational numerical sequences $r = (r\,r\,\ldots\,r\,\ldots)$, to which corresponds the algebraic ordered domain of real algebraic numerical sequences $a = (a\,a\,\ldots\,a\,\ldots)$.

If $F(X)$ is a polynomial in the algebraic ordered domain $\mathfrak{A}$ and all the distinct regular roots of $F(X) = 0$ are $A_1, A_2, \ldots$, then it is the case that

$$F(X) = (X - A_1)^{p_1} \cdot (X - A_2)^{p_2} \cdot \ldots \cdot G(X).$$

And the equation $G(X) = 0$ then has no more regular roots at all since otherwise $G(X)$ would be divisible by a linear factor $(X - A)$. Thus $G(X)$ is definite. Then if $\mathfrak{A}$ is still not a pantachie and if one substitutes for X any real number sequence that does not belong to $\mathfrak{A}$ but that is comparable with each member of $\mathfrak{A}$, then $X - A_1$, $X - A_2$, $\ldots$, $G(X)$, and consequently $F(X)$ have a definite sign $\lessgtr 0$. Hence:

VII. *From an algebraic ordered domain $\mathfrak{A}$ that is not yet a pantachie, there always arises a rational ordered domain $(\mathfrak{A}, X)$ by adjunction to $\mathfrak{A}$ of a further numerical sequence X that is comparable to each element of $\mathfrak{A}$.*

Then again, there is associated to this rational ordered domain $\mathfrak{R}' = (\mathfrak{A}, X)$ an algebraic ordered domain $\mathfrak{A}'$ formed from the regular roots of equations in $\mathfrak{R}'$. Thus $\mathfrak{A}'$ contains $\mathfrak{A}$ as a proper subset since X belongs to $\mathfrak{A}'$ but not to $\mathfrak{A}$. Thus we find:

VIII. *Each algebraic ordered domain that is not yet a pantachie can be enlarged to a more extensive algebraic ordered domain.*

Finally, let a "chain," i.e., (§1) a well-ordered set of algebraic ordered domains

$$\mathfrak{A}_0 \, \mathfrak{A}_1 \ldots \mathfrak{A}_\alpha \ldots$$

be given in such a way that $\mathfrak{A}_\alpha$ is a proper subset of $\mathfrak{A}_\beta$ for $\alpha < \beta$, and let $\mathfrak{A}$ be the union of all these elements, where we assume that the chain has no last term. Then $\mathfrak{A}$ is again an algebraic ordered domain. For each element and consequently for each finite set of elements of $\mathfrak{A}$, there exists an ordered domain $\mathfrak{A}_\alpha$ to which all these elements belong. If Ξ, H are two elements of $\mathfrak{A}$ and both are in $\mathfrak{A}_\alpha$, then $\Xi \pm H$, ΞH, $\Xi : H$ are in $\mathfrak{A}_\alpha$, thus also contained in $\mathfrak{A}$; i.e., $\mathfrak{A}$ is a rational ordered domain (that $\mathfrak{A}$ is an ordered domain was already proved in §1). Moreover, if $F(X) = 0$ is an equation in $\mathfrak{A}$, then its coefficients are already contained in some $\mathfrak{A}_\alpha$, so each regular root of this equation is in $\mathfrak{A}_\alpha$ and, consequently, also in $\mathfrak{A}$; $\mathfrak{A}$ is an algebraic ordered domain.

If we identify the ε-sets (subsets of **3**) of §1 with algebraic ordered domains, then each chain of limit number type is certainly continuable, and it follows, as it did there, that there exists an algebraic ordered domain $\mathfrak{A}_\xi$ that is not contained in a more extensive algebraic ordered domain. According to VIII, such an ordered domain is a *pantachie*.

With this, we have proved the existence of an algebraic and, a fortiori, of a rational pantachie. In this pantachie, the direct and inverse similarity of all pantachie segments to one another and to the whole pantachie can be proved by means of elementary transformations such as

$$X = A + X', \quad X = BX', \quad X = 1 : X',$$

which because of the simplicity needs no explanation. Pantachies of this kind, of which there certainly exist infinitely many, thus display an extensive analogy with the series of real numbers.

§4
Relations with the Second Number Class

The properties of the type H (§2), which is the type of all pantachies and pantachie segments in case the continuum is of the second infinite cardinality, suggest to us the problem of proving, independently of the Continuum Hypothesis, the existence of a pantachie that

(1) is cofinal with Ω (or coinitial with Ω^*),
(2) contains Ω-elements (or Ω^*-elements),
(3) contains $\Omega\omega^*$-gaps (or $\omega\Omega^*$-gaps),
(4) contains $\Omega\Omega^*$-gaps.

The solution of these problems would signify a principal step forward, inasmuch as the scope of the *uncountable*, which the Fundamental Theorem (§1, I) left open, would thereby undergo a restriction in the sense in which one is to be expected from CANTOR's Continuum Hypothesis; with that, the first closer relationship between the continuum and the second number class would be established. We shall actually solve the fourth problem, i.e., we shall prove the existence of an $\Omega\Omega^*$-gap; the dispatch of the first three is still awaited, but it can at least be shown that any one of the three cases 1st, 2nd, or 3rd implies the other two.

Again, we take as basis the domain of all real numerical sequences, graduated according to the eventual sign of the difference (the restriction of §3, where from each class of finally equal numerical sequences only one representative was kept, is again considered as in effect). For two numerical sequences $A < B$, which one has to imagine with universally, i.e., for each n, specified components, it is the case that by definition

$$a_n < b_n \quad \text{for} \quad n \geqq \nu;$$

we are going to call the *smallest* natural number that one may put here for ν the *critical number* for A and B, and we are going to denote it by (AB). If $(AB) = 1$, then without exception $a_n < b_n$ for each n; If $(AB) > 1$, then

$$a_n < b_n \quad \text{for} \quad n \geqq (AB), \quad \text{but certainly}$$
$$a_n \geqq b_n \quad \text{for} \quad n = (AB) - 1.$$

If $A < B < C$, then it follows immediately that

(1) $\qquad (AC) \leqq \max(AB), (BC).$

If $\mathfrak{A} = \{A\}$ is a transfinite set of numerical sequences and $\mathfrak{A} < B$, then the following three cases are possible:

(α) All the critical numbers (AB) remain below a fixed number; we then say that B *uniformly* [[*gleichmäßig*]] surpasses the set $\mathfrak{A}$.

(β) Not all, but a transfinite set of the critical numbers $(A\,B)$ remain below a fixed number; we say that B *half uniformly* [[*halb gleichmäßig*]] (or half non-uniformly) surpasses the set $\mathfrak{A}$.

(γ) No transfinite set of critical numbers $(A\,B)$ remain below a fixed number; we say that B *non-uniformly* [[*ungleichmäßig*]] surpasses the set $\mathfrak{A}$.

If the cases (β) and (γ) should be combined, we say that B *not uniformly* [[*nicht gleichmäßig*]] surpasses the set $\mathfrak{A}$; similarly, for (α) and (β) combined, we say that B *not non-uniformly* [[*nicht ungleichmäßig*]] surpasses the set $\mathfrak{A}$.

Case (γ) is only possible if the set $\mathfrak{A}$ is countable. For if $\mathfrak{A}$ is uncountable of cardinality $\aleph_\alpha$ ($\alpha > 0$), then of the $\aleph_\alpha$ critical numbers $(A\,B)$ that are supposed to be distributed among the $\aleph_0$ values $1, 2, 3, \ldots$, some $\aleph_\alpha$ must take the same value ν, and so B uniformly surpasses a subset $\mathfrak{A}'$ equivalent to $\mathfrak{A}$.

If $\mathfrak{A} < B < C$, then it follows from inequality (1):

I. *If B uniformly surpasses the set $\mathfrak{A}$, so does C; if B half uniformly surpasses the set $\mathfrak{A}$, then C not non-uniformly surpasses it. If C non-uniformly surpasses $\mathfrak{A}$, so does B; if C half uniformly surpasses $\mathfrak{A}$, then B not uniformly surpasses it.*

After this, we prove the following theorem:

II. *If $\mathfrak{A}$ is a countable well-ordered set of numerical sequences and C is a numerical sequence that non-uniformly surpasses each initial segment [[Abschnitt]] of $\mathfrak{A}$, then there is also a numerical sequence B ($\leqq C$) that non-uniformly surpasses the whole set $\mathfrak{A}$.*

Actually, here this ought to be only a question of transfinite segments; the addendum is unnecessary if we agree that each finite set is non-uniformly surpassed by each greater numerical sequence.

Theorem II is obvious when $\mathfrak{A}$ has a last term. Moreover, if $\mathfrak{A}$ is of type ω and if C any sequence $> \mathfrak{A}$, and if $P, Q, R, S, \ldots$ are the elements of $\mathfrak{A}$, then choose a sequence of *increasing* natural numbers $\lambda, \mu, \nu, \ldots$ according to the inequalities

$$\lambda \geqq (P\,Q), (Q\,C); \quad \mu \geqq (Q\,R), (R\,C); \quad \nu \geqq (R\,S), (S\,C); \quad \ldots,$$

so that

$$p_n < q_n < c_n \quad \text{for} \quad n \geqq \lambda,$$
$$p_n < q_n < r_n < c_n \quad \text{for} \quad n \geqq \mu, \quad \text{etc.}$$

If one then chooses a sequence B such that

$$b_n < p_n \quad \text{for} \quad n < \lambda,$$
$$p_n < b_n < q_n \quad \text{for} \quad \lambda \leqq n < \mu,$$
$$q_n < b_n < r_n \quad \text{for} \quad \mu \leqq n < \nu, \quad \text{etc.,}$$

then $\mathfrak{A} < B < C$ and the critical numbers are $(P\,B) = \lambda$, $(Q\,B) = \mu$, $(R\,B) = \nu, \ldots$; hence B non-uniformly surpasses $\mathfrak{A}$.

Now in general, let $\mathfrak{A}$ be of type a limit number of the second number class, and let $\mathfrak{A}_k$ denote the set of those elements A for which $(A\,C) = k$; so $\mathfrak{A}$ decomposes into the subsets $\mathfrak{A}_1\,\mathfrak{A}_2\,\mathfrak{A}_3\,\ldots$. Each set $\mathfrak{A}_k$ is either finite (possibly empty) or it is of type ω and $\mathfrak{A}$ is cofinal with it; for based on the hypothesis that C non-uniformly surpasses each initial segment of $\mathfrak{A}$, only a finite subset of $\mathfrak{A}_k$ can lie in each such initial segment. Accordingly, determine a sequence C_1 so that $\mathfrak{A} < C_1 \leqq C$ and C_1 non-uniformly surpasses the set $\mathfrak{A}_1$. This is possible; for if $\mathfrak{A}_1$ is finite, then one can take $C_1 = C$, and if $\mathfrak{A}_1$ is infinite, then according to the above proved fact, an element C_1 can be found between the ω-sequence $\mathfrak{A}_1$ and C that non-uniformly surpasses $\mathfrak{A}_1$; then $\mathfrak{A}_1 < C_1 < C$ and consequently, $\mathfrak{A} < C_1 < C$. Further, seek a sequence C_2 ($\mathfrak{A} < C_2 \leqq C_1$) that non-uniformly surpasses $\mathfrak{A}_2$, and continue like this. In this way, one obtains a set $\mathfrak{C}$: $C \geqq C_1 \geqq C_2 \geqq C_3 \geqq \cdots$ where $\mathfrak{A} < \mathfrak{C}$ and C_k non-uniformly surpasses the set $\mathfrak{A}_k$. Finally, by the Fundamental Theorem, choose a sequence B between $\mathfrak{A}$ and $\mathfrak{C}$. Then according to I, B non-uniformly surpasses each $\mathfrak{A}_k$. However, from this it follows that B non-uniformly surpasses the entire set $\mathfrak{A}$. For were there a transfinite set $\mathfrak{A}'$ for whose elements $(A'\,B)$ is constant, then it would only have finitely many elements in common with each $\mathfrak{A}_k$; then $\mathfrak{A}'$ would already be non-uniformly surpassed by C and all the more so by B. With this, II is proved.

From this comes the following theorem

III. *One can determine an Ω-sequence $\mathfrak{A} = \{A_\alpha\}$ and an Ω^*-sequence $\mathfrak{B} = \{B_\alpha\}$ of numerical sequences so that $\mathfrak{A} < \mathfrak{B}$ and so that each B_α non-uniformly surpasses the set $\mathfrak{A}_\alpha$ of predecessors of A_α.*

We prove this through transfinite induction by showing that for each index α, the determination of A_α and B_α in accordance with the stated requirements is possible if it has already been done for all smaller indices $\xi < \alpha$. Then for $\mathfrak{A}_\alpha = \{A_\xi\}$, $\mathfrak{B}_\alpha = \{B_\xi\}$, it is the case that $\mathfrak{A}_\alpha < \mathfrak{B}_\alpha$, and B_ξ non-uniformly surpasses the set $\mathfrak{A}_\xi$. If one chooses any C between $\mathfrak{A}_\alpha$ and $\mathfrak{B}_\alpha$, then C non-uniformly surpasses each set $\mathfrak{A}_\xi$, i.e., each initial segment of $\mathfrak{A}_\alpha$. By II, one can then determine a sequence B_α between $\mathfrak{A}_\alpha$ and C that non-uniformly surpasses $\mathfrak{A}_\alpha$; and finally, one can choose any A_α between $\mathfrak{A}_\alpha$ and B_α. Then it is the case that

$$\mathfrak{A}_\alpha < A_\alpha < B_\alpha\ (< C) < \mathfrak{B}_\alpha.$$

And the determination of A_α and B_α is completed. One can initiate the process with any pair of sequences $A_0 < B_0$ or sets of sequences $\mathfrak{A}_\omega < \mathfrak{B}_\omega$.

Moreover, it is easily seen that the sequences $\mathfrak{A}$ and $\mathfrak{B}$ of Theorem III define an $\Omega\Omega^*$-gap, i.e., there does not exist any numerical sequence between $\mathfrak{A}$ and $\mathfrak{B}$. Namely, assuming that such exists, let $\mathfrak{A} < X < \mathfrak{B}$; then, as we noted in the discussion of case (γ), there exists a subset of the second infinite cardinality $\mathfrak{A}' = A_{\xi_0}\,A_{\xi_1}\,\ldots A_{\xi_\alpha}\,\ldots$ such that all the critical numbers $(A_{\xi_\alpha}\,X)$ are $= \lambda$, and there exists a similar subset $\mathfrak{B}' = B_{\eta_0}\,B_{\eta_1}\,\ldots B_{\eta_\beta}\,\ldots$ such that $(X\,B_{\eta_\beta}) = \mu$. Then according to (1)

$$(A_{\xi_\alpha}\,B_{\eta_\beta}) \leqq \max\lambda,\mu;$$

i.e., all these critical numbers remain below a finite bound. If one takes the first ω indices $\xi_0\,\xi_1\,\xi_2\,\ldots$ and if η is one of the indices η_β that is greater than $\xi_0\,\xi_1\,\xi_2\,\ldots$, then B_η uniformly surpasses the set $A_{\xi_0}\,A_{\xi_1}\,A_{\xi_2}\,\ldots$; thus it not non-uniformly surpasses the set $\mathfrak{A}_\eta$ (of which this is a part). This is a contradiction to the construction of Theorem III, and thus it is really true:

IV. *The sequences $\mathfrak{A}$ and $\mathfrak{B}$ of Theorem III have no numerical sequence between them, and so they define an $\Omega\Omega^*$-gap.*

By constructing a pantachie over the ordered domain $\mathfrak{A}, \mathfrak{B}$, one obtains a pantachie that has at least one $\Omega\Omega^*$-gap.

With that, the fourth of the problems stated at the beginning is solved. The first three can be couched in the following questions:

(1) Does there exist a *transcendent* Ω-sequence $\mathfrak{A}$ of numerical sequences, i.e., one that is not surpassed by any numerical sequence?

(2) Does there exist an Ω-sequence $\mathfrak{A}$ with the limit B, i.e., so that no numerical sequence lies between $\mathfrak{A}$ and B?

(3) Does there exist an $\Omega\omega^*$-gap $\mathfrak{A}, \mathfrak{B}$, i.e., two sequences ($\mathfrak{A} < \mathfrak{B}$) of numerical sequences with no further numerical sequence lying between them and with $\mathfrak{A}$ of type Ω and $\mathfrak{B}$ of type ω^*?

By the application of a transformation[1]) that reverses the ordering of numerical sequences (such as $X = -X'$ or $X = 1 : X'$), one observes that these questions could also be stated for corresponding Ω^*-sequences, Ω^*-elements, and $\omega\Omega^*$-gaps. An answer to these questions that is independent of the Continuum Hypothesis seems to offer no less difficulty than the Continuum Hypothesis itself; nevertheless, it can be at least pointed out that all three reduce to the first question.

The identity of the first and second problems is seen by an elementary transformation. If $\mathfrak{B} = \{B_\alpha\}$ is an Ω-sequence of numerical sequences with the limit B, then $A_\alpha = 1 : (B - B_\alpha)$ yields a transcendent Ω-sequence of numerical sequences, and conversely, if $\mathfrak{A} = \{A_\alpha\}$ is a transcendent Ω-sequence of positive numerical sequences, then $B_\alpha = B - \dfrac{1}{A_\alpha}$ runs through an Ω-sequence with limit B. In the last case, if not all the A_α should be positive, then consider instead the sequence $C + A_\alpha$, say, where C is any sequence $> -A_0$.

In order to also demonstrate the identity of the first and the third problem, we first mention the following:

V. *If $\mathfrak{X} = \{X_\alpha\}$ is a transcendent set of numerical sequences (i.e., for which there is no higher element), then there also exists a well-ordered transcendent set $\mathfrak{Y} = \{Y_\beta\}$ of at most the same cardinality as $\mathfrak{X}$.*

Arbitrarily choose
$$Y_0 > X_0,$$
$$Y_1 > X_1, Y_0,$$

and generally

---
[1]) Cf. §3.

$$Y_\alpha > X_\alpha, \mathfrak{Y}_\alpha,$$

where $\mathfrak{Y}_\alpha$ is the set of already constructed Y_β with indices $\beta < \alpha$. Either this choice is possible for each α and $\mathfrak{Y} = \{Y_\alpha\}$ is well-ordered, of the same cardinality as $\mathfrak{X}$, and likewise transcendent because $Z > \mathfrak{X}$ would follow from $Z > \mathfrak{Y}$. Or a first index α exists at which this choice is not possible, and then $\mathfrak{Y}_\alpha$ is transcendent because if there were a $Z > \mathfrak{Y}_\alpha$, then there would also still be an element $Y_\alpha > X_\alpha, Z$ or $Y_\alpha > X_\alpha, \mathfrak{Y}_\alpha$. In this case, $\mathfrak{Y}_\alpha$ is well-ordered and equivalent to a subset of $\mathfrak{X}$.

If $\mathfrak{X}$ is of the second infinite cardinality, then $\mathfrak{Y}$ is too since, according to the Fundamental Theorem, no countable set is transcendent; $\mathfrak{Y}$ is then cofinal with Ω and can thus be replaced by a transcendent Ω-sequence.

Now let $\mathfrak{A}$ and $\mathfrak{B}$ be an $\Omega\omega^*$-gap, i.e., $\mathfrak{A}$ is an Ω-sequence of numerical sequences

$$A^\alpha \equiv (a_1^\alpha \, a_2^\alpha \, \ldots \, a_n^\alpha \, \ldots) \qquad (\alpha = 0, 1, \ldots, \omega, \ldots),$$

and $\mathfrak{B}$ an ω^*-sequence of numerical sequences

$$B^\beta \equiv (b_1^\beta \, b_2^\beta \, \ldots \, b_n^\beta \, \ldots) \qquad (\beta = 1, 2, \ldots);$$

in addition, $\mathfrak{A} < \mathfrak{B}$ and they contain no element between them. We claim that the numerical sequences

$$X_\alpha = ((\alpha\,1)\,(\alpha\,2)\,\ldots\,(\alpha\,n)\,\ldots)$$

formed from the *critical numbers* $(A^\alpha \, B^\beta) = (\alpha\,\beta)$ constitute a transcendent set $\mathfrak{X}$. If this is proved, then according to Theorem V, the existence of a transcendent Ω-sequence follows from the existence of an $\Omega\omega^*$-gap.

If we arrange the critical numbers $(\alpha\,\beta)$ in a tableau with $\aleph_1$ rows and $\aleph_0$ columns so that $(\alpha\,\beta)$ appears in row α and column β, then first of all the following holds:

The numbers in each column cannot remain below a fixed whole number y_β (so that for each α, $(\alpha\,\beta) < y_\beta$ would be the case).

Namely, were this the case, for which we may further assume that $y_1 < y_2 < y_3 < \cdots$, then by the meaning of the critical numbers, it would be that *for each α*

$$a_n^\alpha < b_n^\beta \qquad \text{for} \quad n \geqq y_\beta, \quad \text{thus}$$
$$a_n^\alpha < b_n^1 \qquad \text{for} \quad y_1 \leqq n < y_2,$$
$$a_n^\alpha < b_n^1, b_n^2 \qquad \text{for} \quad y_2 \leqq n < y_3,$$
$$a_n^\alpha < b_n^1, b_n^2, b_n^3 \qquad \text{for} \quad y_3 \leqq n < y_4, \quad \text{etc.}$$

From this, one could determine a numerical sequence C so that for each α

$$a_n^\alpha < c_n \leqq b_n^1 \qquad \text{for} \quad y_1 \leqq n < y_2,$$
$$a_n^\alpha < c_n \leqq b_n^1, b_n^2 \qquad \text{for} \quad y_2 \leqq n < y_3,$$
$$a_n^\alpha < c_n \leqq b_n^1, b_n^2, b_n^3 \qquad \text{for} \quad y_3 \leqq n < y_4, \quad \text{etc.}$$

Then it would be the case that $C > A^\alpha$; furthermore, for each β eventually $c_n \leqq b_n^{\beta+1} < b_n^\beta$, and thus also $C < B^\beta$; therefore $\mathfrak{A} < C < \mathfrak{B}$, contrary to the assumption that $\mathfrak{A}, \mathfrak{B}$ is supposed to be a gap.

Nor can there exist any *subtableau* of $\aleph_1$ rows and $\aleph_0$ columns where the numbers $(\alpha' \beta')$ in each column remain below a number $y_{\beta'}$; for this subtableau contains the critical numbers between two subsets $\mathfrak{A}', \mathfrak{B}'$, whereby $\mathfrak{A}$ is cofinal with $\mathfrak{A}'$ and $\mathfrak{B}$ is coinitial with $\mathfrak{B}'$, and thus $\mathfrak{A}', \mathfrak{B}'$ is also a gap.

Now if the set $\mathfrak{X} = \{X_\alpha\}$ were not transcendent, then there would exist a sequence $Y > \mathfrak{X}$ and this would, as we have noted, uniformly surpass an uncountable subset $\mathfrak{X}' = \{X_{\alpha'}\}$, i.e., it would thus be that $(\alpha' \beta) < y_\beta$ for $\beta \geqq \beta_0$. Hence there would exist, contrary to what is proved above, a subtableau formed from the $\aleph_1$ rows α' and the $\aleph_0$ columns $\beta_0, \beta_0+1, \beta_0+2, \ldots$ with finitely many columns left over. Consequently, the set $\mathfrak{X}$ is transcendent, and the existence of a transcendent Ω-sequence follows from the existence of an $\Omega\omega^*$-gap.

However, the theorem as well as the proof can be reversed. Let $\mathfrak{X} = \{X\}$ be a transcendent Ω-sequence or a transcendent set of the second infinite cardinality, whose elements $X \equiv (x_1\, x_2 \ldots x_n \ldots)$ we assume are sequences of *increasing natural numbers* (if it is not yet the case, this can be done through the enlargement of the original sequences). Then let $\mathfrak{B} = B^1\, B^2 \ldots B^\beta \ldots$ be an arbitrary ω^*-sequence of numerical sequences that, for the sake of simplicity, we may think of as given in such a form that $b_n^\beta > b_n^{\beta+1}$ *throughout* (for each n); furthermore, this can also be accomplished by replacing $B^2, B^3, \ldots$ in turn by finally equal sequences. Now we associate with each numerical sequence X of the given kind a sequence A ($< \mathfrak{B}$) that has the critical numbers $x_1 + 1, x_2 + 1, \ldots$ with respect to $B^1, B^2, \ldots$; e.g., one can let A agree with the first x_1 components of B^1, with the next $x_2 - x_1$ components of B^2, with the next $x_3 - x_2$ components of B^3, and so on. So a set $\mathfrak{A} = \{A\} < \mathfrak{B}$ of the second infinite cardinality corresponds to the set $\mathfrak{X}$, and we claim that no further numerical sequence lies between $\mathfrak{A}$ and $\mathfrak{B}$. Namely, were $\mathfrak{A} < C < \mathfrak{B}$, then according to (1) it would be the case that for each A

$$x_\beta + 1 = (A\,B^\beta) \leqq \max(A\,C), (C\,B^\beta),$$

and if one chooses β_0 so large that $x_{\beta_0} \geqq (A\,C)$, then

$$(C\,B^\beta) > x_\beta \text{ for } \beta \geqq \beta_0,$$

i.e., the numerical sequence

$$Y = ((C\,B^1)\,(C\,B^2) \ldots (C\,B^\beta) \ldots)$$

would be $> X$ for each X; thus $Y > \mathfrak{X}$, and contrary to the assumption, $\mathfrak{X}$ is not a transcendent set. Thus no sequence lies between $\mathfrak{A}$ and $\mathfrak{B}$. One can express this as follows: for the domain $\mathfrak{Z}^\mathfrak{B}$ of all numerical sequences $< \mathfrak{B}$, $\mathfrak{A}$ is a transcendent set of the second infinite cardinality; from this it follows precisely as in Theorem V for the domain $\mathfrak{Z}$ of all numerical sequences that there must also exist in $\mathfrak{Z}^\mathfrak{B}$ a transcendent Ω-sequence (for according to the Fundamental Theorem, a countable set cannot be transcendent in $3^\mathfrak{B}$). Thus if a transcendent Ω-sequence exists for 3, then there also exists one for the domain $3^\mathfrak{B}$ where $\mathfrak{B}$ is an arbitrary ω^*-sequence, i.e., there exists an $\Omega\omega^*$-gap $\mathfrak{A}, \mathfrak{B}$. For example, if one takes for $\mathfrak{B}$ the numerical sequences $1, \frac{1}{2}, \frac{1}{3}, \ldots$, then it follows that in the domain of positive numerical sequences converging to zero there exists a transcendent Ω-sequence in case one exists in the domain of all numerical sequences.

With this, it is proved that the existence of $\Omega\omega^*$-gaps and transcendent Ω-sequences are mutually dependent, and after the three mentioned questions are recognized as identical, further investigation of pantachie types would first have to turn to the main question of the existence of transcendent Ω-sequences.

§5
Convergent and Divergent Sequences

Through our considerations at least some light also falls on the graduation of convergent and divergent sequences. Here we take as our basis the domain 3 of *positive* numerical sequences ($a_n > 0$) with graduation according to $sg(a_n - b_n)$, and we call the sequence $A \equiv (a_1 \, a_2 \ldots a_n \ldots)$ a convergent sequence or a divergent sequence according to the behavior of $\sum a_n$. Along with C, each finally smaller sequence is also convergent, and along with D, each finally larger sequence is divergent; between C and D only one of the two relations $C < D$ and $C \parallel D$ is possible. The Fundamental Theorem (§2, I) now receives the following strengthening:

I. *If $\mathfrak{A}$ is an at most countable set of convergent sequences, then there always exists a convergent sequence $> \mathfrak{A}$. If $\mathfrak{B}$ is an at most countable ordered domain of divergent sequences, then there always exists a divergent sequence $< \mathfrak{B}$. If $\mathfrak{A} < \mathfrak{B}$, then there always exists a convergent sequence as well as a divergent sequence between $\mathfrak{A}$ and $\mathfrak{B}$.*

Proof. (1) Let $\mathfrak{A}$ consist of the convergent sequences $A, B, C, D, \ldots$. Choose increasing integers $\lambda, \mu, \nu, \ldots$ so that

$$\sum_{\lambda}^{\infty} b_n < \tfrac{1}{2}, \quad \sum_{\mu}^{\infty} c_n < \tfrac{1}{4}, \quad \sum_{\nu}^{\infty} d_n < \tfrac{1}{8}, \ldots,$$

and then set

$$\begin{aligned} x_n &= a_n & &\text{for} \quad n < \lambda, \\ x_n &= a_n + b_n & &\text{for} \quad \lambda \leqq n < \mu, \\ x_n &= a_n + b_n + c_n & &\text{for} \quad \mu \leqq n < \nu, \quad \text{etc.} \end{aligned}$$

Then

$$\sum_{1}^{\infty} x_n = \sum_{1}^{\infty} a_n + \sum_{\lambda}^{\infty} b_n + \sum_{\mu}^{\infty} c_n + \cdots$$

is convergent and $X > A, B, C, \ldots$.

(2) If $\mathfrak{B}$ has a least element P, then $\tfrac{1}{2}P$ is a divergent sequence $< \mathfrak{B}$. If $\mathfrak{B}$ does not have such an element, then it is coinitial with an ω^*-sequence

$P > Q > R > S > \cdots$. Choose increasing whole numbers $\lambda, \mu, \nu, \ldots$ so that
$$\sum_{1}^{\lambda-1} p_n > 1 \quad \text{and} \quad q_n < p_n \quad \text{for} \quad n \geq \lambda,$$
$$\sum_{\lambda}^{\mu-1} q_n > 1 \quad \text{and} \quad r_n < q_n \quad \text{for} \quad n \geq \mu,$$
$$\sum_{\mu}^{\nu-1} r_n > 1 \quad \text{and} \quad s_n < r_n \quad \text{for} \quad n \geq \nu, \quad \text{etc.}$$

Then
$$X \equiv (p_1 p_2 \ldots p_{\lambda-1} q_\lambda q_{\lambda+1} \ldots q_{\mu-1} r_\mu r_{\mu+1} \ldots r_{\nu-1} s_\nu \ldots)$$
is a divergent sequence $< P, Q, R, \ldots$; hence $X < \mathfrak{B}$.

The hypothesis that $\mathfrak{B}$ be an *ordered domain* could be replaced by a less restrictive one; however, it recommends itself by its simple formulation, and it suffices for applications. In any event, $\mathfrak{B}$ must not be an entirely arbitrary set since certainly for two divergent sequences P and Q there exists no divergent sequence $< P, Q$ in case, say, $p_1 + p_3 + p_5 + \cdots$ and $q_2 + q_4 + q_6 + \cdots$ are convergent series, or generally, in case there is some partition of the numerical sequence $1, 2, 3, \ldots$ into two complementary subsets $L \equiv (l_1 \, l_2 \, l_3 \, \ldots)$ and $M \equiv (m_1 \, m_2 \, m_3 \, \ldots)$ for which both sequences
$$P_L \equiv (p_{l_1} \, p_{l_2} \, \ldots \, p_{l_n} \, \ldots), \quad Q_M \equiv (q_{m_1} \, q_{m_2} \, \ldots \, q_{m_n} \, \ldots)$$
are convergent. Because of the following, we note that this case is also possible for monotonically decreasing sequences. In order to give an example, we collect in turn groups of $1, 2, 2^3, 2^6, \ldots, 2^{\frac{n(n-1)}{2}}, \ldots$ numbers from the numerical sequence $1, 2, 3, \ldots$; we let these groups alternately belong to L and M, and we arrange the given values of p_n and q_n in the following tableau

	L	M	L	M	L	M
n	1	2 3	4 ... 11	12 ... 75	76 ... 1099	1100 ... 33867
p_n	2^{-1}	2^{-2}	2^{-6}	2^{-7}	2^{-15}	2^{-16}
q_n	2^{-1}	2^{-3}	2^{-4}	2^{-10}	2^{-11}	2^{-21}

$$\underbrace{}_{1} \underbrace{}_{2} \underbrace{}_{2^3} \underbrace{}_{2^6} \underbrace{}_{2^{10}} \underbrace{}_{2^{15}}$$

and we fill the empty positions in some way to obtain monotonic sequences. Then it is the case that
$$\sum p_{l_n} < 1 \cdot 2^{-1} + 2^3 \cdot 2^{-6} + 2^{10} \cdot 2^{-15} + \cdots = 2^{-1} + 2^{-3} + 2^{-5} + \cdots$$
converges,
$$\sum p_{m_n} > 2 \cdot 2^{-2} + 2^6 \cdot 2^{-7} + 2^{15} \cdot 2^{-16} + \cdots = 2^{-1} + 2^{-1} + 2^{-1} + \cdots$$
diverges,
$$\sum q_{l_n} > 1 \cdot 2^{-1} + 2^3 \cdot 2^{-4} + 2^{10} \cdot 2^{-11} + \cdots = 2^{-1} + 2^{-1} + 2^{-1} + \cdots$$
diverges,

$\sum q_{m_n} < 2 \cdot 2^{-3} + 2^6 \cdot 2^{-10} + 2^{15} \cdot 2^{-21} + \cdots = 2^{-2} + 2^{-4} + 2^{-6} + \cdots$
converges.

Thus P and Q are divergent monotonic sequences for which there certainly exists no divergent sequence $< P, Q$.

3) Now if $\mathfrak{A} < \mathfrak{B}$, determine some element Y between $\mathfrak{A}$ and $\mathfrak{B}$ according to the Fundamental Theorem, and further determine a convergent sequence $C > \mathfrak{A}$ and a divergent sequence $D < \mathfrak{B}$; then $\mathfrak{A} < C, Y$ and $\mathfrak{B} > D, Y$. Finally, determine again according to the Fundamental Theorem an element X such that $\mathfrak{A} < X < C, Y$ and an element Z such that $\mathfrak{B} > Z > D, Y$. Then X is convergent and Z is divergent (since $X < C$ and $Z > D$), and $\mathfrak{A} < X < Y < Z < \mathfrak{B}$; thus X and Z are between $\mathfrak{A}$ and $\mathfrak{B}$.

With this, the "strengthened Fundamental Theorem" I is proved. It obviously also holds for the domain $\mathfrak{Z}_{\mathfrak{A}}^{\mathfrak{B}}$ of all numerical sequences between $\mathfrak{A}$ and $\mathfrak{B}$ if $\mathfrak{A}$ is an at most countable set of convergent sequences and $\mathfrak{B}$ is an at most countable ordered domain of divergent sequences and $\mathfrak{A} < \mathfrak{B}$; it holds as well for the domain $\mathfrak{Z}_{\mathfrak{A}}$ of all sequences $> \mathfrak{A}$ and the domain $\mathfrak{Z}^{\mathfrak{B}}$ of all sequences $< \mathfrak{B}$. So it holds for the domain $\mathfrak{Z}^{\mathfrak{E}}$ (cf. §2) of all positive sequences X with $\lim x_n = 0$. And from this, it again follows that it also holds for the domain of all sequences monotonically or strictly monotonically decreasing to 0 since the permutation of an arbitrary sequence to a monotonic one preserves convergence or divergence, as well as the final sign $\gtreqless$ (§2). Theorem I has the well-known consequence that the usual "scales" of convergence criteria and divergence criteria always can break down since they are based on the comparison of an unknown series with a set of known convergent or divergent series where the set is either countable or, with regard to its effect, is replaceable by a countable set.

Our studies can be carried over to the domain of positive functions that are distinguished according to the convergence or divergence of the improper definite integral $\int^\infty f(x)\,dx$ without difficulty.

P. DU BOIS-REYMOND is known to have spoken of a "boundary between convergence and divergence" and to have attempted to describe this boundary in an obscure way by "ideal" elements (series or functions). Since a positive sequence can only be convergent or divergent and not some third sort, there could be the question of such a boundary only in the sense that there would be a last convergent sequence or a first divergent sequence; both of these are excluded in each ordered domain of sequences for which Theorem I holds. This theorem, however, leaves open the possibility that perhaps no further sequences, either convergent or divergent, would lie between an *uncountable* set $\mathfrak{C}$ of convergent sequences C and a divergent sequence D ($> \mathfrak{C}$); so if $\mathfrak{C}$ is an ordered domain, then in a pantachie constructed over $\mathfrak{C}$ and D, the sequence D really would be the first divergent sequence. This remark leads to the investigation of the distribution of convergent and divergent sequences in a *pantachie*. If $\mathfrak{P}_c$ is the set of convergent sequences and $\mathfrak{P}_d$ the set of divergent sequences in the pantachie $\mathfrak{P}$, then it is obviously the case (if neither of the two sets is empty) that $\mathfrak{P}_c < \mathfrak{P}_d$, and in particular, the cut $\mathfrak{P} = \mathfrak{P}_c + \mathfrak{P}_d$ being of whatever form, the question is whether $\mathfrak{P}_c$ can have a last element or $\mathfrak{P}_d$ can have a first element. The answer is that any cut-form that is at all possible can be realized as $\mathfrak{P}_c + \mathfrak{P}_d$, or stated somewhat more precisely:

II. *If $\mathfrak{X}$ is an arbitrary pantachie interval and if $\mathfrak{X} = \mathfrak{X}' + \mathfrak{X}''$ is an arbitrary cut, then a pantachie interval $\mathfrak{W} = \mathfrak{W}_c + \mathfrak{W}_d$ that is similar to $\mathfrak{X}$ can be given in such a way that $\mathfrak{W}_c$ is similar to $\mathfrak{X}'$ and $\mathfrak{W}_d$ is similar to $\mathfrak{X}''$. In particular, one can also get that $\mathfrak{W}$ consists of just monotonic sequences with terms decreasing to zero.*

The proof is quite simple as long as one leaves out the monotonicity requirement. Let the pantachie interval $\mathfrak{X} = \{X\}$ be bounded by the sequences L and M ($L \leq X \leq M$); as a pantachie interval, $\mathfrak{X}$ is not extendible in its interior, i.e., if $\mathfrak{X} = \mathfrak{X}' + \mathfrak{X}''$ is any cut ($\mathfrak{X}' \neq 0$, $\mathfrak{X}'' \neq 0$), then there exists no element between $\mathfrak{X}'$ and $\mathfrak{X}''$. Now choose four sequences $B < C < D < E$, of which C is convergent and D is divergent, and perform (§3) a linear transformation

$$U = PX + Q,$$

by means of which L and M are mapped to B and C; then U runs through an ordered domain $\mathfrak{A}$ ($B \leq U \leq C$) that is similar to $\mathfrak{X}$ and that of course, as is immediately evident, is again a pantachie interval, i.e., its interior is not extendable. Likewise, let V run through the pantachie interval $\mathfrak{B}$ ($D \leqq V \leqq E$) by virtue of a linear transformation

$$V = RX + S.$$

Now if $\mathfrak{X} = \mathfrak{X}' + \mathfrak{X}''$ is a cut in $\mathfrak{X}$, then assign to each sequence X' in $\mathfrak{X}'$ the sequence

$$W' \equiv (u_1', u_1', u_2', u_2', \ldots, u_n', u_n', \ldots)$$

and to each sequence X'' in $\mathfrak{X}''$ the sequence

$$W'' \equiv (u_1'', v_1'', u_2'', v_2'', \ldots, u_n'', v_n'', \ldots).$$

W' runs through an ordered domain $\mathfrak{W}'$ that is similar to $\mathfrak{X}'$ and W'' runs through an ordered domain $\mathfrak{W}''$ that is similar to $\mathfrak{X}''$; furthermore $\mathfrak{W}' < \mathfrak{W}''$ and the ordered domain $\mathfrak{W} = \mathfrak{W}' + \mathfrak{W}''$ is again a pantachie interval since for each W, it is the case that

$$(w_1\, w_3\, w_5\, \cdots) \equiv (u_1\, u_2\, u_3\, \cdots) \equiv U,$$

and thus an element insertible inside $\mathfrak{W}$ would produce an element insertible inside $\mathfrak{A}$. Finally, each U and each W' is convergent, and each V and each W'' is divergent; hence $\mathfrak{W}_c = \mathfrak{W}'$ and $\mathfrak{W}_d = \mathfrak{W}''$.

With this, the first part of Theorem II is proved. One recognizes immediately that here too the case $\mathfrak{P}_c = 0$, thus the case of a pantachie of only divergent sequences, can also be realized; for if by II one produces a pantachie $\mathfrak{P}_c + \mathfrak{P}_d$ in which $\mathfrak{P}_c$ has a last element C and if Y runs through the set $\mathfrak{P}_d$, then $Y - C$ runs through a pantachie of (positive) divergent sequences. On the other hand, the case $\mathfrak{P}_d = 0$ is excluded for the domain 3 since $\lim c_n = 0$ for each convergent sequence, thus $C < \mathfrak{E}$ (see §2) and at best this case could occur in the domain $3^{\mathfrak{E}}$.

The proof of Theorem II is somewhat more complicated when the requirement of monotonicity is placed on the sequences W. We are going to immediately consider the case of *strict monotonicity*, and first we suppose that we have chosen a divergent strictly monotonic sequence D in which not only $\lim d_n = 0$, but actually $\liminf n d_n = 0$ as well (e.g., $d_n = \frac{1}{n \ln n}$ for $n > 1$), and then we suppose that we have chosen a second such sequence $E > D$ for which

(1) $$e_1 > d_1 > e_2 > d_2 > e_3 > d_3 \cdots.$$

Then choose increasing integers $\lambda, \mu, \nu, \ldots$ so that

(2) $$\lambda d_\lambda + \mu d_\mu + \nu d_\nu + \cdots \text{ converges}$$

(e.g., $\lambda d_\lambda < \frac{1}{2}$, $\mu d_\mu < \frac{1}{4}$, $\nu d_\nu < \frac{1}{8}$, ...). In addition, choose sequences $B < C$ according to the following inequalities:

(3) $$\begin{cases} d_\lambda > c_1 > b_1 > e_{\lambda+1}, \\ d_\mu > c_2 > b_2 > e_{\mu+1}, \\ d_\nu > c_3 > b_3 > e_{\nu+1}, \quad \text{etc.} \end{cases}$$

And finally choose a strictly monotonic sequence A corresponding to the inequalities

(4) $$a_1 < d_\lambda - c_1, \quad a_{\lambda+1} < d_\mu - c_2, \quad a_{\mu+1} < d_\nu - c_3, \quad \ldots.$$

Now by linear transformations, we again assign to the given pantachie interval $\mathfrak{X}$ ($L \leq X \leq M$) the pantachie intervals $\mathfrak{A}$ ($B \leq U \leq C$) and $\mathfrak{B}$ ($D \leq V \leq E$), for which we can obviously arrange (if need be, by replacing U and V by finally equal sequences) that *for each* n

(5) $$b_n \leqq u_n \leqq c_n, \quad d_n \leqq v_n \leqq e_n;$$

then it is the case that

$$u_n \geqq b_n > c_{n+1} \geqq u_{n+1},$$
$$v_n \geqq d_n > e_{n+1} \geqq v_{n+1},$$

and thus the U and V are all strictly monotonic with $\lim u_n = \lim v_n = 0$.

Let us then consider the following sequence

$$\Phi \equiv \Phi(X) \equiv (\Phi_1, u_1, \Phi_2, u_2, \Phi_3, u_3, \ldots),$$

where for clarity we understand by $\Phi_1, \Phi_2, \Phi_3, \ldots$ the following groups of $\lambda, \mu - \lambda, \nu - \mu, \ldots$ terms:

$$\Phi_1 \equiv (u_1 + a_1, u_1 + a_2, \ldots, u_1 + a_\lambda),$$

$$\Phi_2 \equiv (u_2 + a_{\lambda+1}, u_2 + a_{\lambda+2}, \ldots, u_2 + a_\mu),$$
$$\Phi_3 \equiv (u_3 + a_{\mu+1}, u_3 + a_{\mu+2}, \ldots, u_3 + a_\nu), \quad \text{etc.}$$

Thus Φ runs through an ordered domain that is similar to $\mathfrak{X}$. Moreover, each Φ is strictly monotonic, which will be proved from the monotonicity of A as soon as we show that u_n is larger than the first term of Φ_{n+1}. Now it is indeed the case that

$$u_2 + a_{\lambda+1} \leq c_2 + a_{\lambda+1} < d_\mu < e_\mu \leq e_{\lambda+1} < b_1 \leq u_1,$$

thus $u_1 > u_2 + a_{\lambda+1}$, likewise $u_2 > u_3 + a_{\mu+1}$, and so on. What is more, each Φ is convergent; since the terms of Φ_1, u_1 are $\leq u_1 + a_1 \leq c_1 + a_1 < d_\lambda$, and likewise the terms of Φ_2, u_2 are less than d_μ, and so on, the sum of Φ is thus less than the sum

$$(\lambda + 1)d_\lambda + (\mu - \lambda + 1)d_\mu + (\nu - \mu + 1)d_\nu + \cdots,$$

which converges by (2).

Furthermore, we consider the sequence

$$\Psi \equiv \Psi(X) \equiv (\Psi_1, u_1, \Psi_2, u_2, \Psi_3, u_3, \ldots),$$

in which

$$\Psi_1 \equiv (v_1, v_2, \ldots, v_\lambda),$$
$$\Psi_2 \equiv (v_{\lambda+1}, v_{\lambda+2}, \ldots, v_\mu),$$
$$\Psi_3 \equiv (v_{\mu+1}, v_{\mu+2}, \ldots, v_\nu), \quad \text{etc.}$$

These sequences, too, form an ordered domain similar to $\mathfrak{X}$, and they are divergent since $V \geq D$ is divergent. They are likewise strictly monotonic, to which end we must furnish a proof that u_n lies between the last term of Ψ_n and the first term of Ψ_{n+1}. It is the case, however, that

$$v_\lambda \geq d_\lambda > c_1 \geq u_1 \geq b_1 > e_{\lambda+1} \geq v_{\lambda+1};$$

thus $v_\lambda > u_1 > v_{\lambda+1}$, likewise $v_\mu > u_2 > v_{\mu+1}$, and so on.

Moreover, the smallest term of Ψ_n is larger than the largest term of Φ_n; for we have

$$v_\lambda \geq d_\lambda > c_1 + a_1 \geq u_1 + a_1;$$

thus $v_\lambda > u_1 + a_1$, likewise $v_\mu > u_2 + a_{\lambda+1}$, and so on. It follows from this that all terms of Ψ_n are larger than the corresponding terms of Φ_n, and that for $X' < X''$ surely $\Phi(X') < \Psi(X'')$.

Now finally, let $\mathfrak{X} = \mathfrak{X}' + \mathfrak{X}''$ be an arbitrary cut, and if X' and X'' run through all elements of $\mathfrak{X}'$ and $\mathfrak{X}''$, we put

$$W' = \Phi(X'), \quad W'' = \Psi(X'').$$

W' runs through an ordered domain $\mathfrak{W}'$ of convergent, strictly monotonic sequences that is similar to $\mathfrak{X}'$, and W'' runs through an ordered domain $\mathfrak{W}''$ of divergent, strictly monotonic sequences that is similar to $\mathfrak{X}''$. It is the case that $\mathfrak{W}' < \mathfrak{W}''$ and that the ordered domain $\mathfrak{W} = \mathfrak{W}' + \mathfrak{W}''$ is a pantachie interval since for each W

$$(w_{\lambda+1}\, w_{\mu+2}\, w_{\nu+3}\, \cdots) \equiv (u_1\, u_2\, u_3\, \cdots) \equiv U,$$

and so any element insertible inside $\mathfrak{W}$ would produce an element insertible inside $\mathfrak{A}$. Thus we have found a pantachie interval $\mathfrak{W}$ of strictly monotonic sequences in which $\mathfrak{W}_c$ is similar to $\mathfrak{X}'$ and $\mathfrak{W}_d$ is similar to $\mathfrak{X}''$, and with that, we have completely proved Theorem II.

Here, we leave undecided whether the case $\mathfrak{P}_c = 0$ is also realizable. The case $\mathfrak{P}_d = 0$, as in the domain $\mathfrak{Z}$, is also excluded in the domain $\mathfrak{M}$ of *all* sequences monotonically converging to zero since for each convergent, monotonic sequence it is well known that $\lim n c_n = 0$; by taking for H the sequence

$$H \equiv (1, \tfrac{1}{2}, \tfrac{1}{3}, \ldots, \tfrac{1}{n}, \ldots)$$

and for $\mathfrak{H}$ the ordered domain $H, \tfrac{1}{2}H, \tfrac{1}{3}H, \ldots$, we can express this by the final inequality $C < \mathfrak{H}$. Thus a pantachie of just convergent sequences could at best exist in the domain $\mathfrak{M}^{\mathfrak{H}}$.

According to II, it is therefore the case that Du Bois Reymond's "boundary," in the form of a last convergent element or a first divergent element, can actually occur *inside a pantachie*; but the cut $\mathfrak{P}_c + \mathfrak{P}_d$ can also represent a gap, and indeed any gap-form that is at all possible inside a pantachie.

[Manuscript received 3/2/1909, declared ready to print 4/18/1909]

Appendix

Sums of $\aleph_1$ Sets[a]

By
F. Hausdorff (Bonn)

Let a separable metric space X be represented as a sum of increasing Borel sets X^ξ ($\xi < \Omega$):

(1) $$X = \sum_\xi X^\xi, \qquad X^0 \subset X^1 \subset \cdots \subset X^\xi \subset \cdots.$$

Let

(2) $$T^\xi = X - X^\xi$$

be the complement of X^ξ; thus

$$0 = \prod_\xi T^\xi, \qquad T^0 \supset T^1 \supset \cdots \supset T^\xi \supset \cdots.$$

Call the sum (1)

k-convergent (convergent in category) if T^ξ is eventually [[schliesslich]] (for $\xi \geq \alpha$) of the 1st category in X,

m-convergent (convergent in measure) if for *each* absolutely additive, finite, non-negative measure function $|A|$ that is defined for at least the Borel sets $\subset X$ (also for X itself), it is the case that eventually $|T^\xi| = 0$.

Both properties are trivial if eventually $T^\xi = 0$. We say that the sum (1) is *uncountable* if it is always the case that $T^\xi \neq 0$, $X^\xi \neq X$; then by passing to a subsequence, the sum can be transformed into one with pairwise distinct terms X^ξ.

In particular, if

(3) $$T^\xi = \sum_n T_n^\xi,$$

where the index n runs through a countable set N, the T_n^ξ are Borel sets, and for each n the sets

(4) $$T_n^0, T_n^1, \ldots, T_n^\xi, \ldots$$

are disjoint, then (1) is *k*-convergent and *m*-convergent. For the sets (4) are

---

[a]*Summen von $\aleph_1$ Mengen* is reprinted in translation with the permission of the publisher of Fundamenta Mathematicae.

eventually (for $\xi \geqq \alpha_n$) of the 1st category[1]), respectively, of measure zero, and for sufficiently large α and $\xi \geq \alpha$, this holds for all n simultaneously.

In place of (3) and (4) one can write (e.g., with $F_n^\xi = X - \sum_{\alpha<\xi} T_n^\alpha$)

(5) $$T^\xi = \sum_n (F_n^\xi - F_n^{\xi+1}),$$

where the F_n^ξ are Borel sets and for each n

(6) $$F_n^0 \supset F_n^1 \supset \cdots \supset F_n^\xi \supset \cdots$$

holds. This subsumes the "Lusin-Sierpiński" space partitions that arise from an analytic set $A \subset X$ or, said more precisely, from a "determining system" of Borel sets that in a well-known way assigns "indices" $< \Omega$ to the points of space[2]); X^ξ is the set of points with indices $\leqq \xi$. Thus these always give k-convergent and m-convergent sums.

In what follows, I intend to show (§1) that a separable, complete, uncountable space X can be represented as a sum of increasing, distinct sets $X^\xi = G_\delta$, as well as to investigate (§2) the partition of the space 2^X of closed sets A of the compact space X according to the index of the first perfect derivative of A; again in both cases these give k-convergent and m-convergent sums.

Already long ago Herr Sierpiński (Fund. Math. (1920), p. 224, Problem 6) asked the questions — of course, to be answered without the Continuum Hypothesis:

(A) Can a sum of $\aleph_1$ sets of 1st category be of 2nd category?

(B) Can a sum of $\aleph_1$ sets of measure zero be of positive outer measure?

Affirmative answers to these questions would produce k-divergent, respectively, m-divergent representations (1). The fact that up to now the known representations that are definable without the Continuum Hypothesis are all k-convergent and m-convergent sheds light on the difficulty of both of Sierpiński's problems.

§1

As usual, among the dyadic number sequences

$$a \equiv (a_1, a_2, \ldots, a_n, \ldots) \qquad (a_n = 0 \text{ or } 1)$$

we define a "final" ordering by the rules:

$[a \leqq b] = [b \geqq a]$ means: *eventually* (for $n \geqq k$) $\ a_n \leqq b_n$.

Denote the smallest natural number k that achieves this by $(a\,b)$. If $a \leqq x$, $x \leqq b$, then it is also the case that $a \leqq b$ and

---

[1]) The conclusion regarding category already holds when the T_n^ξ satisfy the Baire property with respect to the space X (Kuratowski, *Topologie* I (1933), p. 265) or, as we are going to say, are β-sets. A β-set Z is, except for a set of 1st category, identical with a Borel set, i.e., the set $(Z-Y) + (Y-Z)$ is of 1st category, in which one may take Y as open or closed or as $G_\delta \subset Z$ or as $F_\sigma \supset Z$.

[2]) Lusin-Sierpiński, C. R. 175 (1922), p. 357–359 and Journ. de Math. (7) 2 (1923), p. 53–72, in particular p. 61. Sierpiński, Fund. Math. 8 (1926), p. 362–369 and Fund. Math. 21 (1933), p. 29–34. Sélivanowski, Fund. Math. 21 (1933), p. 20–28.

(7) $\qquad (a\,b) \leq \max[(a\,x), (x\,b)].$

Let $\varrho = [a \leq b]$ and $\sigma = [a \geq b]$, and let ϱ' and σ' be the negations of these statements (ϱ' means that $a_n > b_n$ holds infinitely often). Then we have four possible cases:
$$[a = b] = \varrho\,\sigma,$$
$$[a < b] = \varrho\,\sigma',$$
$$[a > b] = \varrho'\,\sigma,$$
$$[a \,\|\, b] = \varrho'\,\sigma'.$$

In the fourth case a and b are said to be incomparable, in the first three they are said to be comparable. If $a < x$ and $x < b$ then $a < b$. The final equality $a = b$ means that eventually $a_n = b_n$; complete identity $a \equiv b$ means that $a_n = b_n$ always.

The $x = a$ that are finally equal to a form a countable class $K(a)$. A set A of pairwise comparable sequences that contains at most one element from each class $K(a)$ is an ordered set in the usual sense.

By $A < B$, it is of course understood that for $a \in A$ and $b \in B$ always $a < b$.

Our intention is to construct an $\Omega\Omega^*$-gap
$$a^0 < a^1 < \cdots < a^\xi < \cdots \,|\, \cdots < b^\xi < \cdots < b^1 < b^0,$$
i.e., two ordered sets $A < B$ of types Ω and Ω^* for which an x with $A < x < B$ does not exist[1]).

First Interpolation Theorem. *If $A < B$ are two at most countable ordered sets of sequences, then there exists a sequence x with $A < x < B$.*

Let it be noted that without the assumption of ordered A and B the statement need not hold. If, for example, A consists of the sequences $(1, 0, 1, 0, 1, 0, \ldots)$ and $(0, 1, 0, 1, 0, 1, \ldots)$, and B consists of the one sequence $(1, 1, 1, 1, \ldots)$, then there is no x with $A < x < B$.

Proof. It suffices to prove that $A' < x < B'$ where $A' \subset A$, $B' \subset B$ and A is cofinal with A' (i.e., there is no $a \in A - A'$ with $a > A'$) and B is coinitial with B'; thus one may assume that A is of type 1 or type ω and B is of type 1 or type ω^*. All four of these cases are proved together by showing that: if
$$a \leqq b \leqq c \leqq \cdots < \cdots \leqq r \leqq q \leqq p,$$
then an x can be interpolated ($a < x < p$, $b < x < q$, $c < x < r$, ...). Now one can partition the sequence N of natural numbers into finite, pairwise disjoint intervals
$$N = N_0 + N_1 + N_2 + N_3 + N_4 + \cdots$$
ordered according to their indices, in such a way that:

---
[1]) I already did this in 1909 for sequences of real or rational numbers (*Graduierung nach dem Endverlauf*, Abh. Sächs. Ges. d. Wiss. 31). Since this work is undoubtedly little known and since the construction for dyadic sequences, which is necessary for the full proof of Theorem 1, needs some modifications, I would like to present the matter completely.

In $N_1 + N_2$ it is the case that $a_n \leqq p_n$, but in N_1 and in N_2 at least once $a_n < p_n$.

In $N_3 + N_4$ it is the case that $a_n \leqq b_n \leqq q_n \leqq p_n$, but in N_3 and in N_4 at least once $b_n < q_n$.

In $N_5 + N_6$ it is the case that $a_n \leqq b_n \leqq c_n \leqq r_n \leqq q_n \leqq b_n$, but in N_5 and in N_6 at least once $c_n < r_n$. And so on.

We let x agree in $N_1, N_2, N_3, N_4, N_5, N_6, \ldots$ with $a, p, b, q, c, r, \ldots$ ($x_n = a_n$ for $n \in N_1$, and so on). Then it is the case that $a_n \leqq x_n \leqq p_n$ holds in $N_1 + N_2 + N_3 + N_4 + \cdots$, while $x_n < p_n$ holds infinitely often in $N_1 + N_3 + \cdots$, and $x_n > a_n$ holds infinitely often in $N_2 + N_4 + \cdots$; thus $a < x < p$. In just this way, it can be shown that $b < x < q$, $c < x < r$, ... Q. E. D.

Now let $A < b$. We distinguish two cases by means of the above defined indices $(a\,b)$:

(γ) b lies *near* to A, briefly $A\,\gamma\,b$, if for each $k = 1, 2, 3, \ldots$ there are only finitely many a in A for which $(a\,b) \leqq k$ holds.

(δ) b lies *far* from A, briefly $A\,\delta\,b$, if for some k there are infinitely many a in A for which $(a\,b) \leqq k$ holds.

If A is finite, then $A\,\gamma\,b$; if A is uncountable, then $A\,\delta\,b$; for countable A both cases can occur.

(8) If $A < x < b$ and b lies near to A, then x also lies near to A.

For if x lies far from A, so that there are infinitely many $a \in A$ with $(a\,x) \leqq k$, then according to (7), for these a it is the case that

$$(a\,b) \leqq \max[k, (x\,b)] = l$$

and b lies far from A.

Second Interpolation Theorem. *Let A be an countable ordered set of sequences, $A < y$, and for each initial segment [[Abschnitt]] $A(a)$ of A let $A(a)\,\gamma\,y$ hold. Then there exists an x with $A < x < y$ and $A\,\gamma\,x$.*

($A(a)$ is the set of all elements of A coming before a).

Proof. First of all, let A be of type ω, $A = \{a, b, c, \ldots\}$, $a < b < c < \cdots < y$. We can partition the set of natural numbers into

$$N = N_0 + N_1 + N_2 + \cdots$$

in such a way that:

In N_1 it is the case that $a_n \leqq b_n \leqq y_n$, at least once $a_n < b_n$ and at least once $b_n < y_n$.

In N_2 it is the case that $a_n \leqq b_n \leqq c_n \leqq y_n$, at least once $b_n < c_n$ and at least once $c_n < y_n$. And so on.

We let x agree in $N_1, N_2, N_3, \ldots$ with $a, b, c, \ldots$. In $N_1 + N_2 + \cdots$ it is the case that $a_n \leqq x_n \leqq y_n$, thus $a \leqq x \leqq y$, and similarly $b \leqq x \leqq y$, $\ldots$; thus $A < x \leqq y$ and certainly $x < y$ since in each $N_1, N_2, \ldots$ a place is found with $x_n < y_n$. On the other hand, in N_1 there is a place with $x_n = a_n < b_n$, in N_2 one with $x_n = b_n < c_n$, and so on, so that the indices $(b\,x), (c\,x), \ldots$ must tend to ∞, i.e., $A\,\gamma\,x$.

In the general case, we can assume that $A\,\delta\,y$ since for $A\,\gamma\,y$ each interpolated element x between A and y would likewise lie near A, and we can assume that A has no last element for if a were the last element of A, then it would follow easily from $A(a)\,\gamma\,y$ that $A\,\gamma\,y$ also. For sufficiently large k ($\geqq h$), the set A^k of $a \in A$ with $(a\,y) \leqq k$ is infinite; however, since only finitely many elements of A^k lie in each segment $A(a)$, A^k is of type ω and cofinal with A (as one recognizes by consideration of an ω-sequence $a^0 < a^1 < \cdots$ cofinal with A). On the strength of what has already been proved, we can then construct dyadic sequences y^h, y^{h+1}, ... in such a way that

$$A < \cdots < y^{h+2} < y^{h+1} < y^h < y, \quad A^k\,\gamma\,y^k.$$

For since A^h is of type ω, an element y^h with $A^h < y^h < y$ and $A^h\,\gamma\,y^h$ can be interpolated between $A^h (\subset A) < y$; after that, since A is cofinal with A^h, it is also the case that $A < y^h$ and $A^{h+1} < y^h$ and an element y^{h+1} can be interpolated with $A^{h+1} < y^{h+1} < y^h$ and $A^{h+1}\,\gamma\,y^{h+1}$, and so forth. Now let an x be interpolated between A and $B = \{y, y^h, y^{h+1}, \ldots\}$: $A < x < B$. Then $A\,\gamma\,x$ holds; for if there were infinitely many a in A with $(a\,x) \leqq l$, then for these a it would also be the case that $(a\,y) \leqq \max[l, (x\,y)] = k$, these a would thus belong to A^k ($k \geq h$) and $A^k\,\delta\,x$ would hold, in contradiction to $A^k < x < y^k$ and $A^k\,\gamma\,y^k$. With this the Second Interpolation Theorem is proved.

Now we can construct a set $A = \{a^0, a^1, \ldots, a^\xi, \ldots\}$ of type Ω and a set $B = \{b^0, b^1, \ldots, b^\xi, \ldots\}$ of type Ω^* according to the requirement that $A < B$ and

(9) $$A^\eta\,\gamma\,B^\eta,$$

where $A^\eta = \{a^0, \ldots, a^\xi, \ldots\}$ ($\xi < \eta$) is the ηth initial segment of A (of type η); correspondingly, let $B^\eta = \{b^0, \ldots, b^\xi, \ldots\}$ ($\xi < \eta$) be of type η^*. We start with an arbitrary pair $a^0 < b^0$ and then show that for each $\eta > 0$ the pair a^η, b^η can be chosen appropriately, if all the pairs a^ξ, b^ξ for $\xi < \eta$ have been chosen appropriately, i.e., if $A^\eta < B^\eta$ and for $\xi < \eta$ it is the case that $A^\xi\,\gamma\,b^\xi$. Next we interpolate an arbitrary y: $A^\eta < y < B^\eta$. For each initial segment A^ξ of A^η ($\xi < \eta$), it is then the case that $A^\xi\,\gamma\,y$ since $A^\xi < y < b^\xi$ and $A^\xi\,\gamma\,b^\xi$. By the Second Interpolation Theorem, there is a b^η with $A^\eta < b^\eta < y$ and $A^\eta\,\gamma\,b^\eta$, and if we again interpolate an a^η with $A^\eta < a^\eta < b^\eta$, then $A^\eta < a^\eta < b^\eta < B^\eta$ and (9) is fulfilled.

The sets A and B so constructed give the sought-after $\Omega\Omega^*$-gap. For if it were the case that $A < x < B$, then there would exist at least one k so that uncountably many a^ξ with $(a^\xi\,x) \leqq k$ would be available; then if η is so large that the initial segment A^η contains infinitely many of these a^ξ, it would be the case that $A^\eta\,\delta\,x$, in contradiction to $A^\eta < x < b^\eta$ and (9). Unfortunately, up to now the existence of $\Omega\Omega^*$-gaps is the only connection of the second number class to the problems of the final ordering that has been established without the Continuum Hypothesis. It is not known whether there can be two ordered sets $A < B$, one of cardinality $\aleph_1$ and the other at most countable, between which no x can be interpolated, e.g., whether $a^0 < a^1 < \cdots < a^\xi < \cdots < b$ and $b = \lim a^\xi$ can hold in the sense that there is no x with $A < x < b$, or even in the stronger sense that for each $x < b$ there exists an a^ξ with $a^\xi > x$.[b] Now we convert the set of dyadic sequences $x = (x_1, x_2, \ldots)$ into a topological space C in which the sets $(a_1, \ldots, a_k)$ (that is the set of $x = (a_1, \ldots, a_k, x_{k+1}, \ldots)$ with k prescribed initial digits) form a basis of open sets; at the same time these basis sets are closed. C can be metrized as a compact space, e.g., as Baire space, by defining the distance between x and $y \neq x$ as $1/k$ if x and y first differ in the kth digit, or as the well-known Cantor set by identifying x with the real number $2\sum_n x_n/3^n$. For $a, b \in C$, it is the case that

(10) $$F_n = \underset{x}{\mathrm{E}}\,[a_n \leqq x_n \leqq b_n]$$

is closed and

(11) $$T = \underset{x}{\mathrm{E}}\,[a \leqq x \leqq b] = \underline{\lim} F_n = \sum_n F_n F_{n+1} F_{n+2} \cdots$$

is an F_σ. Thus our $\Omega\Omega^*$-gap gives us, in the sets

$$T^\xi = \underset{x}{\mathrm{E}}\,[a^\xi \leqq x \leqq b^\xi],$$

a descending sequence of distinct F_σ-sets with $\prod_\xi T^\xi = 0$ because $a^\xi \in T^\xi - T^{\xi+1}$, and we obtain the theorem:

I. *Each separable, complete, uncountable space X can be represented as a sum $\sum_\xi X^\xi$ of increasing, distinct sets $X^\xi = G_\delta$.*

Since X contains C (topologically) as a compact and therefore closed set, the sets T^ξ are also F_σ in X and the $X^\xi = X - T^\xi$ are G_δ in X.[1])

We are now going to show that the representation $X = \sum_\xi X^\xi$, obtained in the indicated way, is k-convergent and m-convergent. The first claim is trivial since a countable set that is dense in C is eventually contained in $C - T^\xi$; thus T^ξ as an F_σ with dense complement is already of 1st category in C and all the more of 1st category in X. Moreover, in our case all X^ξ (except perhaps for $X^0 = 0$) and T^ξ are dense in X.

Furthermore, let $|A|$ be a measure with the properties specified on p. 241; incidentally, it suffices that the measure be defined for at least the Borel sets $\subset C$ and that the $|T^\xi|$ be eventually finite; we are going to show that eventually it must be that $|T^\xi| = 0$. Were it always the case that $|T^\xi| > 0$, these numbers would be eventually constant, say $= 1$, and we can assume that $|T^0| = |T^1| = \cdots = |T^\xi| = \cdots = 1$ by omitting countably many initial terms. Let the intervals $T^0 \underset{x}{\mathrm{E}}\,[x_n = \delta] = \binom{\delta}{n}$, where $\delta = 0$ or 1,

---

[b] F. Rothberger shows that this question, taken in the "stronger sense," has a negative answer; cf. Rothberger, *On some problems of Hausdorff and of Sierpiński*, Fund. Math. **35** (1948), p. 33.

[1]) With the gap construction for sequences of rational numbers one would obtain sets T^ξ that are indeed F_σ in the space of irrational numbers J, but in a space X containing J topologically are only of the form $G_\delta \cdot F_\sigma$ (cf. the remarks on p. 244).

that are open and closed in T^0 have measure $\left|\begin{smallmatrix}\delta\\n\end{smallmatrix}\right|$, whereby $\left|\begin{smallmatrix}0\\n\end{smallmatrix}\right| + \left|\begin{smallmatrix}1\\n\end{smallmatrix}\right| = 1$. Now let $a^0 < a < b < b^0$; thus with the specifications (10) and (11) $T \subset T^0$ and $T = \underline{\lim}(T^0 F_n)$; therefore $|T| \leq \underline{\lim} |T^0 F_n|$, and in case $|T| = 1$, $\lim |T^0 F_n| = 1$. Since $(0,0,0,\ldots) < a < b < (1,1,1,\ldots)$, there are infinitely many p with $a_p = b_p$ (let P be the set of these), thus $F_p = \underset{x}{\mathrm{E}}\,[x_p = a_p]$ and $T^0 F_p = \binom{a_p}{p}$. Hence it is the case that $\lim \left|\begin{smallmatrix}a_p\\p\end{smallmatrix}\right| = 1$. Now if $x \in C - T$ (not necessarily $x \in T^0$), then there is an infinite set $Q \subset P$ for whose elements $q \in Q$ $x_q = 1 - a_q$; for the inequalities $a_q \leqq x_q \leqq b_q$ are infinitely often not satisfied, but rather (when one assumes $q \geq (a\,b)$) either $x_q < a_q = b_q$ or $x_q > b_q = a_q$. Thus $\lim \left|\begin{smallmatrix}x_q\\q\end{smallmatrix}\right| = 0$ follows from $\lim \left|\begin{smallmatrix}a_q\\q\end{smallmatrix}\right| = 1$; so in any case $\underline{\lim} \left|\begin{smallmatrix}x_n\\n\end{smallmatrix}\right| = 0$. This last relation would now have to hold for *each* sequence $x \in C$ since, according to $\prod_\xi T^\xi = 0$, there certainly is a T^ξ with $x \in C - T^\xi$. With this a contradiction is obtained; if for each n we choose $\binom{x_n}{n}$ to be the one of the two intervals $\binom{0}{n}$, $\binom{1}{n}$ that has the largest measure, we obtain a sequence $x = (x_1, x_2, \ldots)$ with $\underline{\lim} \left|\begin{smallmatrix}x_n\\n\end{smallmatrix}\right| \geqq 1/2$. The representation (1) given here is thus m-convergent.

By the way, this proof of m-convergence holds for *any* $\Omega\Omega^*$-gap; for our special construction one can give yet another proof. Let $(k = 1, 2, 3, \ldots)$

$$T_k^\xi = \underset{x}{\mathrm{E}} \prod_{n \geq k}[a_n^\xi \leqq x_n \leqq b_n^\xi];$$

hence $T^\xi = \sum_k T_k^\xi$. In any case, for $\xi < \eta$ it now follows from $T_k^\xi T_k^\eta \neq 0$ that $\prod_{n \geq k}[a_n^\xi \leqq b_n^\eta]$, thus $(a^\xi\,b^\eta) \leqq k$; according to (9), for a given η this can hold for only finitely many ξ. Therefore if T_{kl}^ξ $(l = 1, 2, 3, \ldots)$ is the set of those x that belong to *exactly* l sets $T_k^{\xi_1}, \ldots, T_k^{\xi_l}$ with $\xi_1 < \xi_2 < \cdots < \xi_{l-1} < \xi_l = \xi$, then it is the case that $T_k^\xi = \sum_l T_{kl}^\xi$ and, obviously, $T_{kl}^\xi T_{kl}^\eta = 0$ for $\xi < \eta$. The T_{kl}^ξ are Borel because $\sum_{m \geq l} T_{km}^\xi$ is the set of x that belong to *at least* l sets $T_k^{\xi_1}, \ldots, T_k^{\xi_l}$ with $\xi_1 < \cdots < \xi_{l-1} < \xi_l = \xi$ and thus equal to the sum of the intersections $T_k^{\xi_1} \cdots T_k^{\xi_l}$ or an F_σ; T_{kl}^ξ is the difference of two F_σs. Thus we have $T^\xi = \sum_{kl} T_{kl}^\xi$, and for each pair of numbers k, l the sets $T_{kl}^0, T_{kl}^1, \ldots, T_{kl}^\xi, \ldots$ are disjoint; therefore (3) and (4) hold, which guarantees m-convergence.

If Sierpiński's question (B) can be answered affirmatively and thus in a closed cube of a Euclidean space there exists a set $A = \sum_\xi A^\xi$ of positive

(Lebesgue) outer measure that is the sum of $\aleph_1$ sets A^ξ of measure zero, then one can assume that the A^ξ are ascending G_δ sets (since each set of measure zero can be included in a G_δ set of measure zero). *In case A is of positive inner measure* and thus contains a perfect set X with $|X| > 0$, then $X = \sum_\xi X A^\xi = \sum_\xi X^\xi$ with $X^\xi \neq X$, and as in Theorem I, one obtains a separable, complete, uncountable space X as an uncountable sum of ascending G_δ sets that in this case, however, is m-divergent.

Theorem I suggests the following problem:

(P) *Can a separable, complete, uncountable space X be represented as a sum $\sum_\xi X^\xi$ of distinct ascending sets $X^\xi = F_\sigma$?*

An affirmative answer to this question would solve Sierpiński's problem (A) in the positive sense. This is because the $T^\xi = X - X^\xi$ are distinct descending G_δ sets; if $F^\xi = \overline{T^\xi}$ is the closure [abgeschlossene Hülle] of T^ξ, then T^ξ is dense in F^ξ and $F^\xi - T^\xi = F^\xi X^\xi$ is of 1st category in F^ξ. The descending F^ξ are eventually equal, $F^\xi = F \neq 0$, and then $F = \sum_\xi F X^\xi$ is represented as a sum of $\aleph_1$ sets of 1st category, while F is of 2nd category in itself.

Conversely, if (A) is solved positively and thus in a separable space E, which we can immediately assume is complete, there exists a set $A = \sum_\xi A^\xi$ of 2nd category that is the sum of $\aleph_1$ sets A^ξ of 1st category, then one can assume that the A^ξ are ascending F_σ sets (since each set of 1st category can be included in an F_σ of 1st category). *In case A is of second inner category*, that is to say, in case A contains an X that is G_δ and of 2nd category (in E), then $X = \sum_\xi X A^\xi = \sum_\xi X^\xi$ with $X^\xi \neq X$, and one thus obtains a representation of a topologically complete space X as an uncountable sum of ascending sets $X^\xi = X F_\sigma$, i.e., the question (P) can be answered affirmatively.

In illustration of the just introduced "inner category," which is an analogue for inner measure, note the following. Let $k(A)$ $(= 1, 2)$ be the category of A in the space E and let the inner category $k_i(A)$ be the maximum of $k(Y)$ for the Borel sets $Y \subset A$. (A correspondingly defined outer category $= \min k(Y)$ for $Y \supset A$ coincides with $k(A)$ because each A of 1st category is contained in an F_σ of 1st category). Hereby, one can replace the Y not only by $X = G_\delta$, but also by β-sets Z (which satisfy the property of Baire with respect to E) because each Z contains an X with $k(Z - X) = 1$, thus $k(X) = k(Z)$; for β-sets, it is the case that $k(Z) = k_i(Z)$. If E is separable and complete, P is its perfect kernel, and the interior [offene Kern] $\underline{P}$ of P is nonempty, thus $k(P) = 2$, then there is a set A with $k(A) > k_i(A)$. This is because one can, exactly as in Bernstein's existence proof of totally imperfect sets, partition P into two disjoint sets A and B that both contain points in common with each uncountable $G_\delta \subset P$; then each $G_\delta \subset A$ is at most countable, hence of 1st category; $k_i(A) = k_i(B) = 1$. If it were the case that $k(A) = 1$, then A and $B = P - A$ would be β-sets, hence both of 1st category; consequently, $k(A) = k(B) = 2$. On the other hand, in the present case there are also sets that are not β-sets and yet have $k = k_i (= 2)$; e.g., if $P = P_1 + P_2$ is a splitting of P into two Borel sets that have interior points, then not both AP_1 and AP_2 can be β-sets; if AP_1, say, is not one, then $AP_1 + P_2$ is a set with the desired property. For $\underline{P} = 0$, however, all sets $A \subset E$ are β-sets and $k(A) = k_i(A)$.

§2

Let X be a compact metric space with the distances $\varrho(a,b)$; likewise let $\mathfrak{X} = 2^X$ be the compact space of closed sets $A \subset X$, $A \neq 0$ with the well-known distance definition $\varrho(A, B)$, for which we recall only the characteristic property: $\varrho(A, B) \leqq \delta$ means that for each $a \in A$ there is a $b \in B$ and conversely for each b there is an a with $\varrho(a, b) \leqq \delta$. Uppercase Latin letters denote sets $\subset X$, uppercase German letters denote sets $\subset \mathfrak{X}$; now let $|M|$ denote the *cardinality* of M. A always denotes closed sets $\neq 0$, thus $A \in \mathfrak{X}$.

Let A^ξ ($\xi < \Omega$) be the ξ-th derivative of A, and let

$$\mathfrak{X}^\xi = \mathop{\mathrm{E}}_A [A^\xi = A^{\xi+1}]$$

be the set of A whose ξ-th derivative is perfect (possibly 0); let

$$\mathfrak{T}^\xi = \mathfrak{X} - \mathfrak{X}^\xi = \mathop{\mathrm{E}}_A [A^\xi \neq A^{\xi+1}]$$

be the complement of $\mathfrak{X}^\xi$. We have

(12) $$\mathfrak{X} = \sum_\xi \mathfrak{X}^\xi, \quad \mathfrak{X}^0 \subset \mathfrak{X}^1 \subset \cdots \subset \mathfrak{X}^\xi \subset \cdots ,$$

and we are going to show:

II. *The $\mathfrak{X}^\xi$ are Borel sets; the sum (12) is k-convergent and m-convergent.* 252

If F is closed and G is open, then the following implications hold:

$$[|AF| = \infty] \to [A'F \neq 0], \quad [A'G \neq 0] \to [|A\,G| = \infty].$$

If one represents a closed set $F \neq 0$ as an intersection $\prod_n F_n$ of closed sets F_n in such a way that F_{n+1} is contained in the interior $\underline{F}_n$ of F_n (e.g., F_n = the set of those x whose lower distance $\delta(x, F)$ from F is at most equal to $1/n$), the following hold:

$$[A'F \neq 0] \to \prod_n [A'F_{n+1} \neq 0] \to \prod_n [A'\underline{F}_n \neq 0]$$
$$\to \prod_n [|A\underline{F}_n| = \infty] \to \prod_n [|AF_n| = \infty]$$
$$\to \prod_n [A'F_n \neq 0] \to [A'F \neq 0],$$

thus

(13) $$[A'F \neq 0] = \prod_n [|AF_n| = \infty].$$

After this, we claim: for each closed $F(\neq 0)$ and for each ξ

(14) $$\mathop{\mathrm{E}}_A [A^\xi F \neq 0] \quad \text{is Borel}.$$

This is true for $\xi = 0$: $\mathop{\mathrm{E}}_A [AF \neq 0]$ is closed.

The inference from ξ to $\xi+1$. Let (14) be true. Let $\{U_1, U_2, \dots\}$ be a countable basis of open sets $\neq 0$ of X; let $V_n = \overline{U}_n$ be the closure of U_n, and for a natural number k let

$$\mathfrak{B} = \{V_{n_1}, V_{n_2}, \dots, V_{n_k}\}$$

be a system of k disjoint V_n. Then

$$[|A^\xi F| \geq k] = \sum_{\mathfrak{B}} [A^\xi F V_{n_1} \neq 0] \cdots [A^\xi F V_{n_k} \neq 0]$$

is the statement that $A^\xi F$ contains at least k distinct points. Since $\mathfrak{B}$ runs through only countably many systems, due to (14) $\underset{A}{\mathrm{E}}[|A^\xi F| \geq k]$ is Borel, likewise for the intersection of these sets for $k = 1, 2, \dots$, i.e., the set $\underset{A}{\mathrm{E}}[|A^\xi F| = \infty]$; and according to (13),

$$\underset{A}{\mathrm{E}}[A^{\xi+1} F \neq 0] = \prod_n \underset{A}{\mathrm{E}}[|A^\xi F| = \infty]$$

is Borel, whereby (14) is transferred from ξ to $\xi+1$.

The inference for a limit ordinal η. If (14) is true for $\xi < \eta$, then

$$\underset{A}{\mathrm{E}}[A^\eta F \neq 0] = \prod_{\xi < \eta} \underset{A}{\mathrm{E}}[A^\xi F \neq 0]$$

is Borel.

With this (14) is proved in general.

Furthermore

$$[A^\xi \neq A^{\xi+1}] = \sum_n [A^\xi V_n \neq 0][A^{\xi+1} V_n = 0],$$

and thus if one sets

$$\mathfrak{F}_n^\xi = \underset{A}{\mathrm{E}}[A^\xi V_n \neq 0],$$

then

$$\mathfrak{T}^\xi = \underset{A}{\mathrm{E}}[A^\xi \neq A^{\xi+1}] = \sum_n [\mathfrak{F}_n^\xi - \mathfrak{F}_n^{\xi+1}].$$

With this, it is not only shown that, along with the $\mathfrak{F}_n^\xi$, also $\mathfrak{T}^\xi$ and $\mathfrak{X}^\xi$ are Borel sets, but also that cases (5) and (6) hold here because of

$$\mathfrak{F}_n^0 \supset \mathfrak{F}_n^1 \supset \cdots \supset \mathfrak{F}_n^\xi \supset \cdots,$$

and consequently, (12) is k-convergent and m-convergent.

In the space $\mathfrak{X}$, Herr Hurewicz (Fund. Math. 15 (1930), p. 4–17) constructed a determining system Φ of closed sets and with this showed that for uncountable X the set of uncountable A is analytic but not Borel. Let it be indicated here without further details: if $\mathfrak{X}_\xi$ is the set of A whose index (with respect to the system Φ that is to be somewhat modified) is $\leq \xi$, then

$$\mathfrak{X}^\xi \subset \mathfrak{X}_{\omega(\xi+1)} \subset \mathfrak{X}^{\xi+2}$$

holds, and since $\mathfrak{X} = \sum_\xi \mathfrak{X}_\xi$ is a Lusin-Sierpiński sum (p. 242), thus is k-convergent and m-convergent, it follows that the same also holds for $\mathfrak{X} =$

$\sum_\xi \mathfrak{X}^\xi$ (from the second inclusion just now indicated), whereas the Borel character of $\mathfrak{X}^\xi$ does not follow from this.

Incidentally, more holds with regard to category:

III. *In each case, $\mathfrak{X} - \mathfrak{X}^1$ is of first category; for perfect X, $\mathfrak{X} - \mathfrak{X}^0$ is already of first category.*

For the proof, by $\mathfrak{J}(M)$ we understand the set of A for which AM is dense-in-itself [[insichdicht]] (possibly 0). Then it is true that:

(15) \qquad For open G, $\mathfrak{J}(G)$ is a G_δ.[1]

First of all, the set $\underset{A}{\mathrm{E}}[|AG| \geq 2]$ is open. If A belongs to this set and $a_1 \neq a_2$ are two points in AG, let $\delta < \tfrac{1}{2}\varrho(a_1, a_2)$ be so small that the open spheres around a_1, a_2 with radii δ belong to G. Then for $\varrho(A,B) < \delta$ it is also the case that $|BG| \geq 2$, for there are points $b_1, b_2 \in B$ with $\varrho(a_1, b_1) < \delta$, $\varrho(a_2, b_2) < \delta$, from which it follows that $b_1 \neq b_2$ and $b_1, b_2 \in G$. Furthermore, since $\underset{A}{\mathrm{E}}[AG = 0] = \underset{A}{\mathrm{E}}[A \subset X - G]$ is closed, $\underset{A}{\mathrm{E}}[|AG| \neq 1]$ is of the form $G + F$, thus in any case a G_δ, and since the in-itself-denseness of AG is equivalent to $\prod_n [|AGU_n| \neq 1]$, where again the U_n form a basis of open sets of X, then $\mathfrak{J}(G)$ is a G_δ.

In addition, let F be closed, $G = X - F$ open, and let
$$\mathfrak{F} = \underset{A}{\mathrm{E}}[AF \neq 0], \quad \mathfrak{G} = \mathfrak{X} - \mathfrak{F} = \underset{A}{\mathrm{E}}[AF = 0];$$
$\mathfrak{F}$ is closed and $\mathfrak{G}$ is open. We claim:

(16) \qquad Along with F, $\mathfrak{F}$ is also nowhere dense [[nirgendsdicht]].

It must be shown that $\mathfrak{G}$ is dense in $\mathfrak{X}$ (it can be assumed that $F \neq 0$). Let $A \in \mathfrak{X}$ be arbitrary; we construct a finite set $N \subset A$ (a δ-net) with $\varrho(A, N) < \delta$, and we replace those points $a \in N$ that belong to $F = X - G = \overline{G} - G \subset G^1$ by points $b \in G$ sufficiently nearby. In this way, a finite set B arises with $\varrho(A, B) < \delta$ and $BF = 0$; hence $B \in \mathfrak{G}$.

Now let $P = \prod_\xi X^\xi (= X^\alpha)$ be the perfect kernel of X, $S = X - P$ the scattered part of X, and $Q = \underline{P}$ the interior of P (it is always permitted for one of these sets to vanish). If $\mathfrak{P} = \mathfrak{J}(P)$ is the set of A with perfect AP and $\mathfrak{Q} = \mathfrak{J}(Q)$ is the set of A with dense-in-itself AQ, then

$$\mathfrak{P} \subset \mathfrak{Q}, \quad \mathfrak{P} \text{ is dense in } \mathfrak{X}.$$

The first claim is self-evident since each open subset of a dense-in-itself set is dense-in-itself; hence along with AP, $APQ = AQ$ is also dense-in-itself. For the proof of the second claim, we show that in any neighborhood of $A \in \mathfrak{X}$ there exists a $B \in \mathfrak{P}$, whereby we assume $AP \neq 0$ (otherwise, it would already be that $A \in \mathfrak{P}$). Let $U = U(AP, \delta)$ be an open sphere of radius δ around AP, and let $B = AS + \overline{UP}$. Then it is the case that $UP \supset AP$ and $B = A + \overline{UP}$ is closed, UP is dense-in-itself and $BP = \overline{UP}$

---

[1]) In particular, the set $\mathfrak{J}(X)$ of perfect A is a G_δ (Banach); cf. Kuratowski, Fund. Math. 17 (1931), p. 260.

is perfect, and $B \in \mathfrak{P}$; furthermore, since UP is a sum of finitely many $U(a, \delta)P$ with $a \in AP$, for each point of $\overline{UP}$ there exists one of AP at a distance $\leq \delta$; thus for each $b \in B$ there is an $a \in A$ with $\varrho(a, b) \leq \delta$, hence $\varrho(A, B) \leq \delta$ (because $A \subset B$).

Thus $\mathfrak{Q}$ is now also dense in $\mathfrak{X}$ and according to (15) a G_δ, and $\mathfrak{X} - \mathfrak{Q}$ is of first category.

The closed set $F = X^1 - Q$ is nowhere dense; its complement $G = I + Q$ (I the set of isolated points of X) is dense since $\overline{G} \supset \overline{I} + Q = \overline{S} + \underline{P} = X$. It is the case that
$$\mathfrak{J}(X^1) \subset \mathfrak{Q}, \quad \mathfrak{J}(X^1) \subset \mathfrak{X}^1;$$
for if AX^1 is perfect, then $AX^1 \subset P$, $AX^1 = AP$ is perfect, and $A \in \mathfrak{P} \subset \mathfrak{Q}$; in addition, $A^1 = AA^1 \subset AX^1 = AP = (AP)^1 \subset A^1$, and thus $A^1 = AP$ is perfect, $A^1 = A^2$, and $A \in \mathfrak{X}^1$. According to (16), it is now the case that $\mathfrak{F} = \underset{A}{\mathrm{E}}[AX^1 \neq AQ]$ is nowhere dense, $\mathfrak{Q} - \mathfrak{J}(X^1) \subset \mathfrak{F}$ is nowhere dense, $\mathfrak{X} - \mathfrak{J}(X^1) = (\mathfrak{X} - \mathfrak{Q}) + (\mathfrak{Q} - \mathfrak{J}(X^1))$ is of 1st category, and $\mathfrak{X} - \mathfrak{X}^1$ is of 1st category, q.e.d.

The set $\underset{A}{\mathrm{E}}[AP = 0] = \underset{A}{\mathrm{E}}[A \subset S]$ is open and contains no perfect set $\in \mathfrak{X}$; thus if X is not perfect, this set is a nonempty open part of $\mathfrak{X} - \mathfrak{X}^0$, so that $\mathfrak{X} - \mathfrak{X}^0$ is still not of 1st category. (One can derive the same from $\underset{A}{\mathrm{E}}[AX^1 = 0] = \underset{A}{\mathrm{E}}[A \subset I]$.) However, if $X = P = Q$ is perfect, then $\mathfrak{X}^0 = \mathfrak{J}(X) = \mathfrak{P} = \mathfrak{Q}$ is a G_δ dense in $\mathfrak{X}$, and $\mathfrak{X} - \mathfrak{X}^0$ is already of 1st category.

Bibliography

[BeP 1987] Beckert, Herbert and Purkert, Walter (eds.). *Leipziger mathematische Antrittsvorlesungen*, Teubner Archive zur Mathematik (Leipzig: Teubner), v. 8, 1987.

[Ber 1967] Bergmann, Günter. *Vorläufiger Bericht über den wissenschaftlichen Nachlaß von Felix Hausdorff*, Jahresbericht der Deutschen Mathematiker-Vereinigung **69** (1967), 62–75.

[Bern 1942] Bernays, Paul. *A System of Axiomatic Set Theory, Part III*, Journal of Symbolic Logic **7** (1942), 65–89.

[Berns 1901] Bernstein, Felix. *Untersuchungen aus der Mengenlehre* (Ph.D. dissertation: Göttingen; printed at Halle).

[Berns 1905a] Bernstein, Felix. *Zur Mengenlehre*, Jahresbericht der Deutschen Mathematiker-Vereinigung **14** (1905), 198–199.

[Berns 1905b] Bernstein, Felix. *Über die Reihe der transfiniten Ordnungszahlen*, Mathematische Annalen **60** (1905), 187–193.

[Berns 1905c] Bernstein, Felix. *Zur Mengenlehre*, Mathematische Annalen **60** (1905), 463–464.

[Berns 1905d] Bernstein, Felix. *Untersuchungen aus der Mengenlehre*, Mathematische Annalen **61** (1905), 117–155; reprint of [Berns 1901] with alterations.

[Bo 1967] Bonnet, Hans. *Felix Hausdorff zum Gedächtnis. Geleitwort*, Jahresbericht der Deutschen Mathematiker-Vereinigung **69** (1967), 75–76.

[Br 1996] Brieskorn, Egbert (ed.). *Felix Hausdorff zum Gedächtnis I: Aspekte seines Werkes* (Braunschweig/Wiesbaden: Vieweg), 1996.

[Ca 1978] Campbell, Paul J. *The Origin of "Zorn's Lemma,"* Historia Mathematica **5** (1978), 77–89.

[Can 1879] Cantor, Georg. *Über unendliche, lineare Punktmannichfaltigkeiten*, Mathematische Annalen **15** (1879), 1–7.

[Can 1884] Cantor, Georg. *Über unendliche, lineare Punktmannichfaltigkeiten*, Mathematische Annalen **23** (1884), 453–488.

[Can 1891] Cantor, Georg. *Über eine elementare Frage der Mannigfaltigkeitslehre*, Jahresbericht der Deutschen Mathematiker-Vereinigung **1** (1891), 75–78.

[Can 1895] Cantor, Georg. *Beiträge zur Begründung der transfiniten Mengenlehre I*, Mathematische Annalen **46** (1895), 481–512; translated in [Can 1915], 85 136.

[Can 1897] Cantor, Georg. *Beiträge zur Begründung der transfiniten Mengenlehre II*, Mathematische Annalen **49** (1897), 207–246; translated in [Can 1915], 137–201.

[Can 1915] Cantor, Georg. *Contributions to the Founding of the Theory of Transfinite Numbers* (Chicago: Open Court), translated and edited by Philip Jourdain; reprinted (New York: Dover), 1952.

[Da 1979] Dauben, Joseph W. *Georg Cantor: His Mathematics and Philosophy of the Infinite* (Cambridge: Harvard University Press), 1979; reprinted (Princeton: Princeton University Press), 1990.

[Di 1967] Dierkesmann, Magda. *Felix Hausdorff. Ein Lebensbild*, Jahresbericht der Deutschen Mathematiker-Vereinigung **69** (1967), 51–54.

[Du 1870] Du Bois-Reymond, Paul. *Sur la grandeur relative des infinis des fonctions*, Annali di matematica pura ed applicata **4** (series 2) (1870), 338–353.

[Du 1873] Du Bois-Reymond, Paul. *Eine neue Theorie der Convergenz und Divergenz von Reihen mit positiven Gliedern*, Journal für die reine und angewandte Mathematik **76** (1873), 61–91.

[Du 1879] Du Bois-Reymond, Paul. *Erläuterung zu den Anfangsgründen der Variationsrechnung*, Mathematische Annalen **15** (1879), 282–315, 564–576.

[Du 1882] Du Bois-Reymond, Paul. *Die allgemeine Funktionentheorie I* (Tübingen: Laupp), 1882.

[Ei 1992] Eichorn, Eugen. *Felix Hausdorff–Paul Mongré: Some Aspects of His Life and the Meaning of His Death*, in *Recent Developments of General Topology and its Applications; International Conference in Memory of Felix Hausdorff (1868-1942)*, W. Gähler et al. (eds.) (Berlin Akademie), 85–117, 1992; translated by Mitch Cohen.

[Ei 1994] Eichorn, Eugen. *In memoriam Felix Hausdorff (1868–1942). Ein biographischer Versuch*, in *Vorlesungen zum Gedenken an Felix Hausdorff*, E. Eichorn and E. Thiele (eds.) (Berlin: Heldermann), 1–49, 1994.

[EiT 1994] Eichorn, Eugen and Thiele, Ernst-Jochen (eds.). *Vorlesungen zum Gedenken an Felix Hausdorff* (Berlin: Heldermann), 1994.

[ErH 1962] Erdős, Paul and Hajnal, András. *On a Classification of Denumerable Order Types and an Application to the Partition Calculus*, Fundamenta Mathematicae **51** (1962), 117–129.

[Fe 2002] Felgner, Ulrich. *Die Hausdorffsche Theorie der η_α-Mengen und ihre Wirkungsgeschichte* in *Felix Hausdorff, Gesammelte Werke, Band II, Grundzüge der Mengenlehre*, E. Brieskorn et al. (eds.) (Berlin-Heidelberg-New York: Springer-Verlag), 634–674, 2002; available in pdf format: http://www.aic.uni-wuppertal.de/fb7/hausdorff/Eta_Alpha.pdf.

[Fer 1999] Ferreirós, José. *Labyrinth of Thought: A History of Set Theory and its Role in Modern Mathematics* (Basel: Birkhäuser), 1999.

[Fi 1981] Fisher, Gordon. *The Infinite and Infinitesimal Quantities of du Bois-Reymond and Their Reception*, Archive for History of Exact Sciences **24**, no. 2 (1981), 101–163.

[Fr 1968] Fraenkel, A. A. *Abstract Set Theory* (Amsterdam: NorthHolland), 1968.

[Fre 1981] Frewer, M. *Felix Bernstein*, Jahresbericht der Deutschen Mathematiker-Vereinigung **83** (1981), 84–95.

[Ga 1992] Garciadiego, Alejandro R. *Betrand Russell and the Origins of the Set-theoretic 'Paradoxes'* (Basil: Birkhäuser), 1992.

[GoK 1984] Gottwald, Siegfried and Kreiser, Lothar. *Paul Mahlo—Leben und Werk*, NTM Schriftenreihe für Geschichte der Naturwissenschaften Technik und Medizin **21** (1984) no. 2, 1–22.

[Gr 1970] Grattan-Guinness, Ivor. *An Unpublished Paper by Georg Cantor: Principien einer Theorie der Ordnungstypen. Erste Mitteilung*, Acta Mathematica **124** (1970), 65–107.

[Ha 1894] Hadamard, Jacques. *Sur les caractères de convergence des séries à termes positifs et sur les fonctions indéfiniment croissantes*, Acta Mathematica **18** (1894), 319–336.

[Har 1915] Hartogs, Friedrich. *Über das Problem der Wohlordnung*, Mathematische Annalen **76** (1915), 436–443.

[He 1974] Hechler, Stephen H. *On the Existence of Certain Cofinal Subsets of $^\omega\omega$*, in *Axiomatic Set Theory, Proceedings of Symposia in Pure Mathematics, Vol. XIII, Part 2* (Providence: American Mathematical Society), 1974, 155–173.

[Hes 1906] Hessenberg, Gerhard. *Grundbegriffe der Mengenlehre* (Göttingen: Vandenhoek & Ruprecht); also in *Abhandlungen der Fries'schen Schule* **N.S. 1** Heft 4, 1906.

[Hes 1907] Hessenberg, Gerhard. *Potenzen transfiniter Ordnungszahlen*, Jahresbericht der Deutschen Mathematiker-Vereinigung **16** (1907), 130–137.

[Hu 1905] Huntington, Edward V. *The Continuum as a Type of Order: An Exposition of the Modern Theory.*, Annals of Mathematics (**2**) **6** (1905), 151–184; reprinted in [Hu 1917].

[Hu 1917] Huntington, Edward V. *The Continuum and Other Types of Serial Order* (Cambridge: Harvard University Press); reprinted (New York: Dover), 1955.

[Je 1966] Jensen, Ronald B. *Independence of the Axiom of Dependent Choices from the Countable Axiom of Choice* (abstract), Journal of Symbolic Logic **31** (1966), 294.

[Jo 1905] Jourdain, Philip E. B. *On a Proof That Every Aggregate Can Be Well-Ordered*, Mathematische Annalen **60** (1905), 465–470.

[Ka 1996] Kanamori, Akihiro. *The Mathematical Development of Set Theory from Cantor to Cohen*, Bulletin of Symbolic Logic **2** (1996), 1–71.

[Ka 2003] Kanamori, Akihiro. *The Higher Infinite: Large Cardinals in Set Theory from Their Beginnings* (Berlin: Springer Verlag), 2003; 2nd edition.

[Ka 2004] Kanamori, Akihiro. *Zermelo and Set Theory*, Bulletin of Symbolic Logic **10** (2004), 487–553.

[Kn 1947] Knopp, Konrad. *Theorie und Anwendung der unendlichen Reihen* (Berlin: Springer Verlag), 1947.

[Ko 1904] König, Julius. *Zum Kontinuum-Problem*, in *Verhandlungen des dritten internationalen Mathematiker-Kongresses in Heidelberg von 8. bis 13. August 1904* (Leipzig: Teubner), 1905, 144–147.

[Ko 1905] König, Julius. *Zum Kontinuumproblem*, Mathematische Annalen **60** (1905), 177–180, 462.

[Koe 1996] Koepke, Peter. *Metamathematische Aspekte der Hausdorffschen Mengenlehre*, in *Felix Hausdorff zum Gedächtnis I: Aspekte seines Werkes*, E. Brieskorn (ed.) (Braunschweig/Wiesbaden: Vieweg), 1996, 71–106.

[Kor 1911] Korselt, Alwin. *Über einen Beweis des Äquivalenzatzes*, Mathematische Annalen **70** (1911), 294–296.

[Kow 1950] Kowalewski, Gerhard. *Bestand und Wandel. Mein Lebensweg zugleich ein Beitrag zur neueren Geschichte der Mathematik* (Munich: Oldenbourg), 1950.

[La 2002] Laugwitz, Detlef. *Debates About Infinity in Mathematics Around 1890: the Cantor-Veronese Controversy, Its Origins and Its Outcome*, NTM (N.S) Schriftenreihe für Geschichte der Naturwissenschaften Technik und Medizin **10** (2002), no. 2, 102–126.

[Lo 1967] Lorentz, G. G. *Das mathematische Werk von Felix Hausdorff*, Jahresbericht der Deutschen Mathematiker-Vereinigung **69** (1967), 54–62.

[Ma 1909] Mahlo, Paul. *Über homogene Teilmengen des Kontinuums*, Berichte über die Verhandlungen der Königlich Sächsischen Gesellschaft der Wissenschaften zu Leipzig, Mathematisch-Physische Klasse **61** (1909), 121–124.

[Ma 1911] Mahlo, Paul. *Über lineare transfinite Mengen*, Berichte über die Verhandlungen der Königlich Sächsischen Gesellschaft der Wissenschaften zu Leipzig, Mathematisch-Physische Klasse **63** (1911), 187–225.

[Ma 1912] Mahlo, Paul. *Zur Theorie und Anwendung der ϱ_0-Zahlen*, Berichte über die Verhandlungen der Königlich Sächsischen Gesellschaft der Wissenschaften zu Leipzig, Mathematisch-Physische Klasse **64** (1912), 108–112.

[Ma 1913] Mahlo, Paul. *Zur Theorie und Anwendung der ϱ_0-Zahlen II*, Berichte über die Verhandlungen der Königlich Sächsischen Gesellschaft der Wissenschaften zu Leipzig, Mathematisch-Physische Klasse **65** (1913), 268–282.

[Mo 1982] Moore, Gregory H. *Zermelo's Axiom of Choice: Its Origins, Development, and Influence* (New York: Springer), 1982.

[Mo 1989] Moore, Gregory H. *Towards a History of Cantor's Continuum Problem*, in *The History of Modern Mathematics, Volume I*, David E. Rowe and John McCleary (eds.) (Boston: Academic Press), 1989, 78–121.

[Mo 1996] Moore, Gregory H. *Felix Hausdorff and the Emergence of Order: 1900-1908*, Proceedings of the Canadian Society for the History and Philosophy of Mathematics **9** (1996), 121–135.

[Pe 1990] Peckhaus, Volker. *'Ich habe mich wohl gehütet, alle Patronen auf einmal zu verschießen'. Ernst Zermelo in Göttingen*, History and Philosophy of Logic **11** (1990), 19–58.

[Pe 2002] Peckhaus, Volker. *Pro and Contra Hilbert: Zermelo's Set Theories*, lecture given to PILM International Symposium, October 2002; available in pdf format: http://www-fakkw.upb.de/institute/philosophie/Personal/Peckhaus/.

[Pl 1993] Plotkin, J. M. *Who Put the "Back" in Back-and-Forth*, in *Logical Methods: In Honor of Anil Nerode's Sixtieth Birthday*, John N. Crossley et al. (eds.) (Boston: Birkhäuser), 1993, 705–712.

[Pr 1890] Pringsheim, Alfred. *Allgemeine Theorie der Divergenz und Convergenz von Reihen mit positiven Gliedern*, Mathematische Annalen **35** (1890), 297–394.

[Pu 1995] Purkert, Walter. *Nachlaß Felix Hausdorff: Findbuch*, 1995; available in pdf or html format: http://hss.ulb.uni-bonn.de:90/ulb_bonn/veroeffentlichungen/hausdorff_felix.

[Pu 2002]   Purkert, Walter. *Grundzüge der Mengenlehre–Historische Einführung* in *Felix Hausdorff, Gesammelte Werke, Band II, Grundzüge der Mengenlehre*, E. Brieskorn et al. (eds.) (Berlin-Heidelberg-New York: Springer-Verlag), 2002, 1–89; available in pdf format: http://www.aic.uni-wuppertal.de/fb7/hausdorff/HistEinfuehrung.pdf.

[PuI 1987]   Purkert, Walter and Ilgauds, Hans-Joachim. *Georg Cantor 1845–1918* (Basel: Birkhäuser), 1987.

[Sa 1975]   Sageev, Gershon. *An Independence Result Concerning the Axiom of Choice*, Annals of Mathematical Logic **8** (1975), 1–184.

[Sc 1993]   Scheepers, Marion. *Gaps in ω^ω*, in *Set Theory of the Reals*, Haim Judah (ed.), Israel Math. Conf. Proc., 6, Bar-Ilan University, 1993, 439–561.

[Sch 1898]   Schoenflies, Arthur. *Mengenlehre*, in *Encyklopädie der Mathematischen Wissenschaften mit Einschluss ihrer Anwendungen*, v. I, A5 (Leipzig: Teubner), 1898, 184–207.

[Sch 1900]   Schoenflies, Arthur. *Die Entwickelung der Lehre von den Punktmannigfaltigkeiten*, Jahresbericht der Deutschen Mathematiker-Vereinigung **8**(ii), (1900), 1–251.

[Sch 1905]   Schoenflies, Arthur. *Über wohlgeordnete Mengen*, Mathematische Annalen **60** (1905), 181–186.

[Sch 1908]   Schoenflies, Arthur. *Die Entwickelung der Lehre von den Punktmannigfaltigkeiten, Zweiten Teil*, Jahresbericht der Deutschen Mathematiker-Vereinigung, Erganzungsband **2** (1908), 1–331.

[Sch 1913]   Schoenflies, Arthur. *Entwickelung der Mengenlehre und ihrer Anwendungen, Erster Hälfte: Allgemeine Theorie der unendlichen Mengen und Theorie der Punktmengen* (Leipzig: Teubner), 1913.

[Sch 1922]   Schoenflies, Arthur. *Zur Erinnerung an Georg Cantor*, Jahresbericht der Deutschen Mathematiker-Vereinigung **31** (1922), 90–106.

[Scho 1996] Scholz, Erhard. *Logischen Ordnungen im Chaos: Hausdorffs frühe Beiträge zur Mengenlehre*, in *Felix Hausdorff zum Gedächtnis I: Aspekte seines Werkes*, Egbert Brieskorn (ed.) (Braunschweig/Wiesbaden: Vieweg), 1996, 107–134.

[Si 1949]   Sierpiński, Wracław. *Sur une propriété des ensembles ordonnés*, Fundamenta Mathematicae **36** (1949), 56–67.

[Si 1965]   Sierpiński, Wracław. *Cardinal and Ordinal Numbers*, 2nd ed. revised (Warsaw: Polish Scientific Publishers), 1965.

[SiT 1930]   Sierpiński, Wracław and Tarski, Alfred. *Sur une propriété caractéristique des nombres inaccessibles*, Fundamenta Mathematicae **15** (1930), 293–300.

[Sk 1920]   Skolem, Thoralf. *Logisch-kombinatorische Untersuchungen über die Erfüllbarkeit und Beweisbarkeit mathematischen Sätze nebst einem Theoreme über dichte Mengen*, Videnskapsselskapets skrifter, I. Matematisk-naturvidenskabelig klasse, no. 4 (1920), 1–36; reprinted in [Sk 1970], 103–136.

[Sk 1970]   Skolem, Thoralf. *Selected Works In Logic by Th. Skolem*, J. E. Fenstad (ed.) (Oslo: Universitetsforlaget), 1970.

[Ta 1924]   Tarski, Alfred. *Sur quelques théorèmes qui équivalent à l'axiome du choix*, Fundamenta Mathematicae **5** (1924), 147–154.

[Ta 1925]   Tarski, Alfred. *Quelques théorèmes sur les alephs*, Fundamenta Mathematicae **7** (1925), 1–14.

[Ta 1948]	Tarski, Alfred. *Axiomatic and Algebraic Aspects of Two Theorems on Sums of Cardinals*, Fundamenta Mathematicae **35** (1948), 79–104.
[Ta 1951]	Tarski, Alfred. *A Decision Method for Elementary Algebra and Geometry* (Berkeley and Los Angeles: University of California Press), 1951.
[vdW 1963]	van der Waerden, B. L. *Modern Algebra v. I* (New York: Ungar), 1963; translated from 2nd revised German edition by Fred Blum.
[We 1895]	Weber, Heinrich. *Lehrbuch der Algebra v. I* (Braunschweig: Vieweg), 1895; 2nd ed. reprinted with corrections (New York: Chelsea), 1955, AMS Chelsea 2002.
[Ze 1904]	Zermelo, Ernst. *Beweis daß jede Menge wohlgeordnet werden kann*, Mathematische Annalen **59**, 514–516; translated by Stefan Bauer-Mengelberg in *From Frege to Gödel*, Jean van Heijenoort (ed.) (Cambridge: Harvard University Press), 1967, 139–141.
[Ze 1908a]	Zermelo, Ernst. *Neuer Beweis für die Möglichkeit einer Wohlordnung*, Mathematische Annalen **65** (1908), 107–128; translated by Stefan Bauer-Mengelberg in *From Frege to Gödel*, Jean van Heijenoort (ed.) (Cambridge: Harvard University Press), 1967, 183–198.
[Ze 1908b]	Zermelo, Ernst. *Untersuchungen über die Grundlagen der Mengenlehre* I, Mathematische Annalen **65** (1908), 261–281; translated by Stefan Bauer-Mengelberg in *From Frege to Gödel*, Jean van Heijenoort (ed.) (Cambridge: Harvard University Press), 1967, 199–215.
[Zo 1935]	Zorn, Max. *A Remark on a Method in Transfinite Algebra*, Bulletin of the American Mathematical Society **41** (1935), 667–670.

Titles in This Series

25 **J. M. Plotkin, Editor,** Hausdorff on ordered sets, 2005

24 **Hans Niels Jahnke, Editor,** A history of analysis, 2003

23 **Karen Hunger Parshall and Adrain C. Rice, Editors,** Mathematics unbound: The evolution of an international mathematical research community, 1800–1945, 2002

22 **Bruce C. Berndt and Robert A. Rankin, Editors,** Ramanujan: Essays and surveys, 2001

21 **Armand Borel,** Essays in the history of Lie groups and algebraic groups, 2001

20 Kolmogorov in perspective, 2000

19 **Hermann Grassmann,** Extension theory, 2000

18 **Joe Albree, David C. Arney, and V. Frederick Rickey,** A station favorable to the pursuits of science: Primary materials in the history of mathematics at the United States Military Academy, 2000

17 **Jacques Hadamard (Jeremy J. Gray and Abe Shenitzer, Editors),** Non-Euclidean geometry in the theory of automorphic functions, 1999

16 **P. G. L. Dirichlet (with Supplements by R. Dedekind),** Lectures on number theory, 1999

15 **Charles W. Curtis,** Pioneers of representation theory: Frobenius, Burnside, Schur, and Brauer, 1999

14 **Vladimir Maz'ya and Tatyana Shaposhnikova,** Jacques Hadamard, a universal mathematician, 1998

13 **Lars Gårding,** Mathematics and mathematicians: Mathematics in Sweden before 1950, 1998

12 **Walter Rudin,** The way I remember it, 1997

11 **June Barrow-Green,** Poincaré and the three body problem, 1997

10 **John Stillwell,** Sources of hyperbolic geometry, 1996

9 **Bruce C. Berndt and Robert A. Rankin,** Ramanujan: Letters and commentary, 1995

8 **Karen Hunger Parshall and David E. Rowe,** The emergence of the American mathematical research community, 1876–1900: J. J. Sylvester, Felix Klein, and E. H. Moore, 1994

7 **Henk J. M. Bos,** Lectures in the history of mathematics, 1993

6 **Smilka Zdravkovska and Peter L. Duren, Editors,** Golden years of Moscow mathematics, 1993

5 **George W. Mackey,** The scope and history of commutative and noncommutative harmonic analysis, 1992

4 **Charles W. McArthur,** Operations analysis in the U.S. Army Eighth Air Force in World War II, 1990

3 **Peter L. Duren et al., Editors,** A century of mathematics in America, part III, 1989

2 **Peter L. Duren et al., Editors,** A century of mathematics in America, part II, 1989

1 **Peter L. Duren et al., Editors,** A century of mathematics in America, part I, 1988